Radioanalytical Methods in Interdisciplinary Research

ACS SYMPOSIUM SERIES **868**

Radioanalytical Methods in Interdisciplinary Research

Fundamentals in Cutting-Edge Applications

Carola A. Laue, Editor
Lawrence Livermore National Laboratory

Kenneth L. Nash, Editor
Argonne National Laboratory

Sponsored by the
ACS Division of Nuclear Chemistry and Technology

American Chemical Society, Washington, DC

Library of Congress Cataloging-in-Publication Data

Radioanalytical methods in interdisciplinary research : fundamentals in cutting-edge applications / Carola A. Laue, editor, Kenneth L. Nash, editor.

   p. cm.—(ACS symposium series ; 868)

   Includes bibliographical references and index.

   ISBN 0–8412–3837–5

   1. Radiochemistry—Research—Congresses.

   I. Laue, Carola A., 1966- II. Nash, Kenneth L. III. American Chemical Society. Division of Nuclear Chemistry and Technology. IV. Series.

QD605.R23   2003
541.3'8'072—dc22   2003057855

The paper used in this publication meets the minimum requirements of American National Standard for Information Sciences—Permanence of Paper for Printed Library Materials, ANSI Z39.48–1984.

Copyright © 2004 American Chemical Society

Distributed by Oxford University Press

All Rights Reserved. Reprographic copying beyond that permitted by Sections 107 or 108 of the U.S. Copyright Act is allowed for internal use only, provided that a per-chapter fee of $24.75 plus $0.75 per page is paid to the Copyright Clearance Center, Inc., 222 Rosewood Drive, Danvers, MA 01923, USA. Republication or reproduction for sale of pages in this book is permitted only under license from ACS. Direct these and other permission requests to ACS Copyright Office, Publications Division, 1155 16th St., N.W., Washington, DC 20036.

The citation of trade names and/or names of manufacturers in this publication is not to be construed as an endorsement or as approval by ACS of the commercial products or services referenced herein; nor should the mere reference herein to any drawing, specification, chemical process, or other data be regarded as a license or as a conveyance of any right or permission to the holder, reader, or any other person or corporation, to manufacture, reproduce, use, or sell any patented invention or copyrighted work that may in any way be related thereto. Registered names, trademarks, etc., used in this publication, even without specific indication thereof, are not to be considered unprotected by law.

PRINTED IN THE UNITED STATES OF AMERICA

Foreword

The ACS Symposium Series was first published in 1974 to provide a mechanism for publishing symposia quickly in book form. The purpose of the series is to publish timely, comprehensive books developed from ACS sponsored symposia based on current scientific research. Occasionally, books are developed from symposia sponsored by other organizations when the topic is of keen interest to the chemistry audience.

Before agreeing to publish a book, the proposed table of contents is reviewed for appropriate and comprehensive coverage and for interest to the audience. Some papers may be excluded to better focus the book; others may be added to provide comprehensiveness. When appropriate, overview or introductory chapters are added. Drafts of chapters are peer-reviewed prior to final acceptance or rejection, and manuscripts are prepared in camera-ready format.

As a rule, only original research papers and original review papers are included in the volumes. Verbatim reproductions of previously published papers are not accepted.

ACS Books Department

Contents

Preface..xi

Introduction to the Field

1. One Hundred Years of Development in Radioanalytical
   Chemistry as a Science to Probe the Limits..2
   Carola A. Laue and Kenneth L. Nash

2. Present Status and Future of Academic Radiochemistry.....................12
   Gregory R. Choppin and Andreas Kronenberg

3. Basic Principles of Radiochemistry...22
   Walter D. Loveland

4. Interface of Environmental, Bioassay, and Radioanalytical
   Standard Reference Materials and Traceability Evaluations..............38
   Kenneth G. W. Inn, Zhichao Lin, Michael Schultz,
   Zhongyu Wu, Ciara McMahon, Iisa Outola, Hiromu Kurosaki,
   Svetlana Nour, Linda Selvig, Lisa Karam,
   and J. M. Robin Hutchinson

Developments in Radioanalytical Methods: Radiation Detection Methods

5. Realization and Applications of Collimator-Less Gamma-
   Ray Imaging Systems...52
   K. Vetter

6. Application of the Triple-to-Double Coincidence Ratio
   Method at National Institute of Standards and Technology
   for Absolute Standardization of Radionuclides by Liquid
   Scintillation Counting..76
   B. E. Zimmerman, R. Collé, J. T. Cessna, R. Broda, and P. Cassette

7. Applications of Californium-252 Neutron Irradiations and Other Nondestructive Examination Methods at Oak Ridge National Laboratory...88
   R. C. Martin, D. C. Glasgow, and M. Z. Martin

8. Sequential and Simultaneous Radionuclide Separation–Measurement with Flow-Cell Radiation Detection...........................105
   R. A. Fjeld, J. E. Roane, J. D. Leyba, A. Paulenova, and T. A. DeVol

9. α-Autoradiography: A Simple Method to Monitor the Migration of α-Emitters in the Environment......................................121
   Carola A. Laue and David K. Smith

Developments in Radioanalytical Methods: Separation Techniques

10. Radioanalytical Techniques and the Characterization of New Separations Reagents..140
    Kenneth L. Nash, Renato Chiarizia, Marian Borkowski, Paul G. Rickert, and Emmanuel Otu

11. Recent Progress in the Development of Extraction Chromatographic Methods for Radionuclide Separation and Preconcentration..161
    Mark L. Dietz

12. Radioanalytical Methods in the Discovery and Characterization of Non-Pertechnetate (^{99}Tc) Species in Hanford Tank Wastes...177
    Rebecca M. Chamberlin, Kenneth R. Ashley, Jason R. Ball, Eve Bauer, Jonathan G. Bernard, Douglas E. Berning, Norman C. Schroeder, and Paul Sylvester

13. Thorium-229 for Medical Applications..193
    Miting Du, Fred Peretz, Rose A. Boll, and Saed Mirzadeh

Interdisciplinary Applications

14. Radioanalytical Chemistry in the Courtroom....................................206
    J. David Robertson

15. Study of Traces of Tritium at the World Trade Center.................218
    Thomas M. Semkow, Ronald S. Hafner, Pravin P. Parekh,
    Gordon J. Wozniak, Douglas K. Haines, Liaquat Husain,
    Robert L. Rabun, and Philip G. Williams

16. Tritium-Helium-3 Age-Dating of Groundwater in the
    Livermore Valley of California...235
    Gail F. Eaton, G. B. Hudson, and J. E. Moran

17. Automation of Radiochemical Analysis: From Groundwater
    Monitoring to Nuclear Waste Analysis.....................................246
    Oleg B. Egorov, Matthew J. O'Hara, R. Shane Addleman,
    and Jay W. Grate

18. Rapid Actinide Column Extraction Methods for
    Bioassay Samples..271
    S. L. Maxwell, III and D. J. Fauth

19. Determination of Isotopic Thorium in Biological
    Samples by Combined Alpha Spectrometry and Neutron
    Activation Analysis..286
    S. E. Glover

20. Applications of Radioanalytical Chemistry to Alzheimer's
    Disease..298
    J. D. Robertson and M. A. Lovell

21. Recent Developments in Neutron Activation Analysis:
    1997–2002...307
    S. Landsberger

Author Index..337

Subject Index...339

Preface

This book's creation was triggered by our symposium entitled *Radioanalytical Methods at the Frontier of Interdisciplinary Science*, which was held at the 223rd National Meeting of the American Chemical Society on April 8–10, 2002 in Orlando, Florida. The symposium's goal was to bring together scientists whose diverse research interests in one way or another rely on and probe the limits of radioanalytical chemistry. This volume based on the Orlando symposium carries the same philosophical objective, but goes beyond the symposium's scope with the inclusion of a basic introduction to radiochemistry, its history and its principles, and a siren's call for advancing radioanalytical and radiochemical education to meet the future (and present) needs of the scientific community for trained radiochemists. With these additions, it is our intention that this volume will serve a novice in the field of radioanalytical methods as a text or valuable reference book while exploring the possible applications of radioanalytical techniques to solving new problems. We hope that this volume will encourage our colleagues (the practicing radiochemist, nuclear scientist, and medical radiation research professional) to expand their horizons by considering how other experts have employed the various existing techniques (and have sought to develop new ones). By combining this diverse collection of investigations in a single location, we hope to stimulate more interdisciplinary interactions.

From the first accidental detection of radioactivity (Becquerel's observation of film exposure in the dark), radioanalytical methods have been at the center of the 20th century's expansion of the periodic table from about 90 known *elements* in the 1890s to the more than 2500 *isotopes* that are known today. Though radioanalytical chemistry has been practiced for more than a century, recent decades have seen continued improvements in detection methods, separations procedures to single out the radioactive analyte of interest, and the application of radioanalytical science to thoroughly modern problems needing modern solutions. Sadly, most scientists, who do not deal with radioanalytical chemistry on a day-to-day basis, are in the main unaware of the fantastic advancements the field has made and the opportunities the field offers.

In recent years radioanalytical methods have been applied in perhaps not so obvious fields as tumor targeting in cancer treatment, tracking the origins of Alzheimer's disease, or in the support of the criminal justice system. The chapters in the interdisciplinary applications section of this book highlight some of these applications. Most notable is a contribution on the fate of tritium in the rubble of the destroyed World Trade Center (WTC) buildings. Though elevated tritium was observed in only a few WTC water samples, the tritium that was found most likely originated in such ordinary items as the emergency exit signs that are found in many buildings around the world, emphasizing that radioactivity (and radioactive materials) represent an integral (and often invisible) part of our daily lives. Medical and material sciences also frequently apply radioanalytical methods and contribute to the advancement of those methods, as seen in the contribution on neutron activation analysis and others. Separations sciences have been a driver in advancing radioanalytical methods from the earliest days of this science. This role continues today in two particular areas: first in the development of new, easier to handle, and more reliable separation procedures, including the design of new separation reagents; and secondly in pushing for lower and lower detection limits in radiation measurement technologies. Both of those aspects are well demonstrated by various contributions to this book.

We believe we have created a quite unique volume that is both less and more than a textbook. In a sense, we have incorporated the entire history of radioanalytical chemistry herein ranging from the origins of radioanalytical chemistry to some of the most recent developments in the field. We believe that practicing radiochemists and other scientist involved in nuclear sciences, radio-pharmaceutical and radiation-medical research, students (or prospective students), and even the casual reader will find something of interest here.

Carola A Laue
Nuclear & Radiochemist/Research Staff
C&MS Enivronmental Services
Lawrence Livermore National Laboratory
7000 East Ave, L-231
Livermore, CA 94550
Phone   (925) 422–3192
Fax   (925) 422–3160
Email   laue1@llnl.gov

Kenneth L Nash
Chemist/Group Leader
Chemistry Division
Argonne National Laboratory
9700 S. Cass Avenue
Argonne, Illinois 60439–4831
(630) 252–3581
(630)252–7501
klnash@anl.gov

Introduction to the Field

Chapter 1

One Hundred Years of Development in Radioanalytical Chemistry as a Science to Probe the Limits

Carola A. Laue[1] and Kenneth L. Nash[2]

[1]Analytical and Nuclear Chemistry Division, Lawrence Livermore National Laboratory, 7000 East Avenue, L-231, Livermore, CA 94550
[2]Chemistry Division, Argonne National Laboratory, 9700 South Cass Avenue, Argonne, IL 60439

The development of radioanalytical chemistry as a science to probe the limits of our understanding of matter and the universe is tightly bound to the evolution of nuclear chemistry, nuclear physics and nuclear technology. The advancements that have been achieved in all of those areas as well as in nuclear medicine, a field presently attracting significant attention, would not have been possible without the steady progress that has been made in radioanalytical chemistry. The objective of this chapter is to highlight the important historic milestones that have led to the present state of the art in radioanalytical chemistry.

Before the turn to the 20th century, chemistry had advanced from the black art of alchemy to a creditable field of scientific inquiry, largely due to significant growth in understanding achieved during the 19th century. Though ten chemical elements had been "known" and widely used since prehistoric times (carbon, sulfur, iron, copper, zinc, silver, tin, gold, mercury and lead), only three new elements were added during the middle ages (when alchemistry enjoyed its heyday). Nineteen new elements were discovered during the 18th century while thirty were identified in the 19th century. As the numbers of elements increased, Mendelejev and Meyer were able to simultaneously and independently (in 1869/70) develop the basic concept of chemical periodicity and the basic organization of the periodic table, as we understand it today. The pace of discovery of new elements through the end of the 19th century is illustrated in Figure 1. By the end of the 19th century, 3 naturally occurring elements (polonium, radium, and actinium) had been identified on the basis of their radioactivity and chemical properties. As with all previously discovered elements, the position of these elements within the periodic table was guided by the trends of chemical properties specific to their periodic groups. Unlike their stable congeners, these elements would not have been discovered without their radioactive characteristics. The discovery of these three elements represents the birth of radiochemistry, as we know it today.

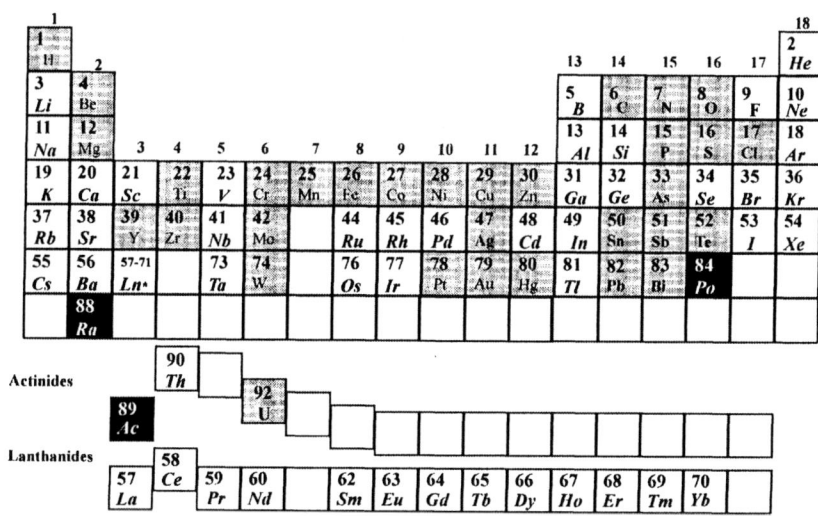

Figure 1: All elements known at 1900 are shown in today's format of the periodic table, as established by Mendelejev, Meyer and Seaborg. The elements with shaded background were known by 1800, all others were identified in the 19th century, including the first three radioactive elements highlighted in black.

In truth, a historical overview of radioanalytical chemistry cannot be separated from the development of nuclear chemistry. More than a century ago, nuclear chemistry evolved on the basis of several scientific achievements, coincidences, fortunate circumstances, and the persistence of the researchers creating this new field. The 19th century had been the century of chemistry — of the atom and molecule — and at the turn to the new century, renowned scientists freely predicted that only a few great discoveries were left to be made. Looking back we note that, although chemistry had gained a substantial knowledge base and was increasing in sophistication, basic knowledge of chemistry was largely build on an empirical foundation while the fundamental principles still remained mysterious.

At the same time, classical physics had developed to an advanced state and a common thought in the physics community, likewise, was that there were no new fundamental principles to be uncovered. Like the chemists, physicists failed to recognize that there was no understanding of the fundamental structure of matter, which could be used, for example, to explain the principles behind the periodic law and chemical behavior displayed by the elements. The existence of atoms was not even conjectural. Furthermore, an important aspect of the classical thermodynamic problem of black body radiation had not been explained. At the end of the 19th century, it became obvious that physics within its existing rigid framework could not explain those phenomena.

Development of the physical and chemical sciences in the 20th century relied on some very basic experiments and discoveries made in the last half of the 19th century. In 1859, Plucker had experimented with cathode rays, but Lorentz, in 1892, first formulated the theory that electricity is due to charged particles. After an initially slow period, scientific observations and breakthroughs appeared at an accelerated pace. The following events (not intended to be comprehensive) were particularly important for the development of the new field of nuclear science:

- In 1895 Roentgen discovered X-rays and Perrin stated that cathode rays are negatively charged particles.
- In 1896, Becquerel observed for the first time an unexpected property of natural matter —radioactivity— in an uranium ore.
- In 1898, Marie and Pierre Curie separated and identified the first naturally occurring radioactive elements, Po and Ra, from pitchblende. The circumstances that led to the identification of those new elements are particularly noteworthy. The Curie's employed common chemical separation procedures, but tracked the new elements in the separation process using the radioactive decay properties of the elements.

Shortly thereafter, Becquerel found that at least some of the observed radiated emissions are electrically charged. Rutherford successfully identified two different types of those radiations and labeled them α and β radiation. Villard determined that γ rays are not charged.

In 1899, Max Planck started the revolution in physics by introducing a theory, which postulates that matter only absorbs or emits energy in arbitrary units or 'quanta'. In the same year, Soddy uncovered the half-life property associated with radioactivity. Although Becquerel suggested in 1900 that β particles were electrons, the Goldhabers proved this in 1948 settling a long standing argument between the peers. Rutherford determined in 1903 that α particles have a positive charge. During his work with Soddy, they formulated the theory on the transmutation by radiation, radioactive decay, which led to coining of the term atomic energy in 1902. Harriet Brooks, a woman not often mentioned, working under the Curies and later with Rutherford, contributed substantially to the formulation of the transmutation theory of one element into another with her work.

As early as 1904, radioactivity and radioanalytical chemistry began making contributions to the advancement of science in general. Rutherford estimated the age of the earth by the radioactive dating technique, which had evolved by then from the U/He to the U/Pb technique. By 1905, the still standing dogmatic castle of classical physics crumbled lastly and forever when Einstein formulated the Special Theory of Relativity, the Law of Mass-Energy Equivalence, the Brownian Theory of Motion, and the Photon Theory of Light. In the same year, Bragg and Kleeman determined that α particles have discrete energies. Thompson revealed the relation between the scattering of X-ray photons and the number of electrons in an atom in 1906. Two years later, Geiger designed a simple device for detecting radiation, and in collaboration with Royds and Rutherford, they successfully identified α particles as He nuclei. In 1910, atmospheric radiation was recognized by Wulf, and in 1911, Hess attributed the radiation increase with altitude as originating in extraterrestrial space. Also in 1911, Rutherford proposed that atomic mass is concentrated in the nucleus, a direct conclusion from his α scattering experiments. Wilson constructed his cloud chamber in 1912 allowing for the first time the detection of protons and electrons.

The year of 1913 produced numerous scientific advancements. Geiger formulated the relationship between atomic number and nuclear charge that year. Soddy coined for the first time the term 'isotopes' ("same place" – meaning that they inherit the same position in the periodic table but differed distinctly in their radioactive properties; chemically they are the same but have a different mass), and Moseley utilized X-rays to confirm the relationship between nuclear charge and atomic number. Although Nagaoka proposed a 'Saturnian' model of the atom in 1904, this postulate did not capture the attention of the scientific community. It was left to Bohr in 1913 to revolutionize the understanding of the atom by proposing a "solar system" as a model of the atom, the quantum theory of atomic orbits, and radioactivity as a property of the nucleus. Rutherford and da Costa revised the nature of γ rays as high-energy photons, which had been previously identified as neutral by Villard.

By 1919, the alchemist's dream of transforming matter was achieved. Rutherford's group artificially transmuted nitrogen into oxygen and hydrogen. Rutherford proved, furthermore, the existence of protons in the nucleus. At the same time Aston, working in Thompson's lab, designed a mass spectrometer that allowed him to identify several isotopes of neon, among others. In 1923 de Broglie demonstrated the wave nature of particles on which basis Schroedinger postulated the particle wave equation three years later. In 1929, Bothe and Kolhorster found that cosmic rays are charged particles.

The year 1929 is also one of major developments of scientific equipment that proved crucial to the further advancement of the young field of nuclear science. Lawrence designed the first cyclotron and van de Graaff devised the Van de Graaff generator. In the early 1930's a new phenomenon was observed. Although Rutherford had speculated on the existence of neutrons as early as 1920, Becker and Bothe first observed an unusual penetrating neutral radiation in 1930. Two years later, Chadwick successfully identified the neutron, which led Heisenberg to state that the nucleus is composed of neutrons and protons. In 1933, Szilard conceived the idea of using a chain reaction of neutron collisions with atomic nuclei to release energy — the possibility of "atomic energy" as a source of human controlled power thus came into existence as a concept.

In 1934, Joliot and Curie-Joliot were able to artificially induce radioactivity. By bombarding a sheet of aluminium-27 with α particles, they created of a new radioactive isotope, or radioisotope, phosphorus-30. In the same year, indications exist that Fermi and Hidin did observe fission, though neither recognized the nature of their results at this time. The first man-made element (i.e., not existing naturally), technetium, was produced and identified by Perrier and Segre in 1937. One year later, at the dawn of the Second World War, Hahn and Strassman induce fission by exposing uranium to neutrons. By chemical separation they are able to prove that barium was formed rather than the anticipated transuranium elements, hence, they concluded that fission of the uranium target had occurred. Meitner, a member of the Hahn team, theoretically explains the fission process together with Frisch. Joliot, Curie-Joliot, and Szilard introduced the theory predicting the possibility of a nuclear chain reaction. Those stunning findings, in the climate of the emerging world war, sparked more than scientific curiosity. Along side with fission, fusion was also recognized in those years; Bethe postulated in 1939 that the fusion of hydrogen nuclei forming deuterium releases energy. He suggested further that the energy release of the sun and stars originates from the energy-releasing fusion of four hydrogen nuclei uniting and forming one helium nucleus. One year later, in 1940, Flerov discovered spontaneous fission, the 'natural' break up of a nucleus into two more stable fragments, in uranium.

In 1940 Corson, MacKenzie, and Segre synthesized element 85, astatine. Scientific publications were delayed for several years during the wartime (and postwar) in the interest of security. New elements such as neptunium (the last published in 1940), plutonium and americium were artificially produced and studied in great detail for military proposes. This chapter of nuclear science has been thoroughly described previously, and so will not be discussed in detail here. Nevertheless, the reader should note that, in addition to producing the world's first two nuclear weapons, the Manhattan project added much to the advancement of the field of nuclear science and to the development of radioanalytical techniques.

Starting with the identification of neptunium, the first transuranium element, a new era in the nuclear sciences began. Probing constantly the limits by designing new irradiation techniques, new separations methods and increasing sensitivities to detect unknown radioactive species, the surge to discover elements heavier than americium began. By 1961, the last of the transuranium (5f) elements, lawrencium (named after E. O. Lawrence, inventor of the cyclotron), was discovered. Chemical separation methods were crucial in the discovery of the new elements because the limited energy discrimination of the detection devices used at that time (to identify the energy of the emitted particles/γ-rays or fission fragments) and the interfering activities produced at the same reaction would have prevented the detection of the element of interest.

The elements from neptunium through einsteinium can be produced in weighable amounts. However, for heavier elements only small numbers of atoms could be created making the identification of new elements by its chemical properties challenging. With the ability to create only small amounts of the new elements, innovative approaches to conducting "one-atom-at-a-time" chemistry were developed to enable the identification of the remaining actinide elements. The trans-actinides that have been discovered since have all relied on these techniques for their creation and characterization.

With the development of the first silicon detection devices, the search for "Superheavy" elements was revitalized in the 70's and culminated in the 1999 and 2001 discoveries of elements 114 and 116, respectively. The Superheavy elements, or trans-actinides, can be produced only in an one-atom-at-a-time fashion today and the longest lived of them survives barely one minute. Chemical characterization is made extremely difficult by such short life-times thus discovery of new elements today is based solely on the isotopic decay signature. The elements that are thought to occupy a region of enhanced nuclear stability (formerly called "Island of Stability") are expected to live considerable longer, although theoretical predictions on the expected half-lives differ by orders of magnitude. Reaching the proton/neutron ratios necessary for attaining this nuclear stability is the primary obstacle to validation of the existence of this region of increased nuclear stability.

Continued development in detection technologies has proven crucial to the advancement of the nuclear and radioanalytical sciences and for the development of technological applications of the science. Those developments dominated the second half of the century of nuclear sciences. Applications such as Neutron Activation Analysis (NAA) would have been impossible without the development of nuclear reactors and semiconductor based gamma ray detection techniques. Detecting contaminants in the environment at below ppt levels would be unthinkable without the advances achieved in chemical separation methods combined with the constantly improving detection techniques such as:

- scintillation counting, which was in its entity first introduced by Kallmann in 1947 (photomultiplier were known and the phenomenon of scintillating material had already been observed in 1903 by Pierre Curie, who had taken a ZnS coated glass plate that had been bombarded with alphas into the dark to see the sparks);
- gas-proportional counting, the logical successor to the earlier electroscopes, ion chambers, and the Geiger-Mueller counters, although the latter are still very valuable for fast survey measurements to ensure safe laboratory practice;
- the high resolution γ- and α- detection based on the still advancing semi-conductor techniques. Such devices have revolutionized the field with its sensitivity and capability to be very isotope selective.

This semi-conductor based instrumentation vitalized not only the identification of new elements, the trans-actinides, but also the identification of most of the more than 2500 isotopes known today. These achievements equal the sudden increase in element discoveries that resulted when Bunsen and Kirchhoff designed a spectroscope in 1859 that enabled identification of elements by their characteristic spectral lines.

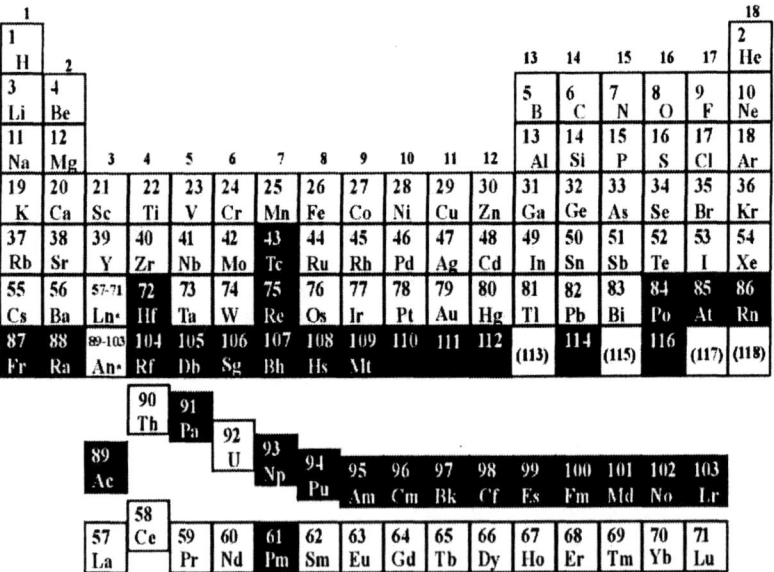

Figure 2: Periodic Table of the Elements as known at the turn of the 21st century, with highlighting of the elements found or identified on the basis of their nuclear properties (black background with white lettering).

Figure 2 shows the Periodic Table of the Elements of today including the additions of the last century. This figure serves to illustrate the role played by nuclear and radiochemistry in advancing our understanding of chemical periodicity. Though more new elements were discovered during the 19th century, in the 20th century a significant number of new, man-made elements were added to the periodic table. Furthermore, we learned that each of the elements in the periodic table was composed of a significant number of chemically identical isotopes, lending a new three-dimensional perspective to our previously recognized two-dimensional picture of the periodic table. The vast majority of 20th century additions to the periodic table are based on the properties of the atomic nuclei rather than properties resulting from their electronic shell or ionic radii. Perhaps the most important development in chemical periodicity of the 20th century is Seaborg's postulation that the actinides represented a 5f transition series. This discovery follows from the important work done in nuclear and radiochemistry.

Our brief outline of the sequential history of nuclear and radiochemistry is designed to demonstrate that this "new" field of science was born at the point when classical physics was not able to explain the fundamental structure of matter and when chemistry had recognized the periodic law of the elements but neither science was able to explain why the elements obeyed the periodic law. Only through the combined efforts of chemists and physicists was this dramatic transformation in science made possible. The discovery of natural and artificial radioactivity led to a revolution in scientific thinking and created an upheaval in our understanding of the universe and in the evolution of knowledge. Moreover, nuclear science has, through its applications, profoundly influenced the development of 20th (and now 21st) century human society in ways that were unimaginable a century ago (with all due respect to Jules Verne).

Present day technologies and scientific spin-offs initiated by 20th century developments in the nuclear sciences include:

- *Age dating* – used in earth science and in forensics. The earth grew a lot older with the introduction of this technique.
- *Nuclear Medicine* – Madame Curie took X-rays out to the battlefields to combat the misery, Seaborg introduced iodine-131 for medical uses, and today we are designing isotope-enhanced agents that specifically target the destruction of cancer cells.
- *Element and isotope discovery* – Most of the elements (> 30) and isotopes were artificially produced and helped refine periodic law of the elements while promoting a more thorough understanding of matter, and the evolution processes that created our earth, stars and solar system.
- *New analytical tools* – The use of tracers made it possible to significantly decrease detection limits and also to uncover reaction mechanisms,

benefiting many fields, including perhaps most notably the biological sciences.
- *New detection techniques* – their development made many new applications possible such as non-destructive analysis of material, etc.
- *Energy generation (civilian and military)* – The development of nuclear energy for peaceful uses (power generation and propulsion) and for defense-related purposes was and is strongly coupled. The two headed-legacy of this application of nuclear science is the ability to produce power without the emission of greenhouse gases (in 1908 Arrhenius argued that the greenhouse effect from coal and petroleum burning was warming the globe – it is becoming more apparent today that Arrhenius was correct), and the need to sequester the confined waste products from the biosphere for an extended period of time. The existence of nuclear weapons has "preserved the peace" during the second half of the 20th century, while the potential for proliferation of these terrible weapons represents a significant threat to our collective future on this planet.

However, the coexistence of positive and negative aspects of the applications of nuclear science should not blind mankind to the potential benefits derived from its application, some of which we have noted above. We must recognize that no human activity is without costs and benefits. Advances in nuclear and radiochemistry made during the last century have taught us that we in fact live in a radioactive world. We are surrounded by natural and artificial terrestrial and extraterrestrial sources of radiation, and have been throughout the history of mankind. The knowledge we have gained, if wisely applied, should serve us well in the future as we continue to occupy this planet and to venture out into space. The future should hold many opportunities for nuclear science, and the nuclear and radiochemists who practice this art to continue to contribute to the advance of human civilization.

Selective Bibliography:

Adloff, J.-P.; MacCordick, H.J. *Radiochim. Acta* **1995**, *70/71*, 13-22.
Adloff, J.-P. *Radiochim. Acta* **1997**, *77*, 1-7.
Burchfield, J.D. *Lord Kelvin and the Age of the Earth*; MacMillian: London, GB, 1975.
Draganic, I. G., Draganic, Z. D., Adloff J.-P. *Radiation and Radioactivity on Earth and Beyond*; CRC Press, Inc., Boca Raton, Florida, 1990.
Genet, M. *Radiochim. Acta* **1995**, *70/71*, 3-12.
Halliday, A.N. *Contemp. Phys.* **1997**, *38*, 103-114.
Rhodes, R. *The Making of the Atomic Bomb*; Simon & Schuster: New York, NY, 1986.

Seaborg, G.T. *Man-Made Transuranium Elements*; Prentice Hall, Inc.: Englewood Cliffs, NJ, 1963.

Seliger, H.H. *Phys. Today* **1995**, *November*, 25-31.

Sime, R.L. *Lise Meitner, A Life in Physics*; University of California Press: Berkeley, CA, 1996.

Rayner-Canham, M.F., Rayner-Canham, G.W. *Harriet Brooks, Pioneer Nuclear Scientist*; McGill-Queen's University Press: Montreal, CA, 1992

Rodden, C.J. *Analytical Chemistry of the Manhattan Project*; McGraw-Hill: New York, NY, 1950.

http://nuketesting.enviroweb.org/hew/
http://www.pafko.com
http://www.weburbia.com
http://www.accessexcellence.org/
http://www.physics.ucla.edu
http://www.aip.org/history/
http://fangio.magnet.fsu.edu/~vlad/pr100/100yrs

Acknowledgement

This work was performed under the auspices of the U.S. Department of Energy by Lawrence Livermore National Laboratory under Contract W-7405-Eng-48. The author greatly acknowledges the generosity of the Chemistry and Material Sciences Directorate Postdoctoral Program for sponsoring such activities outside the job assignments that greatly contribute to the overall career development.

Chapter 2

Present Status and Future of Academic Radiochemistry

Gregory R. Choppin and Andreas Kronenberg

Department of Chemistry and Biochemistry, Florida State University, Tallahassee, FL 32306–4390

After 1980 there has been a steady decrease in the number of university chemistry department offering graduate study in nuclear/radiochemistry and in the number of Ph.D. graduates per year in this field. The need for such scientists is discussed as is the growing severity of the supply/demand problem. Since this problem exists worldwide, it is not possible to import such skills. The result could be a major decrease in activities in nuclear medicine, the nuclear power industry and the US Department of Energy laboratories which would have significant effects to the country.

Introduction

Chemists have played a major role in the study of nuclear phenomena. They have been involved in many areas of nuclear/radiological research and have clarified and added greatly to our understanding of nuclear science. In turn, the use of nuclear science and radioisotopes in the study of many facets of chemistry has been very significant, as have the applications of nuclear science to probe molecular structures and reaction mechanisms. Chemical separation techniques have been key aspects in synthesis of over a dozen new transuranium elements which, in some cases, involved isolating and characterizing a few atoms (*1*).

Nuclear science and its applications evolved steadily through the first forty years of the 20th century. In 1945 the use of atomic weapons to end World War II greatly accelerated research in nuclear science. For the first 20 to 30 years after World War II, there was a constant growth in interest in nuclear and radiochemistry with some change in emphasis over time in the direction of those interests. Between 1945 and 1960 academic radiochemists chemists were involved largely in the study of nuclear reactions and nuclear spectroscopy rather than in separations, etc. From 1960 to 1980, research in nuclear power and in the utilization of radionuclides in nuclear medicine began to play a strong role. With the ending of the Cold War era, the attention of many radiochemists turned to the development of more specific nuclear separations and the use of radionuclides in analysis as well as continued strong interest in nuclear medicine. There was also a great deal of interest in the use of radionuclides in space exploration. As the Cold War ended, a strong need developed to decontaminate the large quantities of materials and areas of land that had been utilized for the preparation and testing of atomic weapons (*2*). These latter activities still constitute a major focus of many radiochemists and nuclear scientists in the laboratories of the U.S. Department of Energy.

Present Situation

Radionuclides continue to be used extensively in the areas of isotope dilution analysis, activation analysis, radioimmunoassay, and in tracer applications in biological, agricultural, environmental, chemical, physical, geochemical and hydrological studies. For example, in nuclear medicine ^{99m}Tc is used in lung, bone, kidney and blood pool imaging, isotopes of palladium, iodine, iridium and gold are used in the treatment of cancer, ^{131}I is used in the diagnosis and treatment of thyroid disease and ^{117m}Sn and ^{223}Ra are used to reduce pain in terminally ill bone cancer patients. At present, approximately 15 million nuclear medicine procedures, either diagnostic or therapeutic treatments, are performed annually (*3*).

A significant challenge remaining from the days of the Cold War that requires the work of nuclear and radiochemists involves the dismantlement and disposition of materials of nuclear weapons (*4*). Important in this work are the fissile cores of ^{235}U and ^{239}Pu weapons as well as tritium and other radionuclides that may have been included in the weapons assembly for tracking purposes. The quantities of these materials are quite large and represent both a challenge in safe handling and disposition as well as an opportunity for use in the future, in nuclear reactors, and in uses of radioisotopes in industry, medicine, research, etc.

It is necessary to decontaminate and to remediate large areas of former weapons production and assembly sites from the Cold War that have significant levels of contamination in the remaining buildings and in the environment. This decontamination has been estimated to require 30-50 years to accomplish and on the order of several hundred billions of dollars to accomplish (5). A product of this decontamination and remediation work will be very large quantities of contaminated materials and other forms of nuclear wastes. This waste material will require permanent disposal since the lifetimes of some of these radioisotopes are in the range of tens of thousands to hundreds of thousands of years. The United States has opened one nuclear waste repository for the disposition of materials from weapons laboratories and plants which are contaminated by plutonium and other nuclides (the Waste Isolation Pilot Plant) in New Mexico. More recently, Congress gave approval for the Department of Energy to proceed with the completion and certification of a high level waste disposal site in Nevada (Yucca Mountain Repository) where the waste from civilian power plants, etc. can be disposed permanently.

Long Term Needs

Due to the pervasive nature of the contamination in many of the Cold War weapons sites, complete decontamination of all these sites is not achievable with present technologies. Those sites that can be completely decontaminated would be released for unrestricted use. However, the areas which would have residual radioactivity contamination must have restrictions on their future use. In these latter areas, it will be necessary to establish stewardship control of these sites (5). This will involve controlled access and long term monitoring of the radioactivity to ensure that it is not being dispersed beyond the controlled zone. Such stewardship must also include a provision that development of new remediation technologies, which can be expected to occur over the next decades and centuries, will initiate reactivation of a remediation program to clean the site, and eventually, bring it to a state where there could be unrestricted use. This approach will require technical personnel well trained in the environmental behavior of the contaminant radioactivity as well as in the status of potential new remediation technologies over the length of time of the stewardship control.

Another important area that does, now, and will require for a long period of time well trained nuclear and radiochemists is international monitoring of compliance with nuclear nonproliferation agreements. In many nations an increase the number of nuclear power stations is expected with the resultant production of significant quantities of radioactive fission products as well as an increase in the production of plutonium as a byproduct. In these nations, it is necessary that there be international validation, by inspection, that none of these nuclear materials are diverted to clandestine weapons production.

Nuclear Power

The Greenhouse Effect problem may be confirmed over the next 10 – 20 years as a major threat to our civilization (6). At present, nuclear power may be the only technology that can provide an adequate and sufficiently large base for power production to replace carbon fuel. Energy produced from solar or wind power is too limited to provide the quantities needed for the expected increases in world population. The major environmental problem related to nuclear power is uranium mining. However, with the large amounts of excess weapons uranium and plutonium now available for burning in reactors to provide power, it has been estimated that no further uranium mining would be needed in the United States for a minimum of 100 and, probably, closer to 300 years. Moreover these new reactors would be "passive", i.e., they would be designed to avoid the operator misunderstanding and mistakes, which caused the serious nuclear power accidents in the past such as those at Three-Mile Island and Chernobyl. The international program of the 4th generation of nuclear power plants, which includes the USA, Japan, France, Great Britain and others is directed to new reactor concepts and new spent fuel treatment technologies (6). There will be a continuing, and probably, increasing need for well trained nuclear and radiochemists in the nuclear power industry if this expected increase in nuclear power occurs.

Associated with a large increase in nuclear power production would necessarily be a corresponding increase in the production of radioactive wastes. There is strong international interest presently in the use of non-aqueous processes to treat such radioactive waste (7). These processes would have the advantages of higher radiation resistance in the processing materials, use of more compact operational equipment (since the net volume of the treatment systems would be much smaller) and production of much smaller amounts of secondary waste. A considerable advantage is that such systems could be designed to have greater resistance to clandestine – either individual or national – diversion of material produced in the reactors for potential use in weapons. A number of these non-aqueous systems are being investigated; however, they have a significant disadvantage in the greater difficulty in conducting the separations when compared to systems using aqueous methods. Also, the decontamination factors in the separations from such non-aqueous systems is usually smaller than that obtained from multi-plate aqueous systems. This latter may not be as great a disadvantage as it would seem since a smaller decontamination factor leads to a higher residual contaminant radioactivity which could reduce the likelihood of diversion of the materials without being detrimental to the recycling of the fissile materials for further burning in a reactor.

In connection with a possible large increase in use of nuclear power and in the quantity of associated fission product "wastes", there is a strong interest in changing the concept of spent fuel treatment by using "efficient" separations (*8*). For example, such separations could isolate the longest lived nuclides such as ^{237}Np, ^{129}I, ^{99}Tc, ^{241,243}Am etc. from the shorter-lived fission products and, rather than placing them in a repository which would demand their continued isolation for a millennia, they would be destroyed by nuclear transmutation to shorter lived nuclides (*9*). In addition to the separation of these long-lived products, other radionuclides produced in reactors could be separated and utilized in space, industrial, medical and other applications and the unburned ^{235}U, ^{239}Pu recovered for recycle and further burning in reactors. This efficient separation approach would result in a much smaller amount of shorter lived radionuclides being sent to permanent repositories. Again, development of such efficient, multi-faceted separation system will require considerable research by nuclear and radiochemists over the next several decades.

Radiation Dose Effect Research

An important problem to which relatively little attention has been paid is the question of the validity of the present policy of linear extrapolation of dose effects (*10*). This assumes that the effects on health, etc. of radiation doses have a linear relationship to dose amount as the dose decrease to a zero value. There is no evidence to justify this assumption and, in fact, for low doses, there is some evidence for a dose threshold below which there is no effect. Unfortunately, at very low doses, the effects are so very small that the uncertainties become much greater than the measured effects and, consequently, it has not been possible to provide definite evidence either for or against a threshold. This is a very important question since, if a threshold can be confirmed for dose effects rather than a linear correlation to zero dose, much of the materials for which very large sums of money would be spent to encapsulate and dispose of in permanent repositories, could be recovered and released for recycle into industrial use. The result would have a very positive effect on the economics of nuclear power. The resolution of this issue will require the work of many bright and diligent radiochemists, probably over several decades, but the lack of trained scientists would be a serious handicap to such research at present.

Areas Needing Nuclear Scientists

The preceding discussion has reviewed areas in which an adequate number of nuclear and radiochemists are needed for the indefinite future. From that brief review, the areas needing a continued supply of such scientists can be listed as:
- basic research ;
- validation of nuclear nonproliferation;
- disposal of nuclear materials from dismantled weapons;
- permanent disposal of waste;
- remediation of environmental problems at former DOE sites;
- development of new, passive nuclear reactors to address the Greenhouse Effect as well as to burn the excess weapons uranium and plutonium so that it cannot be recycled in the into nuclear weapons.
- validation of the relation between radiation effects and dose amounts at very low levels.

In addition, the need for such scientists in nuclear medicine will continue, at least at the present level but may increase with population growth and other areas can be considered for the more distant future. As society moves from fission power to fusion power in the future, there will be a great need for nuclear scientists in the development of valid, useful fusion power reactors. In the near future, it is very likely there will be increased nuclear applications in medicine and in industry. Finally, it is very possible that in the future some of the radioactive fission products from power reactor operation which are now labeled for disposal as nuclear wastes may be worth recovering for use; e.g., as catalysts, in medicine, etc. This becomes increasingly attractive with time due to the decrease in radioactivity of the shorter-lived nuclides.

Studies of Personnel Needs

Unfortunately, the trend over the past 25 years in the United States raises a very serious question of, not whether there will be an adequate supply, but whether there will be any continued education of nuclear/radiochemists. This problem was recognized and recommendations to remedy the situation were given in a National Academy Report in 1988 (*11*). In this Report, it was noted that as nuclear faculty in chemistry departments retire, they are rarely replaced by new faculty. In the past this was not such a problem as it was possible to bring foreign scientists to the US for employment in the nuclear industry and USDOE laboratories. This is no longer a solution since, in almost all nations, the same situation exists as in the United States: a sharp decrease in educational programs available for interested students in nuclear science and engineering (*12, 13*). While the US Department of Energy has been aware of the growing problem, there has not been adequate attention to it as reflected in the fact that in 1990 the Office of Education of the Department of Energy was closed. There has been some response to this problem by DOE by establishment of programs in nuclear/radiochemical education (e.g., the REAP and INIE programs).

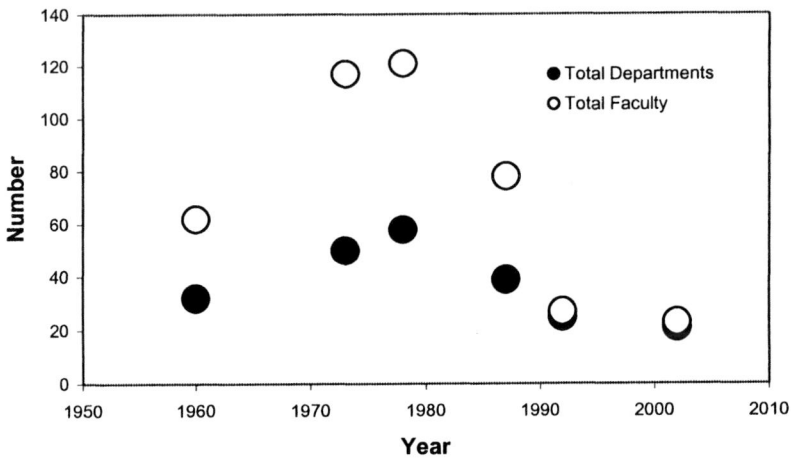

Figure1: Number of faculty and departments offering nuclear/radiochemistry graduate programs

A further problem that has existed for three decades in academic chemistry is the policy of the National Science Foundation to not fund radiochemical research as a result of an agreement with DOE in which the latter assumed responsibility for such funding. This not only barred an important source of funding to academic scientists, but resulted in many chemistry departments viewing radiochemistry as a minor topic since funding by the National Science Foundation is used as a criterion of the importance of a scientific area.

This problem in the United States was recognized a paragraph in an OECD/NEA report (*15*) published in 1992:

"In the United States there is a concern about the adequacy of supply of new graduate scientists and engineers to meet technical employment needs. This concern is particularly acute within the nuclear field because of declines in the number of education programs and number of students in nuclear engineering, health physics, and radiochem-istry. The decline in the number of new graduates is assessed in comparison to current and projected future employment needs. Currently, supplies of new graduates are just meeting employment needs in nuclear engineering and are less adequate in health physics and radiochemistry. If the number of graduates does not increase, these inadequacies of supply are likely to become more severe in the future. "

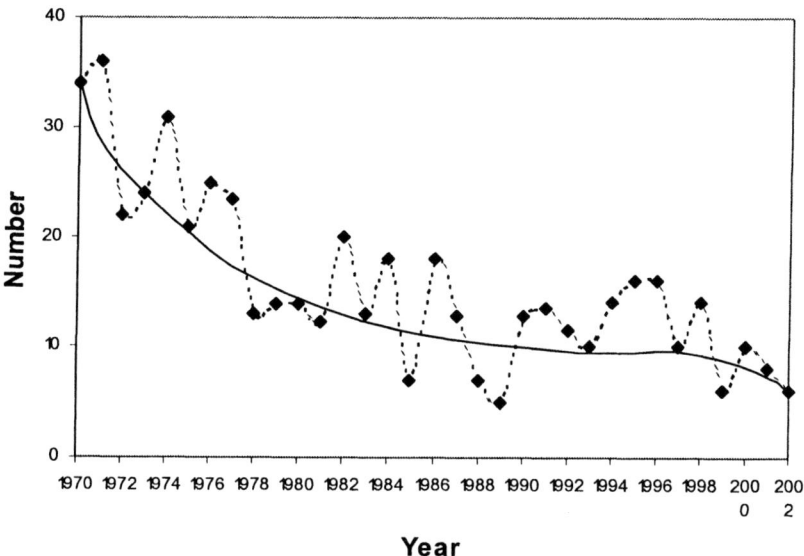

Figure2: Number of doctoral degrees in nuclear/radiochemistry awarded by U.S. universities from 1970-200

The critical nature of the present situation in the United States is confirmed by the results of a recent study conducted for the IAEA (*13*). A national survey of US universities for that study provided the data for Figures 1 and 2. Figure 1 shows the number of departments which provide education in nuclear/radiochemistry and in the number of faculty in the area from the years 1960 to 2002. There were increases in both numbers from 1960 to 1978, followed by a sharp decline after that time. With so few faculty and programs, there is little opportunity for students to study nuclear and radiochemistry even if there is sufficient student interest. This is reflected in Figure 2 in which we see that the number of doctoral degrees, while showing an uneven pattern, has declined steadily from 30+7 to the early 1970's about 5 per year at present. Five new Ph.D.'s per year in nuclear/radiochemistry at the present is certainly not adequate for the United States to fulfill the needs in nuclear power, nuclear medicine, industry, and the Department of Energy programs. As stated earlier, it is not possible to rely on the recruiting of foreign faculty or foreign students, as we have in the past, to fill that need. This was reflected already in a 1987 study

by the International Atomic Energy Agency (*12*). In the report of this study, the major conclusions were:
- · increasingly fewer universities are offering courses in nuclear and radiochemistry;
- · the number of graduates and teachers are declining.

Further, recommendations included:
- · establishment of "regional" centers at an adequate number of national universities;
- · adequate funding of these regional centers would for faculty, instrumentation, research and teaching programs in nuclear and radiochemistry.

The goal would be to ensure that these regional centers would provide a "critical mass" to ensure education of an adequate number of well trained scientists to meet the needs of the nations.

Presently, chemists with preparation in areas of research of chemistry other than nuclear/radiochemistry are being hired and retrained to fill the needs in this field. Due to unique aspects of handling radioactivity properly and safely, several years are required to gain the experience to work with radioactive materials without contamination of the facilities and personnel. Consequently, this is usually an inefficient and often inadequate approach to fulfilling the personnel needs. If these needs increase significantly, such retraining will be unlikely to be able to provide an adequate supply of properly trained nuclear scientists to fulfill our national needs.

Summary

Unfortunately, not only in the United States, but also in Europe and Asia, the picture today grows ever more grim, as these recommendations not only have not been adopted as national policy but they have also failed to persuade national universities to attempt to meet the needs in this area. Unless the federal government, through the USDOE, moves very soon to initiate a program to support adequately nuclear and radiochemistry in U.S. universities, there may be no base for education of trained scientists to complete the decontamination of ex-weapon sites, to ensure proper use of radionuclides in medicine, industry and research and to allow expanded use of nuclear power to meet the threat of the Greenhouse Effect.

Preparation of this paper was supported by a grant from the USDOE-OBES Division of Chemical Sciences.

References:

1. Choppin G.R.; Liljensgen J.–O.; Rydberg, J. *Radiochemistry and Nuclear Chemistry*, 3rd ed.; Butterworth and Heinemann: Burlington, MA, 2002.
2. Fryberger, T. In *Chemical Separations in Nuclear Waste Management*; Choppin G.R., Khankhasayev, M.K., Plendl, H., Eds.; Department of Energy/EM-0591, Battelle Press: Columbus, OH, 2002; pp 17.
3. Arlington Medical Resources, Malvern, PA, 2002; private communication.
4. *Report to the Congress and Secretary of Energy;* The Commission on Maintaining United States Nuclear Weapon Expertise, National Defense Authorization Acts of 1997 and 1998; March 1, 1999.
5. *From Cleanup to Stewardship,* U.S. Department of Energy, Office of Environmental Management: Washington, D.C., 1999.
6. Lake, J.A.; Bennett, R.G.; Kotek, J.F. *Scient. American,* **2002**, *286,* 70-80.
7. Choppin, G.R. In *"Chemical Separations in Nuclear Waste Management: The State of the Art and a Look to the Future;"* Battelle: Columbus, OH, 2002; pp 3.
8. *Efficient Separations and Processing Program: Technology Summary*; U.S. Department Of Energy, Office of Environmental Management: Washington, D.C., 1995.
9. *Nuclear Wastes – Technologies for Separations and Transmutation*; U.S. National Research Council, National Academy Press: Washington, DC, 1996.
10. *The Disposition Dilemma: Controlling Release of Solid Materials from Nuclear Regulatory Commission Licensed Facilities*, U.S. National Research Council, National Academic Press: Washington, D.C., 2002.
11. *Training Requirements for Chemists in Nuclear Medicine, Nuclear Industry and Related Areas*, U.S. National Research Council Report, National Academic Press: Washington, D.C., 1988.
12. *Consultants Meeting on Training Requirements in Modern Aspects of Radiochemistry, Munich, Germany, Sept 21-24, 1987*, Ruiz, H.V., Ed.; IAEA: Vienna, Austria, 1987.
13. *Assessment of the Teaching and Applications in Radiochemistry: Report of a technical meeting, Antalya, Turkey, June 10-14, 2002*; M. Rossbach, Ed., IAEA: Vienna, Austria, 2002.
14. *Nuclear Education and Training: Cause of Concern?*; Organization For Economic Co-Operation And Development/Nuclear Energy Agency: Paris, France, 2000.
15. *Supply of Science and Engineering Graduates for United States Nuclear Industry*; Baker, J.G. and Blair, L.M. Eds.; Organization For Economic Co-Operation And Development/Nuclear Energy Agency; U.S. Department Of Energy, ORISE: Oakridge, TN, 1992.

Chapter 3

Basic Principles of Radiochemistry

Walter D. Loveland

Department of Chemistry, Oregon State University, Corvallis, OR 97331

What is radiochemistry?

Radiochemistry is defined as "the chemical study of radioactive elements, both natural and artificial, and their use in the study of chemical processes"(*1*). Operationally radiochemistry is defined by the activities of radiochemists, i.e., (a) nuclear analytical methods (b) the application of radionuclides in areas outside of chemistry, such as medicine (c) the physics and chemistry of the radioelements (d) the physics and chemistry of high activity level matter and (e) radiotracer techniques. Radiochemistry is closely allied with nuclear chemistry, a three-pronged endeavor made up of: (a) studies of nuclear properties such as structure, reactions, and radioactive decay by people trained as chemists (b) studies of macroscopic phenomena (such as geochronology or astrophysics) where nuclear processes are intimately involved and (c) the application of measurement techniques based upon nuclear phenomena (such as activation analysis or radiotracers) to study scientific problems in a variety of fields. All radiochemists are, by definition, nuclear chemists, but not all nuclear chemists are radiochemists. Many nuclear chemists use purely non-chemical techniques to study nuclear phenomena, and thus, their work is not radiochemistry. Neither nuclear chemistry nor radiochemistry should be confused with radiation chemistry, the study of the chemical effects of ionizing radiation.

Nuclear chemists and radiochemists must possess knowledge of radioactivity, radioactive decay and nuclear reactions, the interaction of radiation with matter and nuclear radiation detectors. These subjects are treated in a number of fine textbooks (*2-5*). Radiochemistry is treated in these textbooks as well as a number of excellent textbooks that focus primarily on radiochemistry (*6-8*).

II. What makes radiochemistry unique?

Radiochemistry involves the application of the basic ideas of inorganic, organic, physical and analytical chemistry. However, the need to manipulate radioactive materials imposes some special constraints (and features) upon these endeavors. The first of these involves the number of atoms involved and the concentration of solutions. The range of activity levels in radiochemical procedures ranges from pCi to MCi, but for the sake of argument, let us assume an activity level, A, typical of radiotracer experiments, 1 µCi (= 3.7×10^4 dis/s = 3.7×10^4 Bq), of a nucleus with mass number ~ 100. If we assume a half-life for this radionuclide of 3 days, the number of nuclei involved can be calculated from the equation

$$N = \frac{A}{\lambda} = \frac{(1\mu Ci)(3.7 x 10^4 \, dps/\mu Ci)(3 days)(24 hr/day)(3600 s/hr)}{\ln 2}$$

where λ is the decay constant of the nuclide (= $\ln 2 / t_{1/2}$). Then

$$N \sim 1.4 \times 10^{10} \text{ atoms}$$

$$\text{Wt. of sample} = 2.3 \times 10^{-12} \text{ g}$$

This quantity of material, if prepared as an aqueous solution of volume 1 L would have a concentration of 10^{-14} mol/L. This simple calculation demonstrates a number of the important features of radiochemistry, i.e., (a) the manipulation of samples involving infinitesimal quantities of material (b) the ability to conduct tracer studies since the quantity of radioactive material added to a system is so small as to not affect the system and (c) the power of nuclear analytical techniques. In the previous example, the quantity of radioactivity, 1 µCi is a significant, easily detectable quantity of radioactivity. Thus it is possible to speak of detection sensitivities in terms of pg or less. In fact, since the decay of a single atom (such as a heavy element) can lead to the emission of an alpha-particle that can be detected with 100% efficiency, it is possible to speak of detecting single atoms—or doing chemistry on an atom by atom basis.

The small numbers (< 100) of atoms involved in some radiochemical procedures can cause problems. The extreme dilution in some solutions can mean that equilibrium is not reached due to kinetic limitations. One must be sure that all phase space is appropriately sampled in every chemical procedure.

Conventional analytical techniques generally operate at the ppm or higher levels. Some techniques such as laser photoacoustic spectroscopy are capable of measuring phenomena at the 10^{-8}-10^{-6} mol/L level. The most sensitive conventional analytical techniques, time-resolved laser-induced fluorescence and ICP-MS are capable of measuring concentrations at the part per trillion level, i.e., 1 part in 10^{12}, but rarely does one see detection sensitivities at the single atom level as routinely found in some radioanalytical techniques. While techniques such as ICP-MS are replacing the use of neutron activation analysis in the routine measurement of ppb concentrations, there can be no doubt about the unique sensitivity associated with radioanalytical methods.

Along with the unique sensitivity and small quantities of material associated with radiochemistry, there are certain problems associated with the use of radioactivity. The most important of these is the need to comply with the regulations governing the safe use and handling of radioactive material. This task is a primary focus in the design and execution of radiochemical experiments and is often a significant factor in the cost of the experiment. Because so many of these rules are site specific, they are not treated in this short review.

There are some chemical effects that accompany high specific activities that are unique to radiochemistry and worth noting. Foremost among these are the chemical changes accompanying radioactive decay. The interaction of ionizing radiation from a radioactive source with air can result in the generation of ozone and the nitrous oxides, which can lead to corrosion problems. Sources containing radium or radium (Ra) produced from the decay of heavier elements, such as uranium, will emanate radon (Rn) gas as the decay product of Ra. The decay products of Rn are particulates that deposit on nearby surfaces, such as the interior of the lungs, leading to contamination problems. In high activity aqueous solutions, one can make various species such as the solvated electron, e_{aq}^-, hydroxyl radicals, OH•, as well as the solvated proton, H_3O^+. The hydroxyl radical, OH•, is a strong oxidizing agent with

$$OH\bullet + e^- \rightarrow OH^- \quad E_0 = 2.8 \text{ V}$$

while the solvated electron, e_{aq}^-, is a strong reducing agent

$$e_{aq}^- + H^+ \rightarrow \tfrac{1}{2} H_2 \quad E_0 = 2.77 \text{ V}$$

Solutions involving high activity levels will change their redox properties as a function of time. For example, a solution of 100 Bq/mL (10^{-7} mol/L) of ^{239}Pu will have all the atoms undergo a redox change in a period of one year. In general, it is hard to keep high specific activity solutions stable. Reagents, column materials, etc. can suffer radiation damage also. In radiotracer studies, the self-decomposition (radiolysis) of ^{3}H or ^{14}C-labeled compounds can lead to a variable concentration and number of products. In addition to the usual concerns about the purity of chemicals, one must be concerned with *radiochemical purity*, the presence of other radioactive species in low concentration. Some tracers (usually cations) in solution behave as colloids rather than true solutions. Such species are termed radiocolloids and are aggregates of $10^3 - 10^7$ atoms, with a size of the aggregate of 0.1 – 500 nm. They are quite often formed during hydrolysis, especially of the actinides in high oxidation states. One can differentiate between real colloids and pseudocolloids, in which a radionuclide is sorbed on an existing colloid, such as humic acid or $Fe(OH)_3$. The chemical behavior of these radiocolloids is difficult to predict, as the systems are not at equilibrium.

III. Radiochemical Separation Techniques

Radiochemical separations involve the conventional separation techniques of analytical chemistry adapted to the special needs of radiochemistry. Radiochemical purity is generally more important than chemical purity. When dealing with short-lived nuclides, speed may be more important than yield or purity. The high cost of radioactive waste disposal may require unusual waste minimization steps. A recent review summarizes some newer developments of relevance to radiochemistry (9) along with the chapters by Dietz and Nash et al. in this book.

Precipitation

The oldest, most well established chemical separation technique is *precipitation*. Because the amount of the radionuclide present may be very small, *carriers* are frequently used. A carrier is usually a stable or long-lived radioisotope of the radionuclide in question or a chemically homologous element. The carrier is added in macroscopic quantities and insures the radioactive species will be part of a kinetic and thermodynamic equilibrium system. It is important that there is an isotopic exchange between the carrier and the radionuclide. There is the related phenomenon of *co-precipitation* wherein the radionuclide is incorporated into or adsorbed on the surface of a precipitate that does not involve an isotope of the radionuclide. Examples of this behavior are the sorption of radionuclides by Fe(OH)$_3$ or the co-precipitation of the actinides on LaF$_3$. Co-precipitation can be minimized by the use of *holdback carriers*, inactive materials that are isotopic with the radionuclide and replace it in co-precipitates. Separation by precipitation frequently involves the chemical process of scavenging, which is the removal of unwanted activities from solution by precipitation. In a typical application, the desired element is put in a soluble oxidation state, a precipitation of some scavenger compound such as Fe(OH)$_3$ is performed, removing unwanted activities, the element of interest is oxidized or reduced to a different oxidation state and removed. Separation by precipitation is largely restricted to laboratory procedures and apart from the bismuth phosphate process used in WWII to purify Pu, has little commercial application.

Solvent Extraction

Separation by liquid-liquid extraction (solvent extraction) has played an important role in radiochemical separations. Ether extraction of uranium was used in early weapons development, and the use of tri-butyl phosphate (TBP) as an extractant for U and Pu was recognized in 1946, resulting in the commercial PUREX process for reprocessing spent reactor fuel. In recent years, there has been a good deal of activity in the development of solvent extraction processes for the removal of transuranic elements, ^{90}Sr and ^{137}Cs from acidic high level waste. Laboratory demonstrations of the TRUEX process that uses the neutral extractant CMPO (octyl(phenyl)-N,N-diisobutylcarbamoyl-methylphosphine oxide) to separate the transuranium elements from acidic high-level waste have been successful. More recently crown ethers have been used as specific extractants for strontium and cesium.

In solvent extraction, a metal ion is distributed between two immiscible phases, usually an aqueous phase and an organic phase. To achieve sufficient solubility in the organic phase, the metal ions must be in the form of a neutral, non-hydrated species. The distribution ratio D is defined as

$$D = [M]_{org}/[M]_{aq}$$

where $[M]_i$ is the metal ion concentration in the ith phase. The relevant equilibria to describe the extraction in systems of lipophilic acidic chelating agents are

$$HL_{org} \rightleftarrows HL_{aq}$$

$$HL_{aq} \rightleftarrows H^+ + L_{aq}^-$$

$$M^{3+}_{aq} + 3HL_{org} \rightleftarrows (ML_3)_{org} + 3H^+_{aq}$$

where K_e is the equilibrium constant for the last reaction. The distribution coefficient D is given as

$$D=[ML_3]_{org}/[M^{3+}]_{aq} = K_e [HL]^3_{org} / [H^+]^3_{aq}$$

If one introduces a water-soluble complexing agent into the system, the $[M^{3+}]_{aq}$ becomes $[M^{3+}] + [MX^{2+}] + [MX_2^+] + ...$ and the measured distribution ratio will include these species as well. The separation factor between two ions, S, is given by the ratio of their distribution coefficients

$$S=D_A/D_B$$

Thus, the most effective separations will involve cases where the target ion interacts strongly with the extractant but is less strongly complexed by the aqueous ligand X.

Besides chelating agents, extractant molecules can be of other types. Cation transfer from the aqueous to organic phase can be accomplished by solvation of a neutral metal complex (TBP, for example) or formation of an ion-association complex (quaternary amines) or by a surfactant that forms extractable aggregates. Cation selectivity is best for the lipophilic chelating agents. Recent developments have suggested the utility of using supercritical fluids, such as CO_2 as solvents, mainly for lesser costs in waste disposal.

Ion Exchange

Ion exchange is one of the most popular radiochemical separation techniques due to its high selectivity and the ability to perform separations rapidly. In ion exchange, a solution containing the ions to be separated is brought into contact with a synthetic organic resin containing specific functional groups that selectively bind the ions in question. In a later step the ions of interest can be removed from the resin by elution with another suitable solution that differs from the initial solution. Typically the solution containing the ions is run through a column of resin. The resins are typically cross-linked polystyrenes with attached functional groups. Most cation exchangers (such as Dowex 50) contain free sulfonic acid groups, SO_3H, where the cation displaces the hydrogen ion. Anion exchangers (such as Dowex 1) contain quaternary amine groups, such as $CH_2N(CH_3)_3Cl$ where the anion replaces the chloride ion. The resin particles have diameters of 0.08 – 0.16 mm and exchange capacities of 3-5 meq per gram of dry resin.

It is common to absorb a group of ions on the column material and then selectively elute them. Complexing agents, which form complexes of varying solubility with the absorbed ions, are used as eluants. There exists a competition between the complexing agent and the resin for each ion and each ion will be exchanged between the resin and the complexing agent several times as it moves down the column. This is akin to a distillation process. The rates at which the different ions move down the column vary causing a spatial separation between "bands" of different ions. The ions can be collected separately in successive eluant fractions (see Figure 1).

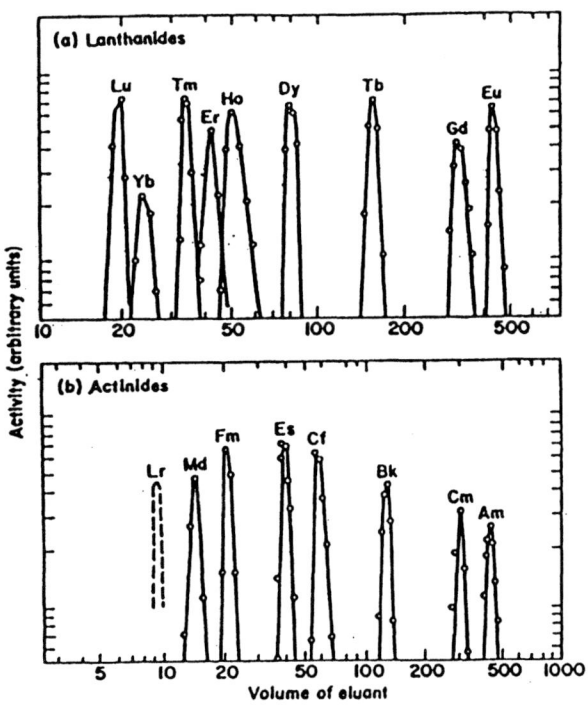

Figure 1. Elution of tripositive lanthanide and actinide ions on Dowex-50.

The most widely cited application of ion exchange techniques is to the separation of the rare earths or actinides from one another. This is done with cation exchange using a complexing agent of α-hydroxyisobutyric acid ("α-but"). The order of elution of the ions from a cation exchange column is generally in order of the radii of the hydrated ions with the largest hydrated ions leaving first; thus lawrencium elutes first and americium last among the tripositive actinide ions (see Figure 1). In the case of the data of Figure 1, the separation between adjacent cations and the order of elution is derived from the comparative stability of the aqueous actinide or lanthanide complexes with αhydroxyisobutyrate. As shown in Figure 1, there is a strikingly analogous behavior in the elution of the actinides and lanthanides that allowed chemists to prove the identity of new elements in the discovery of elements 97-102 (Bk-No). For cation exchange, the strength of absorption goes as $M^{4+} > M^{3+} > MO_2^{2+} > M^{2+} > MO_2^{+}$.

The anion exchange behavior of various elements has been extensively studied. For example, consider the system of Dowex 1 resin and an HCl eluant. Typical distribution ratios for various elements as a function of [Cl$^-$] are shown in Figure 2. One usually sees a rise in the distribution coefficient D until a maximum is reached and then D decreases gradually with further increases in [Cl$^-$]. The maximum occurs when the number of ligands bonding to the metal atom equals the initial charge on the ion. Figure 2 or similar data can be used to plan separations. For example, to separate Ni(II) and Co(II), one needs simply to pass a 12 M HCl solution of the elements through a Dowex 1 column. The Co(II) sticks to the column while the Ni(II) is not absorbed. A mixture of Mn(II), Co(II), Cu(II), Fe(III), and Zn(II) can be separated by being placed on a Dowex 1 column using 12 M HCl, followed by elutions with 6M HCl (Mn), 4 M HCl (Co), 2.5 M HCl (Cu), 0.5 M HCl (Fe) and 0.005 M HCl (Zn). Du et al., for instance, prove in their chapter the viable application of the ion exchange technique to separate and purify thorium.

In addition to the organic ion exchange resins, some inorganic ion exchanges, such as the zeolites, have been used. Inorganic ion exchangers are used in situations where heat and radiation might preclude the use of organic resins although the establishment of equilibria may be slow.

Newer developments have emphasized the preparation of more selective resins. Among these are the chelating resins (such as Chelex-100) that contain functional groups that chelate metal ions. Typical functional groups include iminodiacetic acids, 8-hydroxyquinoline or macrocyclic units such as the crown ethers, calixarenes or cryptands. The bifunctional chelating ion exchange material, Diphonix® resin—a substituted diphosphonic acid resin, shows promise in treating radioactive waste. Important newer resins include those with immobilized phosphorus ligands. (*9*)

Exchange Chromatography

Extraction chromatography is an analytical separation technique that is closely related to solvent extraction. Extraction chromatography is a form of solvent extraction where one of the liquid phases is made stationary by adsorption on a solid support. The other liquid phase is mobile. Either the aqueous or the organic phase can be made stationary. Extraction chromatography has the selectivity of solvent extraction and the multistage character of a chromatographic process. It is generally used for laboratory scale experiments although some attempts have been made to use it in larger scale

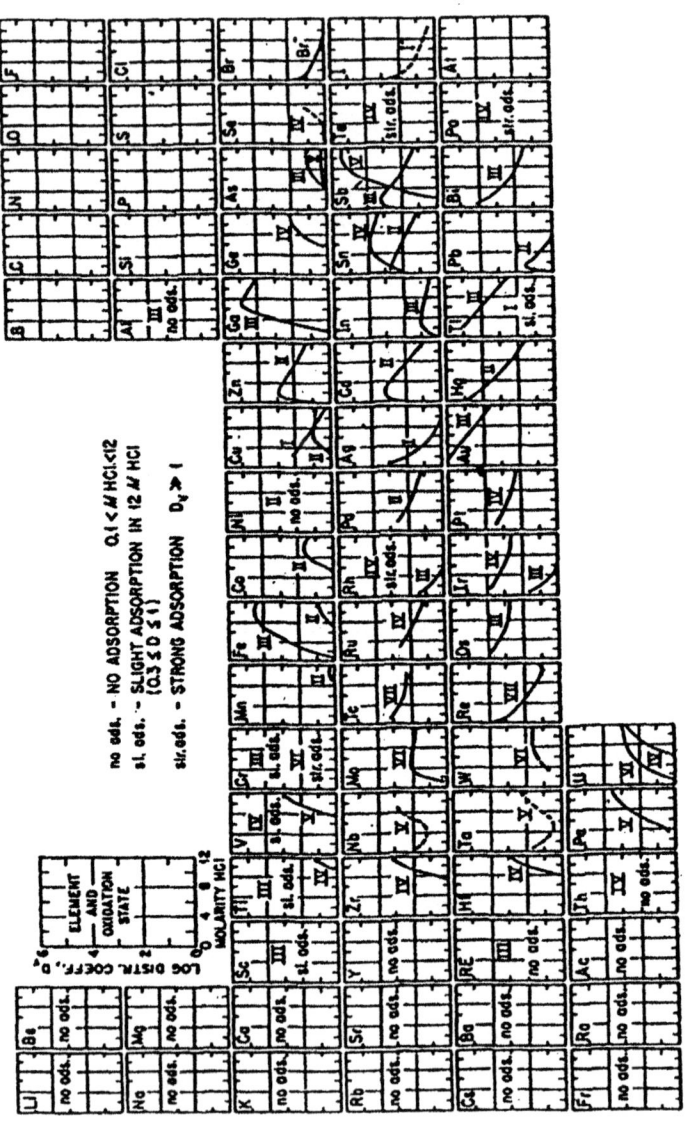

Figure 2. Elution of elements from anion exchange resin. (10).

operations. The common applications involve the adsorption of an organic extract ant onto a variety of inorganic substrates such as silica or alumina or organic substrates such as cellulose or styrene-divinyl benzene copolymers.

The same extracting agents as used in solvent extraction can be used in extraction chromatography. Early applications of extraction chromatography have employed various traditional extractants such as the acidic organophosphorus compounds (di- (2-ethylhexyl) phosphoric acid, HDEHP) or TBP as extractants for the actinide elements. Recent advances have led to a variety of new solvent exchange extractants such as the crown ethers, cryptands or bifunctional organophosphorus compounds. A particularly successful application is the selective sorption of actinides on TRU resins, involving solutions of CMPO in TBP sorbed on Amberlite XAD-7. This resin has found a number of applications in the determination of the actinides in complex matrices.

Rapid Radiochemical Separations

Many of the separation techniques we have described take hours to perform. Many interesting nuclei, such as the heavier actinides, the transactinides, or the light nuclei used in PET have much shorter half-lives. Thus, we review the principles of rapid radiochemical separations (procedures that take seconds to minutes) and refer the reader to (*11, 12*) for details.

In most chemical separation procedures, the goal is to selectively transfer the species of interest from one phase to another, leaving behind any unwanted species. The phase-to-phase transfer is rapid, but the procedures to place the species in the proper form for transfer are slow. The goal of rapid radiochemical separations is to speed up existing chemical procedures or to use new, very fast chemical transformations.

Two procedures are commonly used for rapid radiochemical separations, the batch approach and the continuous approach. In the batch approach, the desired activities are produced in a short irradiation, separated and counted with the procedure being repeated many times to reduce the statistical uncertainty in the data. In the continuous approach, the production of the active species is carried out continuously and the species is isolated and counted as produced.

One of the most widely used techniques for rapid chemical separation is that of gas chromatography, which has been developed for use with the transuranium elements by Zvara and co-workers (*13*). In gas chromatography, volatile elements or compounds are separated from one another by their differences in distribution between a mobile gas phase and a stationary solid phase. Thermochromatography involves passing a gas through a column whose temperature decreases continuously with distance from the entrance. Thus the less volatile species condense on the column walls first with the more volatile species depositing last. This technique was used to show the chemical properties

(*14-16*) of the transactinides Rf-Hs and their behavior relative to their chemical congeners.

IV. Applications of Radiochemistry

The discipline of radiochemistry has been actively pursued since the early 1900s. Because of this, there are a very large number of successful applications of radiochemistry. In this short review, we can only describe a few of the most important of these applications and refer the reader to standard textbooks that catalog many of these applications (*2-8*).

Radioanalytical Techniques

Activation Analysis

As indicated earlier, the exquisite sensitivity of radiochemistry lends itself to the development of Radioanalytical techniques. One of the most important of these is *activation analysis*. Activation analysis is an elemental analysis technique that allows one to determine the amount of an element X contained in some material Y. The basic steps in activation analysis are: (a) Irradiate Y with ionizing radiation, so that X will change into a radioactive isotope of X, X*. (b) Using chemical or instrumental techniques, "isolate" X and X* from all other elements in Y (not necessarily quantitatively) and measure the activity of X*. "Isolation" can involve a chemical separation, followed by counting X* or use of a technique like γ-ray spectroscopy that allows the detection of γ-rays associated with the decay of X* without any intervening chemistry. (c) Calculate the amount of X present. This last step (c) is usually done by irradiating and counting a known amount of X under the same conditions used for Y. Then the mass of X in Y = (known mass of X)(activity of X* in Y)/activity of X* in known).

The most common choice for the irradiation of samples is to irradiate them with reactor thermal neutrons. In this case, the activation reaction is $^{A}X(n,\gamma)^{A+1}X$. This choice is motivated by the high capture cross sections for thermal neutrons, the copious fluxes of neutrons available in reactors ($> 10^{12}$ n/cm^{2}sec), the ease with which neutrons penetrate matter allowing analysis of rocks and other solids and the small activation of C, N, and O, major elements in biological samples, by these neutrons. Activation analysis using neutrons is called NAA. If the "isolation" method involves a radiochemical separation, it is termed RNAA, while instrumental "isolation" of the activity of interest is INAA. Other activating particles include photons, fast (~ 14 MeV) neutrons and charged particles, which are used primarily for surface analysis.

Typical detection sensitivities for NAA are ~ ppb with some elements being detected at parts-per-trillion levels. NAA is a multi-elemental analysis technique, allowing the simultaneous determination of 30-40 elements in most samples. It is a non-destructive technique, a fact of some importance for analysis of very valuable specimens such as lunar samples, forensic samples, art artifacts, etc. The drawbacks of NAA include the need to handle radioactivity, the availability of nuclear reactors for irradiations, the long times and high labor costs frequently associated with assaying longer-lived ($t_{1/2}$ > 1 day) radionuclides, and the lack of chemical speciation information. While activation analysis continues to be an important tool for the analysis of geological samples, some biological materials and environmental samples, and in art and archeology, it is being replaced in many applications by ICP-MS. This latter technique has similar or better sensitivities than NAA, and can be used to get chemical speciation information and is generally more cost effective.

The chapter by Landsberger in this book contains more information about this subject.

PIXE

PIXE (particle induced x-ray emission) is an elemental analysis technique based upon the bombardment of thin samples with charged particles, usually 1-3 MeV protons. The charged particles interact with the inner shell electrons of the atoms causing ionization and subsequent x-rays that follow the rearrangements of the atomic electrons. The x-ray energies are characteristic of the atomic number Z of the ionized atom and the number of emitted x-rays can be used to determine the number of atoms present. Either absolute or comparative measurements can be used for quantitative determinations. Because of the limited penetrating power of both the exciting protons and the emitted x-rays, PIXE samples should be thin. Sensitivities for PIXE analysis are typically at the ppm level and are good for elements with Z=20-90. PIXE is a non-destructive, multi-elemental, rapid analysis technique that has found widespread use in measuring elemental abundances in environmental samples such as filters loaded with particulate matter. Because the detected radiation is the result of atomic rather than nuclear transitions, PIXE is really a non-nuclear technique, but because of the use of nuclear particle accelerators and x-ray detectors, it is usually classified as a radioanalytical method. The chapters by Robertson in this book describe applications of PIXE.

Radioimmunoassay (RIA)

There are a large number of nuclear analytical techniques based upon the use of radiotracers. One technique that is most widely used and utilizes immensely radiotracer is *radioimmunoassay* (RIA). RIA is a sensitive quantitative method for determining trace amounts of biological molecules based upon their ability to displace a radioactively labeled form of the molecule from combination with its antibody. A known amount of the biomolecule, such as a hormone, is labeled with a radiotracer and mixed with a sample containing an unknown, unlabeled amount of that hormone and a limited amount of its antibody. The radiolabeled hormone and the unlabeled hormone compete for binding sites on the antibody. If the antigen(hormone)-antibody complex is isolated and its activity determined, the amount of unlabeled hormone can be determined using standard calibration curves. This technique can be used to detect concentrations of hormones, drugs, vitamins enzymes, viruses, serum proteins, etc. at levels as low as 10^{-10} mol/L.

Nuclear Medicine

Perhaps the most important and most rapidly developing application of radiochemistry today is in nuclear medicine. The use of radionuclides in medicine can be for either diagnosis or therapy. An estimated 10 to 12 million nuclear medicine imaging and therapeutic procedures are performed each year in the U.S and over 100 million laboratory test using radionuclides are performed. Therapeutic procedures may involve direct irradiation of target organs using x-ray machines, particle accelerators, etc. and *in-situ* irradiation by physically implanted radioactive material or chemically administered radionuclides like ^{131}I. Of the diagnostic procedures, over 90% use either ^{99}Tc^m or one of the iodine isotopes (^{123,125,131}I). Common nuclear medicine diagnostic applications include diagnosis of hyperthyroidism (Grave's Disease), cardiac stress tests to analyze heart function, bone scans for detecting tumors, etc. The radionuclides used in these procedures are generated using reactors (^{131}I), radionuclide generators (^{99}Tc^m), medical (in hospital) cyclotrons (the PET nuclides ^{11}C and ^{18}F), or ordinary cyclotrons (^{123,125}I, ^{201}Tl).

^{99}Tc^m is a favored radionuclide for diagnosis because it only emits 141 keV γ-rays, which are very convenient for imaging. ^{99}Tc^m ($t_{1/2}$=6.0 h) is produced using radionuclide generators or "cows". The nuclide ^{99}Mo is produced in reactors using the ^{98}Mo(n,γ) reaction or fission. The ^{99}Mo is isolated, oxidized to MoO$_4^{2-}$, loaded on an alumina column to which it binds. The ^{99}Mo decays to ^{99}Tc^m, producing pertechnetate ions, TcO$_4^-$. The TcO$_4^-$ is eluted off the column with 0.9% NaCl. The column is a "generator" or "cow" in that the ^{99}Mo is continually decaying to ^{99}Tc^m, replacing the removed species and the column can be "milked" repeatedly to produce the ^{99}Tc^m.

Another important application of radiochemistry in nuclear medicine is in PET (positron emission tomography). Positron emitters emit β$^+$ particles which annihilate in matter producing two 0.511 MeV photons that travel in opposite directions from the point of annihilation. If both of these γ-rays are detected in coincidence, one can locate the point of emission. The two most important PET nuclides are ^{11}C ($t_{1/2}$ = 20.3 m) and ^{18}F ($t_{1/2}$ = 1.83 h). These nuclides can be attached to glucose molecules and used for real time imaging of organs like the brain that take up glucose.

Physics and Chemistry of the Heaviest Elements

The actinides and transactinides are all radioactive and thus studies of their physics and chemistry will necessarily involve radiochemistry. Among the active areas of research are the study of the fate, transport and speciation of the actinides in the environment, the physics and chemistry associated with nuclear waste management and site remediation, stockpile stewardship, the fundamental chemistry and physics of the actinide elements (thermodynamics, crystal structure, etc.) and the characterization of the basic chemical and physical properties of the transactinides.

This latter area dealing with the properties of the transactinides is among the most challenging (and high profile) aspects of radiochemistry. These elements have short half-lives and the typical production rates are about one atom/experiment. The experiments must be carried out hundreds of times and the results summed to produce statistically meaningful results.

The elements Lr - 112 are expected (non-relativistically) to be d-block elements because they are expected to involve the filling of the 6d orbital. However relativistic calculations have shown that rutherfordium prefers a 6d 7p electron configuration rather than the $6d^2$ configuration expected non-relativistically and predicted by a simple extrapolation of periodic table trends. This prediction also implies that $RfCl_4$ should be more covalently bonded than its homologue $HfCl_4$ and $ZrCl_4$. In particular, the calculations show $RfCl_4$ to be more volatile than $HfCl_4$ which is more volatile than $ZrCl_4$ with bond dissociation energies in the order $RfCl_4 > ZrCl_4 > HfCl_4$. (The periodic table extrapolations would predict the volatility sequence $ZrCl_4 > HfCl_4 > RfCl_4$).

The first aqueous chemistry of rutherfordium showed that it eluted from liquid chromatography columns as a 4 + ion consistent with its position in the periodic table as a d-block element rather than a trivalent actinide. Gas chromatography of the rutherfordium halides has shown the volatility sequence $ZrCl_4 > RfCl_4 > HfCl_4$ with a similar sequence for the tetra bromides. Thus rutherfordium does not follow the expected periodic table trend nor is its behavior in accord with relativistic calculations.

The aqueous chemistry of hahnium has also shown unexpected trends. Hahnium does not behave like its homologue tantalum in aqueous solutions but similar to niobium or the pseudo-group 5 element, protactinium, under certain conditions. Gas phase thermochromatography of $NbBr_5$, $TaBr_5$, and $HaBr_5$

shows $NbBr_5$ and $TaBr_5$ to behave similarly while $HaBr_5$ is less volatile. Just the opposite trend was predicted by relativistic calculations.

Thus the chemistry of hahnium and rutherfordium deviates significantly from periodic table trends, a fact that is partly explained by relativistic calculations.

The study of the chemistry of seaborgium (*14*) is remarkable for its technical difficulty as well as the insight offered. In an experiment carried out over a two-year period, 15 atoms of seaborgium were identified in a thermal chromatography experiment. From this experiment, one concluded that the volatility sequence $MoO_2Cl_2 > WO_2Cl_2 > SgO_2Cl_2$ was followed. This observation agreed with both the extrapolations of periodic table trends and relativistic calculations. In an aqueous chemistry experiment, three atoms of seaborgium were detected, showing seaborgium to have a hexavalent character expected of a Group 6 element. The most stable oxidation state of seaborgium is +6 and like its homologues molybdenum and tungsten, seaborgium forms neutral or anionic oxo - or oxohalide compounds.

More recently, bohrium (element 107) has been characterized (*15*), in a thermochromatography experiment involving the oxychloride. BhO_3Cl was found to be less volatile than the oxychlorides of the lighter members of group VII, in agreement with relativistic calculations. Hassium (element 108) has also been characterized (*16*) in the form of its tetraoxide, HsO_4, in a thermochromatography experiment. HsO_4 appears to be less volatile than its group VIII congener OsO_4.

Radiotracers

The basic idea behind the use of radiotracers is that all the isotopes of a given element will behave the same chemically. Thus the atoms of radioactive ^{24}Na will behave in the same way as those of stable ^{23}Na in a chemical system. So what? The point is that it is easier to follow the radioactive ^{24}Na atoms than it is to detect the non-radioactive ^{23}Na atoms, due to the exquisite sensitivity associated with the detection of radioactivity. The other unique advantages of radiotracer experiments include their simplicity and small expense (compared to competing technologies such as mass spectrometry). In a well-designed experiment, the presence of radiotracers does not affect the system under study and any analysis is non-destructive. Interference from other species that may be present is not important (as compared to conventional methods of analysis where interferences may thwart the analysis.) Perhaps the most outstanding advantage of the use of radioisotopes is the opportunity offered to trace dynamic mechanisms. Such biological phenomena as ion transport across cell membranes, turnover, intermediary metabolism, or translocation in plants could, before the advent of radiotracer methods, be approached only indirectly. Medical imaging with radiotracers is another example of the ability to study dynamic processes.

There are literally countless applications of radiotracers that have been summarized in elementary textbooks. (*2,4,5*). Radiotracers have important applications in the biological, environmental and physical sciences along with many industrial applications. In many physical applications, radiotracers are used to track the large-scale motion of fluids, as in ground water, estuaries or pollutant emissions. (see the chapter by Eaton et al.) (A useful variant on the use of radiotracers for uncontrolled environmental releases is the use of *stable isotope tracers*, where a stable material is used as the tracer and then made radioactive in the lab to enhance detection). In chemistry, radiotracers have found widespread use in the study of chemical reaction mechanisms and the testing of separation procedures.

References

1. *The Random House College Dictionary, Revised Edition*; Random House: New York, NY, 1984.
2. Wang, C.H.; Willis, D.; Loveland, W. *Radiotracer Methodology in the Biological, Environmental and Physical Science;* Prentice-Hall: Englewood Cliffs, NJ, 1975.
3. Loveland, W; Morrissey, D.J.; Seaborg, G.T. *Modern Nuclear Chemistry*; Wiley: New York, NY, 2002.
4. Friedlander, G.; Kennedy, J.W.; Macias, E.S.; Miller, J.M. *Nuclear and Radiochemistry;* Wiley: New Orkney, 1981.
5. Choppin, G.R.; Liljenzin, J.O.; Rydberg, J. *Radiochemistry and Nuclear Chemistry; 3rd Edition;* Butterworth-Heinemann: Woburn, MA, 2001.
6. Adloff, J.P.; Guillaumont, R. *Fundamentals of Radiochemistry;* CRC Press: Boca Raton, FL, 1993.
7. Lieser, K *Nuclear and Radiochemistry: Fundamentals and Applications;* VCH: New York, NY, 1997.
8. Ehmann, W.D.; Vance, D.E. *Radiochemistry and Nuclear Methods of Analysis;* Wiley: New York, NY, 1991.
9. *Metal Ion Separations and Preconcentration;* Bond, A.H.; Dietz, M.L.; Rogers, R.D., Eds.; ACS Symposium Series 716; ACS: Washington, DC, 1999.
10. Kraus, K.; Nelson, D. Paper 837, Geneva Conference, Vol. 7. (1956).
11. Herrman, G.; Trautman, N. *Annu. Rev. Nucl. Part. Sci.* **1982**, *32*, 117-122.
12. Meyer, R.A.; Henry, E.A. *Proc. Workshop Nucl. Spectrosc. Fission Products, Grenoble1979;* Bristol, London, 1979; pp 59-103.
13. Zvara, I., *et al. Sov. Radiochemistry* **1972**, *14*, 115-117.
14. Schädel, M., *et al. Nature* **1997**, *388*, 55-57.
15. Eichler, R., *et al. Nature* **2000**, *407*, 63-65.
16. Düllmann, Ch. E., *et al. Nature* **2002**, *418*, 859-862.

Chapter 4

Interface of Environmental, Bioassay, and Radioanalytical Standard Reference Materials and Traceability Evaluations

Kenneth G. W. Inn[1], Zhichao Lin[2], Michael Schultz[3], Zhongyu Wu[2], Ciara McMahon[4], Iisa Outola[1], Hiromu Kurosaki[1], Svetlana Nour[1], Linda Selvig[1], Lisa Karam[1], and J. M. Robin Hutchinson[1]

[1]Ionizing Radiation Division, National Institute of Science and Technology, Gaithersburg, MD 20899-8462
[2]WEAC, Food and Drug Administration, Winchester, MA 01890
[3]Ametek, 801 South Illinois Avenue, Oak Ridge, TN 37831
[4]Radiological Protection Institute of Ireland, 3 Clonskeagh Square, Clonskeagh Road, Dublin 14, Ireland

Radiochemistry is the cornerstone for certifying environmental natural-matrix Standard Reference Materials (SRMs), and for verifying environmental and radiobioassay test samples for the National Institute of Standards and Technology (NIST) traceability evaluation programs. Natural-matrix SRMs provide the basis for radiochemical methods validation, demonstration of traceability for environmental radionuclide measurements, and the basis for measurement comparison over time within and among laboratories. The certifications of eight natural-matrix radionuclide SRMs are conducted through the efforts of intercomparisons of experienced laboratories from around the world. While the SRM program has been successful in certifying many radionuclides in environmental matrices, there have also been several unresolved radiochemical issues left to future efforts. The evaluation programs provide the program participants with a direct and unambiguous evaluation of the traceability of their radioanalytical capabilities by NIST. Because the participants in the evaluation programs depend on absolute confidence in the results of the evaluation to assess their radioanalytical methods for bias from the national physical standard, Bq, NIST must verify the production of the test samples to approximately 1 percent (relative combined standard uncertainty) with radiochemical measurements that depend on extraordinary laboratory practices.

Introduction

Environmental radiochemistry has been an ongoing sub-discipline that has found applications in understanding the dispersion of fallout in the atmosphere, on and below the land surface, and through the water systems on land and in the oceans. Additionally, radiochemical studies have been applied to dating ages of meteorites, characterizing glacial and land erosion, subduction of Earth-surface plates, and elucidating the formation of minerals from molten masses and the diagenesis of deep-sea sediments. More recently, environmental radiochemistry has also been applied to the characterization, remediation, and long-term monitoring of cleaned up nuclear weapon production-related sites and waste storage. On the other hand, radiobioassay applications of radiochemistry have focused on health and safety protection of occupationally exposed employees working on the production of nuclear weapons. In this case, non-invasive whole-body counting of occupational workers and selected sectors of the general public are conducted; radiochemical measurements are made on urine, feces, nasal swipes, tissue, and blood samples for indirect evaluations of internal contamination; and autopsy radiochemical evaluations are conducted on partial and whole-body donors for reevaluation of bio-kinetic models and reassessment of previous whole-body counting and in-vitro measurements histories.

Over the past 20 years, NIST has incorporated radiochemical analysis at the interface of environmental, radiobioassay, and radioanalytical traceability evaluations to provide the quality and metrology infrastructure upon which the environmental and radiobioassay communities rely. NIST's efforts result in the certification of low-level radionuclide natural-matrix SRMs for the validation of radiochemical methodologies, and also serve as the basis for data comparability over time and between laboratories. Additionally, NIST provides environmental and radiobioassay measurement evaluation programs that demonstrate participating laboratories' traceability to the international unit of radioactivity, the Bq.

Because the massic activity of radionuclides in the environment or bioassay samples is usually low, and the matrices involved are complex, certification of environmental radionuclide SRMs and traceability evaluation samples require extraordinary approaches. The challenge to NIST is to extend and demand excellence from its radiochemical metrology expertise to provide the quality radiochemical measurements that are expected for its programs.

Natural-Matrix Radionuclide SRMs

In 1977 the Low-level Working Group of the International Committee for Radionuclide Metrology met in Woods Hole, MA to discuss Standard Reference Materials for environmental and radiobioassay measurements. One of the major needs identified by the meeting was for natural-matrix SRMs. These materials are ones in which naturally occurring or anthropogenic radionuclide species are naturally incorporated into the matrix (*1, 2*). The National Bureau of Standards (now NIST) launched a program to develop the requested suite of benchmark natural-matrix radionuclide SRMs for measurements along the major pathways from environmental radioactive sources to man, and within man (see Table 1). At least a 10-year supply of each SRM is developed as long-term bases for comparisons and quality control. These SRMs are used by laboratories to: a) develop and test radiochemical procedures; b) calibrate instruments; c) conduct internal quality assessments; d) compare laboratories; e) evaluate the competency of analysts and laboratories; and f) serve as a basis for measurement comparability over time and between laboratories. To date, eight SRMs have been issued: River Sediment (SRM 4350B), Human Lung (SRM 4351), Human Liver (SRM 4352), Rocky Flats Soil - I (SRM 4353), Freshwater Lake Sediment (SRM 4354), Peruvian Soil (SRM 4355), Ashed Bone (SRM 4356) and Ocean Sediment (SRM 4357) (*3-13*). NIST is currently working on the certification of radionuclides in Shellfish, Seaweed and Rocky Flats Soil - II. Additional matrices such as food and vegetation will be considered for future SRM development.

Each of the SRMs is certified through strictly controlled intercomparisons among a number of experienced national and international laboratories. Each laboratory submits the results of multiple measurements for each reported radionuclide using the methods for which they have the most confidence and experience. The results are carefully screened using radiochemical experience and statistical tools (*13-15*), then certified for massic activity and uncertainty. Participating laboratories are acknowledged in the SRM certificates, and their results are revealed in the certified and uncertified lists and in the open literature. Certificates can be viewed by accessing the NIST SRM webpage: (http//patapsco.nist.gov/srmcatalog/tables/view_table.cfm?table=205-11.htm).

Table 1. NIST Natural-Matrix Radionuclide SRMs

SRM	Matrix	Certified Radionuclides	Massic Activity $(Bq \cdot g^{-1})$
4350B	Columbia River Sediment	^{226}Ra, ^{238}Pu, ^{239}Pu/^{240}Pu, ^{60}Co, ^{137}Cs, ^{152}Eu, ^{154}Eu	10^{-5} to 3×10^{-2}
4351	Human Lung	^{232}Th, ^{234}U, ^{238}U, ^{239}Pu/^{240}Pu	10^{-4} to 10^{-3}
4352	Human Liver	^{238}Pu, ^{239}Pu/^{240}Pu	5×10^{-5} to 2×10^{-3}
4353	Rocky Flats Soil I	^{226}Ra, ^{228}Th, ^{230}Th, ^{232}Th, ^{234}U, ^{238}U, ^{238}Pu, ^{239}Pu/^{240}Pu	10^{-4} to 7×10^{-2}
		^{90}Sr	8×10^{-3}
		^{40}K, ^{137}Cs, ^{228}Ac	10^{-2} to 7×10^{-1}
4353A	Rocky Flats Soil II	In Process	In Process
4354	Freshwater Lake Sediment	^{228}Th, ^{232}Th, ^{235}U, ^{238}U, ^{238}Pu, ^{239}Pu/^{240}Pu	2×10^{-4} to 3×10^{-2}
		^{90}Sr	1
		^{60}Co, ^{137}Cs	6×10^{-2} to 3×10^{-1}
4355	Peruvian Soil	^{228}Th, ^{230}Th, ^{232}Th, ^{239}Pu/^{240}Pu	8×10^{-6} to 4×10^{-2}
		^{60}Co, ^{125}Sb, ^{137}Cs, ^{152}Eu, ^{154}Eu, ^{155}Eu	10^{-5} to 3×10^{-4}
4356	Ashed Bone	^{226}Ra, ^{230}Th, ^{232}Th, ^{234}U, ^{238}U, ^{238}Pu, ^{239}Pu/^{240}Pu, ^{243}Cm/^{244}Cm	10^{-4} to 10^{-2}
		^{90}Sr	4×10^{-2}
4357	Ocean Sediment	^{226}Ra, ^{228}Th, ^{230}Th, ^{232}Th, ^{238}Pu, ^{239}Pu/^{240}Pu	2×10^{-3} to 10^{-2}
		^{90}Sr	4×10^{-3}
		^{40}K, ^{137}Cs, ^{228}Ra	10^{-2} to 2×10^{-1}
4358	Shellfish	In Process	In Process
4359	Seaweed	In Process	In Process

Table 2: Natural Matrix Radionuclide SRM Measurement Discrepancies

Nuclide	Matrix	Discrepancy	Cause
^{90}Sr	Rocky Flats Soil –I Columbia River Sediment	300% Range	Not Resolved
^{137}Cs	Peruvian Soil	30%	^{87}Rb contamination
^{210}Pb	Freshwater Lake Sediment	10%	Not Resolved
	Rocky Flats Soil – I	20%	Sample Dissolution
	Columbia River Sediment	20%	Sample Dissolution
^{232}Th	Rocky Flats Soil –I	20%	Sample Dissolution
	Human Lung	200% Range	Blank and Heterogeneity
^{234}U	Human Lung and Liver	200% Range	Heterogeneity
	Freshwater Lake Sediment	15%	Not Resolved
^{238}U	Rocky Flats Soil – I	20%	Sample Dissolution
	Columbia River Sediment	20%	Sample Dissolution
	Ocean Sediment	20%	Sample Dissolution
^{238}Pu	Human Lung	1000% Range	Heterogeneity
^{239}Pu/^{240}Pu	Rocky Flats Soil - I	10%	Sample Dissolution
^{241}Am	Human Lung	50%	Not Resolved

Occasionally, radionuclides cannot be certified because of discrepancies among the reporting laboratories that can not be resolved (Table 2). Many times, such discrepancies are not resolved because of technical or resource limitations. An example of an unresolved interlaboratory discrepancy was ^{90}Sr in SRM 4353 (Rocky Flats Soil - I) and SRM 4350B (Columbia River Sediment) where the range of reported values was 300 percent. Another example of an unresolved interlaboratory discrepancy was for ^{210}Pb in SRM 4354 (Freshwater Lake Sediment) where the difference between gamma-ray spectroscopy and radiochemical beta counting methods differed by 10 percent. Likely causes for interlaboratory discrepancies include: a) incomplete sample dissolution of all of the radionuclide species, b) interferences; and c) insufficient measurement statistics.

Frequently, laboratory discrepancies are resolved and the radionuclide certified. For example, the results for the ^{137}Cs in the Peruvian Soil were resolved when it was learned that the ^{87}Rb carried in radiochemical separations made a significant contribution to the low-level gas-flow proportional counting because the radio-cesium content in this material was extremely low. Another example was a 10 percent discrepancy between acid leaching and total dissolution of the plutonium in SRM 4353 (Rocky Flats Soil - I) which was discovered to be due to refractory phases of plutonium in the sample that was not assessable without considerable quantities of fluoride ions in the sample dissolution step. NIST continues to work on resolving the discrepancies as the opportunity and resources arise, and as measurement technology improves.

Traceability Evaluations

As NIST's radiometrology capabilities grew, it was approached by communities that were interested in establishing direct assessment by NIST through traceability evaluations (measurement assurance programs or MAPs). Three major aspects addressed by the MAPs are a) the radionuclide concentration, b) the uncertainty of the measurements being routinely conducted, and c) the complexity of the evaluation matrix. Examples of these three aspects for various NIST radionuclide MAPs are described in Table 3. The applications for radiochemistry become dominant as the radionuclide concentration becomes lower and the matrix becomes more complex, i.e., for radiobioassay and environmental applications.

NIST has adopted for its use the definition of traceability by the International Vocabulary of Basic and General Terms in Metrology: "property of the result of a measurement or the value of a standard whereby it can be related to stated references, usually national or international standards, through an unbroken chain of comparisons all having stated uncertainties." (ISO VIM, 2nd ed., 1993, definition 6.10). The key phrases, as interpreted by the NIST Ionizing Radiation Division (IRD) are: comparison with national or international standards; unbroken chain of comparisons; and stated uncertainties. In addition, the IRD: collaborates with its constituent communities to develop acceptance criteria on performance evaluation (test) of measurement comparisons, e.g., ANSI N42.23, ANSI N42.22, ANSI N13.30, and DOE/EM National Analytical Management Program (NAMP) Acceptance Criteria (*16-20*); suggests that comparisons be conducted on a periodic basis; strives to assure that the evaluations provide appropriate evaluations; and expects that the constituent communities operate under a quality system, such as ISO 17025 (21). The IRD finds that while many may claim traceability to NIST, the most direct and clear evidence of measurement traceability is through performance evaluations where "blind" NIST traceable samples are sent to a participating laboratory for measurement, and the returned results are compared with the NIST values. Alternatively, the participating laboratory could certify its own material and send it along with its certificate to NIST for measurement comparison.

NIST currently conducts a traceability evaluation program for the environmental community (DOE/EM NAMP) where the two DOE reference laboratories are evaluated, and with the *in vitro* radiobioassay community through the NIST Radiochemistry Intercomparison Program (NRIP). The NAMP program is focused on the analysis of ^{90}Sr, plutonium, americium and uranium isotopes, and gamma-ray emitting radionuclides in soil, water and air filters at 0.1 to 40000 Bq per sample. The NAMP Radiological Traceability Program (RTP) acceptance criteria require that their Reference Laboratory results on samples prepared by NIST show less than an eight to 12 percent difference from the NIST value, depending on matrix and radionuclide. And, for samples prepared by the NAMP reference laboratories and sent to NIST for measurement, the RTP acceptance criterion is defined by ANSI N42.22 where the difference from the NIST value should be less than or equal to three times the combined standard uncertainties of the reference laboratory and NIST. The NRIP program offers ^{90}Sr, and plutonium, americium and uranium isotopes in air filters, synthetic feces, soil, water and synthetic urine at 0.03 to 0.3 Bq per sample. For NRIP, the acceptance criterion for results from samples sent by NIST to the reporting laboratory is defined by the ANSI N42.22 guidance. In addition, for synthetic urine and fecal samples, ANSI N13.30 provides guidance criteria for bias (-25 to 50 percent difference from the NIST value), and for precision of the bias ($< \pm 40$ percent).

For both the NAMP and NRIP programs, the process of developing and verifying the NIST evaluation samples have been previously described by Wu, et al. (22). The challenges that NIST must address are: 1) production of appropriate test samples with certified values that are traceable to the NIST SRMs from which they were derived, and with sufficiently small uncertainty to not affect the results of the evaluations, and 2) verification measurements of the dilution factors between the SRMs and the final test samples of sufficiently small uncertainty to confidently conclude that the dilutions were executed accurately. The key to meeting these challenges lies in the quality of the radiochemical verification measurements of the master solutions that are spiked into the individual samples. The NIST target criteria for adequate verification of the dilution factors are $< \pm 1$ percent agreement between the measured and gravimetric values, with a $< \pm 1$ percent relative combined standard uncertainty for the verification measurements. The adequacy of these criteria was assessed and confirmed by McCurdy, et. al. (*23*). To meet these strict criteria, a number of replicate samples of the master solutions are assayed for each target radionuclide, along with reagent blank samples, under Class 100 hoods, scrupulously clean reagents and disposable lab ware are used. Multiple and redundant precipitation, and anion or extraction chromatographic separation steps are employed in the radiochemical schemes. Additionally, to assure appropriate statistics, counting times are prolonged using well shielded low-background alpha spectroscopy, proportional counting and HPGe gamma-ray spectroscopy detectors while individual spectra are carefully evaluated for effects from impurities and peak overlap, and Evaluated Nuclear Structure Data File reference values are used as the source for decay probabilities, half lives and their associated uncertainties.

Table 3. NIST radionuclide measurement assessment (traceability evaluation) program characteristics

MAP	Relative Standard Uncertainty (u_c, %)	Massic Activity	Matrix
Commercial Standards	0.1 - 1	Bq - kBq	Solutions and Gases
Radiopharmaceuticals	0.1 - 1	kBq	Solutions
Nuclear Power Plant	0.5 - 2	Bq - kBq	Solution, air filter, sand
Military Applications	0.5 - 2	Bq - kBq	Large Area Sources
Environmental Remediation	< 30	mBq - Bq	Air filters, soil and water
Radiobioassay	< 40	mBq - Bq	Thorax and whole-body counting Urine and Feces

All of these precautions are taken only to verify the traceability of the master solutions to the starting SRMs. When the verification results do not meet the target acceptance criteria, they are evaluated for fitness-for-use in our programs, and, if judged sufficient, the samples are still used for the traceability evaluation program. An example of the results we have obtained for the verification measurements on the 1999 master solution is displayed in Table 4. The next step is verification, with relative wide-window NaI measurements, that the master solutions were correctly added to the individual samples. Each sample is counted for sufficient time to accumulate more than 100,000 counts between 50 and 1400 keV. The integrated spectrum is then divided by the mass of the master solution that was spiked into the sample, and plotted on a normal probability curve to see if the data come from one population. "Outliers" are recounted to verify that they are different from the rest of the sample population. To date, we have never confirmed an "outlier," and suspicious samples are never sent to evaluate a participating laboratory.

The results for both the NAMP radiological traceability program and the NRIP traceability evaluation programs for production laboratories have successfully demonstrated radiochemical capabilities of both the participating laboratories and NIST's radiochemical capabilities. To date, the performances of the NAMP reference laboratories are well within the program's acceptance criteria. Likewise, when averaged over all matrices for all participating NRIP laboratories, the mean values for each analyte nuclide were very close to the NIST values, and the relative standard deviations of the results were on the order of $\pm$ 14%. Results for ^{238}Pu, ^{239}Pu, and ^{241}Am are shown in Table 5 and provide an estimate of the state-of-the-art at production radiochemistry laboratories. Apparent mean values and standard deviations for the data sets for each of the nuclides are found in the second and third columns. Normal probability plots for these three nuclide data sets, however, indicate "outliers" which skew the data distributions. This "skewness" of the data sets is not surprising because the data were generated by the different analytical methods being used by the participating laboratories. When the "outliers" were removed from the data sets, the relative standard deviations in column five decreased significantly for the plutonium isotopes, indicating a potential improvement in laboratory traceability when analytical methods gain better control. On the other hand, removing the "outliers" for the ^{241}Am data set did not improve the distribution's standard deviation significantly, indicating that the current analytical capabilities for this nuclide may be more difficult to improve than for plutonium.

Conclusions

The NIST Low-level Radiochemistry Project has developed many radiochemical measurement applications to provide SRMs and evaluation programs to the environmental and radiobioassay communities for validating their measurement methodologies, supporting the accuracy of their measurement traceability to the Bq, and establishing a basis for data comparability over time and between laboratories.

The high quality radiochemical measurements by NIST take advantage of the Institute's unique metrology situation. The work is done in a non-production environment, the majority of the radiochemical manipulations are carried out under class 100 air to minimize sample contamination, most lab ware are used once then discarded to minimize sample contamination, bulk supplies of chemicals are selected from single lots for each project to minimize reagent blank variability, relatively long counting times are used to minimize the uncertainty from measurement statistics, the responsibility of most projects is assigned to a single analyst who would execute allphases of the project, and, most importantly, the responsible analyst has a wide network of very experienced colleagues from whom advice is sought to develop carefully laid out plans before the project is executed.

Table 4: 1999 master solution verification measurement results

Nuclide	Massic Activity (Bq/g)	Number of Replicate Measurements	Difference from Gravimetric Value (%)	Relative Standard Uncertainty (u_c, %)
^{241}Am	3.8665	3	-0.63	0.6
^{238}Pu	3.9001	3	0.57	1.4
^{234}U	3.7746	3	0.12	1.2
^{238}U	3.9187	3	0.12	1.2
^{90}Sr	169.34	3	0.98	0.8
^{137}Cs	1633.6	4	1.0	1.1
^{60}Co	1567.9	4	0.50	0.8

It is our experience that both classical and recently developed radiochemistry techniques have proven themselves reliable and sensitive when used appropriately and within their limitations, and will have continued and important uses for the environmental and radiobioassay communities well into the future, particularly with the new emphasis on Homeland Security (interdiction, consequence management, attribution, remediation and waste management). It is most important, however, that the analyst has the experience and expertise that lead to sound judgment for selecting the best radiochemical methods for the analytical situation.

Table 5: Performance of Laboratories Participating in NRIP

Nuclide	Apparent Mean Difference from NIST Value (%)	Apparent Relative Standard Deviation (1s, %)	Mean Difference from NIST Value without "Outliers" (%)	Relative Standard Deviation without "Outliers" (1s, %)
^{238}Pu	-2.0	18.6	-2.7	4.8
^{239}Pu	2.3	12.0	-1.7	4.3
^{241}Am	-1.6	12.2	-0.1	11.1

References

(1) Hutchinson, J.M.; Mann, W.B., Eds., *Env. Int.* **1978**, *1*, 7-116.
(2) Hutchinson, J.M.R. *Env. Int.* **1978**, *1*, 11-14.
(3) H.L. Volchok, H.L.; et. al. *Env. Int* **1980**, *3*, 395-398.
(4) Inn, K.G.W.; Noyce, J.R. *NBS SP 609* **1982**, 117-127.
(5) Inn, K.G.W.; et. al., *Env. Int* **1984**, *10*, 91-97.
(6) Liggett, W.S.; et. al., *Env. Int* **1984**, *10*, 143-151.
(7) Inn, K.G.W. *J. Radioanaly. Nucl. Chem.* **1987**, *115*, 91-112.
(8) Inn, K.G.W.; et. al. *J. Radioanaly. Nucl. Chem.* **1990**, *138*, 219-229.
(9) Krey, P.W.; et. al. *J. Radioanal. Nucl. Chem.* **1994**, *177*, 5-18.
(10) Liggett, W.S.; Inn, K.G.W. In *Environmental Sampling*, Keith, Larry H., Ed., American Chemical Society: Washington, D.C. **1996**, Second Edition, 185-202.
(11) Inn, K.G.W.; et. al. *Int. J. Appl. Rad. Iso.* **1996**, *47*, 967-970.
(12) Lin., Z.; et. al. *Appl. Radiat. Isot.* **1998**, *49*, 1301-1306.
(13) Inn, K.G.W.; et. al. *J. Radioanaly. Nucl. Chem.* **2001**, *248*, 227-231.
(14) Inn, K.G.W.; et. al. *Nucl. Inst. Meth. in Physics Res.* **1984**, *223*, 443-450.
(15) Lin, Z.; Inn, K.G.W.; Filliben, J.J. *J. Radioanaly. Nucl. Chem.* **2001**, *248*, 163-173.
(16) *American National Standard Measurement and Associated Instrument Quality Assurance for Radioassay Laboratories*; ANSI N42.23-1996, Institute of Electrical and Electronics Engineers, Inc., New York, NY, 1996.
(17) *American National Standard – Traceability of Radioactive Sources to the National Institute of Standards and Technology (NIST) and Associated Instrument Quality Control*; ANSI N42.22-1995, Institute of Electrical and Electronics Engineers, Inc., New York, 1995.

(18) *Performance Criteria for Radiobioassay*; ANSI N13.30-1996, Health Physics Society, McLean, VA, 1996.
(19) *National Analytical Management Program - Radiological Traceability Program*, U.S. Dept. Energy, Idaho Falls, ID, April **2002**.
(20) Morton, J.S.; et. al. *J. Radioanaly. Chem.* **2001**, *248*, 175-177.
(21) *General Requirements for the Competence of Testing and Calibration Laboratories*; International Standards Organization/International Electrotechnical Commission (ISO/IEC) 17025, Geneva, Switzerland, **1999**.
(22) Wu, Z.; et. al. *J. Radioanaly. Chem.* **2001**, *248*, 155-161.
(23) McCurdy, D.E; et. al. *J. Radioanaly. Chem.* **2001**, *248*, 185-190.

Developments in Radioanalytical Methods

Radiation Detection Methods

Chapter 5

Realization and Applications of Collimator-Less Gamma-Ray Imaging Systems

K. Vetter

Glenn T. Seaborg Institute, Lawrence Livermore National Laboratory, 7000 East Avenue, Livermore, CA 94550

Due to advances in manufacturing large and highly segmented semi-conductor detectors, it is now possible to build efficient and compact gamma-ray imaging systems for gamma-ray energies up to several MeV. These new devices allow one to obtain positions and energies of individual gamma-ray interactions and enable the use of "Compton" tracking algorithms not only to determine the scattering sequence of the interactions, but also to determine the origin of the incident gamma rays. The main gain in this approach for localizing and characterizing gamma-ray sources is the increased sensitivity due to the removal of the collimator, which allows to significantly increase the efficiency while maintaining the spatial resolution even of ultra-high resolution collimators. The basic concept of gamma-ray imaging based on tracking and different possible approaches to optimize the sensitivity for different gamma-ray energies will be presented. Applications in nuclear medicine to enhance cancer diagnostic but in particular therapeutic capabilities, in nuclear nonproliferation as well as in nuclear physics will be addressed.

Introduction

In this first chapter on "Radioanalytical Methods at the Frontier of Interdisciplinary Science: Trends and Recent Achievements", which is based on presentations at the 2002 American Chemistry Society Meeting, we will neither touch upon chemistry to synthesize or extract nor to identify or characterize by means of chemical methods, as it will be done in many of the following chapters. Instead, this chapter on the realization of new concepts of gamma-ray imaging emphasizes the interdisciplinary character of this symposium on radioanalytical methods. Inter- or multidisciplinary research has become a very important aspect in modern science to maximize synergistic effects by combining technologies and concepts from different fields. In this chapter, the multi-disciplinary character of radioanalytical methods is reflected by the combination of advances in physics instrumentation and, for example, the synthesis and characterization of new radiopharmaceuticals by means of chemical methods and biological validation. The combination of this research as it is currently being realized at the Glenn T. Seaborg Institute will potentially allow for earlier cancer detection and for an improved cancer treatment planning.

However, the intention of this paper is to focus – and therefore contrast the main body of this book – on recent advances in the detection of gamma radiation as part of physics instrumentation. Advances in manufacturing three-dimensional position sensitive gamma-ray detectors and in analog as well as digital electronics enable one to track gamma-radiation similar to charge particles in high-energy experiments and to determine the incident direction of gamma rays. This capability allows for the detection of gamma rays without using a collimator, which potentially can increase the sensitivity of gamma-ray imaging significantly. The ability to image and characterize known as well as unknown gamma-ray sources can be utilized not only in nuclear medicine but in a variety of other fields such as national security, e.g. nuclear nonproliferation, in nuclear waste control and monitoring as well as in basic sciences such as nuclear and astrophysics.

Table I shows different categories of applications and corresponding modes of gamma-ray detection. In the "emission" mode, natural radioactive sources can be detected and identified by their characteristic gamma-ray decay. Here, the imaging capability not only enables localization or to measurement of the distribution of gamma-ray sources, but also the enhancement of the signal-to-noise ratio between the object of interest and background radiation, thus extending applications to weak sources or sources embedded in relatively high external radiation fields. The second, "particle-induced emission", mode is applied in situations in which γ-ray emission is induced by external particle beams such as neutrons or protons. Applications can be found in so-called active interrogation procedures to find and identify contraband, such as drugs and explosives, or generally material which is not radioactive or emits only radiation which can easily be shielded. Again, gamma-ray imaging devices can be employed to localize suspicious material in large objects or to increase the signal-to-background ratio in the induced complex radiation field. The third, "transmission", mode is applied as a standard X-ray device which is based on the variation of densities and atomic numbers of objects. With different photon energies the contrast can be tuned to different densities. The last, "scattering", mode uses the detection of scattered incident photons for monitoring the integrity of a material, for following the course of corrosion, observation of fractures, etc.

Table I: Gamma-ray detection modes and their corresponding applications. The number of crosses indicates importance of each mode.

	Emission	Particle-Induced Emission	Transmission	Scattering
Nonproliferation	XXX			
Stockpile Stewardship	XXX	XXX	XXX	X
Waste management	XXX			
Contraband/Sensitive Material Detection		XXX	XX	
Material Integrity and Properties			XXX	XXX
Nuclear Medicine	XXX	X(BNCT)		
Nuclear/Astrophysics	XXX	XXX		

In Table II, nuclear nonproliferation and nuclear medicine are taken as examples to show two different classes of scenarios. A "qualitative" scenario aims at the detection, identification and localization while the "quantitative" scenario aims at the measurement of the precise two- or potentially three-dimensionally distribution of radioactivity.

Recent advances in the manufacture of two-dimensional segmented semiconductor detectors enable the determination of energies and three-dimensional positions of individual gamma-ray interactions in the detector (1). Using tracking algorithms, which are based in the underlying interaction processes such as the Compton scattering and the photoelectric effect, it is possible to determine the time sequence of the interactions (2). Knowing the energies and positions of the first interactions allows the determination of the incident angle of the incoming gamma ray without the use of a collimator which will significantly increase the efficiency a gamma-ray imaging system. However, to realize the full potential of gamma ray tracking and Compton-tracking based gamma-ray imaging, particularly for low gamma-ray energies, the excellent position and energy resolution of a semi-conductor detector is needed.

Table II Examples of applications and imaging goals.

	Search and separate "qualitative"	*Characterize "quantitative"*
Nuclear Nonproliferation	Improve signal-to-background to find nuclear material	Measure distribution of nuclear material
Nuclear Medicine	Identify smaller cancer lesions Early diagnosis	Measure distribution of activity uptake Diagnosis/therapy

In the following, we will briefly discuss conventional – collimator-based - gamma-ray imaging systems. We will illustrate the limitations and efforts to build more sophisticated collimator-based systems, which try to minimize the drawbacks of such systems. In chapter 3, we will extend the concept to image gamma-ray distribution without using a collimator. Chapter 4 summarizes applications in basic science such as nuclear physics, astrophysics as well nuclear medicine, with the focus on early cancer detection or the improvement in cancer treatment planning will be illustrated. While gamma-ray imaging has many more application in nuclear medicine, e.g. to study drug addiction etc., we focus on cancer related applications. The final chapter will also discuss a major effort of the Glenn T. Seaborg Institute to increase not only the gamma-ray imaging capabilities for molecular targeted radiation therapy but to design, develop, and synthesize new radiolabed agents with potentially higher affinity to cancerous tissue than currently available targeting agents.

Collimator-Based Gamma-Ray Imaging Systems

The requirement of a collimator arises from the fact that it is very difficult to focus gamma rays as in optical lens systems. Multilayer technologies have been developed and already utilized in astrophysics that allow focusing of x-rays. This technology will enable focusing of photons up to 30 keV, potentially even up to energies above 100 keV (see for example (*3*)). These systems, if built, will be characterized by excellent angular resolution but very small field-of-view (FOV). Gamma-ray collimators in a large variety have been developed particularly for nuclear medicine applications. Figure 1 illustrates a parallel hole collimator as part of a conventional Anger camera, which is used for planar imaging or single-photon emission computer tomography (SPECT) (see for example (*4,5*)). In addition to the collimator, which can be built in fan-beam or pinhole geometries, such systems consist of large NaI(Tl) scintillator detectors with attached photomultipliers to read out the light which is generated by the interacting gamma ray. The position of the incident gamma ray can be deduced by the light distribution across the photomultipliers.

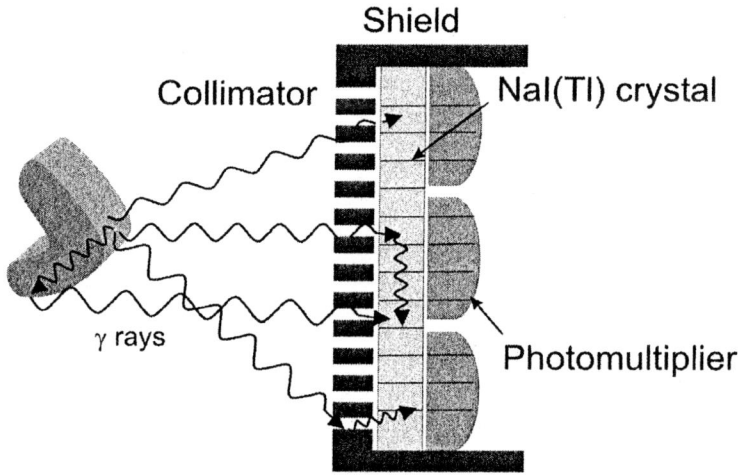

Figure 1: Illustration of a collimator-based gamma-ray detection system. Gamma rays from the object are supposed to penetrate only the holes in the collimator, however, due to the trade-off between thickness and transmission through the collimator, gamma rays penetrate the absorber or can scatter in the absorber, both leading to blurring in the image. In addition, multiple interaction of the gamnma ray in the detector prevents the determination of the proper incident directions.

While conventional devices were built with single large crystals and a few large photomultipliers to obtain the position information, newer and more compact devices are built of many optically insulated scintillation crystals with position sensitive photomultipliers as indicated in Figure 1.

Figure 1 also indicates the limitations and illustrates the trade-offs that characterize a collimator-based gamma-ray imaging system or gamma camera. Increasing the resolution by decreasing the size of the holes reduces the transmission probability and therefore the geometrical efficiency. Also the thickness of a collimator represents a compromise between the probability of a gamma ray penetrating a hole as expected or penetrating the absorber, which is normally built of lead or hevimet. Another problem associated with a collimator is the scattering of gamma rays in the collimator, particularly forward, the scattering process in which most of the energy remains in the gamma ray while the incident direction is changed. Also, scattering processes in the scintillation crystal, which are becoming more likely for higher gamma-ray energies, such as 369keV from ^{131}I, degrade the possible resolution. However, these scattering processes as well as the scattering in the environment or the patient are not related to the collimator. The degradation of scattering processes, e.g. in the patient or in the collimator, can be reduced by measuring the full energy of the emitted gamma ray, assuming the energy is known. Gamma rays which underwent large angle scattering before entering the collimator can be suppressed very efficiently with NaI detectors, which are characterized by an energy resolution of about 10%. Small-angle scattering processes cannot be separated and induce a blurring in the images. However, the new generation of room-temperature semi-conductor detectors, such as CdZnTe, are able to reduce this effect due to their improved energy resolution, as well as conventional semi-conductor detectors built of silicon or germanium (*6,7,8*). Finally, collimator-based systems are not only equipped with heavy collimators at the front, but also heavy absorber surrounding the rest of the detector to prevent gamma rays from entering from elsewhere, which makes the whole instrument heavy and less compact.

In Figure 2, typical efficiency values are plotted for conventional cameras with different parallel-hole dimensions. The efficiencies shown here represent the geometrical efficiencies of the used collimators. The spatial resolution varies between 7 mm and 13 mm, the efficiencies between 6×10^{-4} and 2×10^{-4}, both dependent on the radioisotopes and the corresponding gamma-ray energy used. The trade-off between efficiency and spatial resolution is clearly visible. In addition, the significant drop in sensitivity for higher gamma-ray energies can be seen.

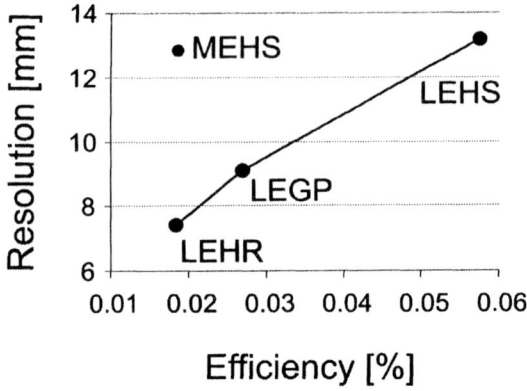

Figure 2: Spatial resolution and geometrical efficiencies for low-energy and high-resolution (LEHR), low-energy and general purpose (LEGP), low-energy and high-sensitivity (LEHS), and medium energy and high sensitivity (MEHS) collimators, which can be found in conventional clinical gamma cameras. Low energy reflects the gamma-ray energy of 140 keV from ^{99m}Tc, medium energy 369 keV from ^{131}I. Both are radioisotopes, that are widely used for cancer diagnosis and treatment (see (9)).

Collimator-Less Gamma-Ray Imaging Systems

The limitations and the collimator-driven trade-off between efficiency and spatial resolution of collimator-based gamma-ray imaging system motivate the question for ways to image gamma rays without the use of collimators. One approach to build an "optical" lens system for gamma rays, as discussed briefly above, is currently under development but will be limited to very small fields-of view. In addition, the focal length and therefore the dimensions of these systems are expected to be substantial in comparison with state-of-the-art systems, particularly for gamma-ray energies above 100 keV (3). However, one of the fundamental interaction processes of gamma rays with matter, the Compton scattering, enables an alternative: based on energy and momentum conservation it is possible to relate the incident angle of a gamma-ray to energies and positions of interactions in a detector without using a collimator. More specifically, as illustrated in figure 3 for two interactions, the energy of the first interaction and the total gamma-ray energy and the positions of the first two interactions allow for the determination of the incident angle. The Compton-scattering formula (1) shows the relation between the incident angle θ and the energies. E_γ represents the incident gamma-ray energy, E_γ' the gamma-ray energy after the first scattering and E_e the deposited gamma-ray energy. The incident angle can only be defined to a cone since the direction of the emerging Compton electron is very difficult to determine. However, observing three gamma rays from the same location enables the measurement of the incident direction, as will be demonstrated below. This concept of the so-called Compton camera is not new and was first discussed in the 70's (*10*) while first experiments have been performed in the 80's (*11,12*) and were refined and extended to different applications in the 90's (*13,14,15*).

$$\cos(\theta) = 1 + \frac{511}{E_\gamma} - \frac{511}{E_\gamma'}; \quad E_\gamma' = E_\gamma - E_e \qquad (1)$$

The potential advantages of a Compton camera or collimator-less imaging system is obvious: (1) no collimator-based trade-off between efficiency and spatial resolution and (2) no scattering in or penetration of the collimator. However, now the question arises: How efficient and to what accuracy can we measure individual gamma-ray interactions? Ultimately, how do expected sensitivities compare with a collimator-based system? We will in the following discuss gamma-ray detectors and concepts, which are able to provide the necessary information, how to translate positions and energies of individual interactions into a gamma-ray path and ultimately, what efficiencies and sensitivities we can anticipate.

Gamma-Ray Tracking

In this chapter, we will discuss concepts to realize collimator-less gamma-ray imaging devices, which are based on the measurement of energies and positions of individual gamma-ray interactions. We will briefly discuss possible detectors and detector configurations and how to obtain the required information. It will be shown how to translate this information into the gamma-ray path by employing gamma-ray tracking.

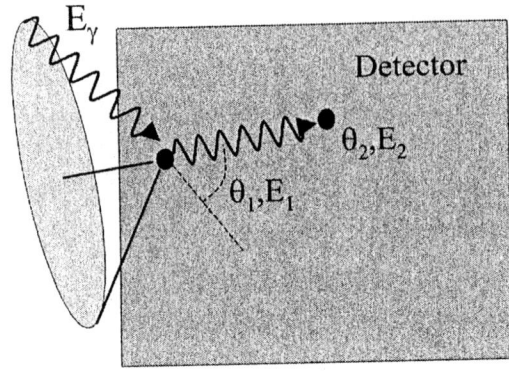

Figure 3: An incident gamma ray interacts at first by Compton scattering and is then fully absorbed by the photoelectrical effect. Based on the energies and the positions of the individual interactions it is possible to determine a cone of allowed incident angles.

Gamma-ray imaging detectors

Recent advances to manufacture highly segmented semi-conductor detectors with sufficiently large dimensions enable to realize efficient collimator-less gamma-ray imaging systems. Particularly, high-purity germanium detectors (HPGe) in planar or coaxial geometry and CdZnTe detectors in planar geometry can now be produced in sufficiently large sizes and fine, two-dimensional segmentation. In a semi-conductor, the electron that is produced via Compton scattering or the photoelectrical effect is directly converted into electron-hole pairs. These charge carriers are driven by the applied electrical field and induce signals on the segmented electrodes. Since the electrodes are segmented orthogonally in two dimensions, the electrode(s) that collect the charge(s) determine(s) two dimensions of one interaction. The two-dimensional segmentation scheme can either be realized in a pixelation of one electrode or in strips on both electrodes, which are arranged orthogonally (double-sided strip detector, or DSSD) (*7,16*). The advantage of the DSSD geometry is the lower number of electrodes and therefore electronics channels which have to be processed. To obtain the third dimension of the gamma-ray interactions, pulse-shape analysis can be applied, since the shape of the measured signals reflects the origin of the charge carriers. The pulse-shapes can even be used to improve the position resolution beyond the segment sizes by analyzing signals from segments, which are adjacent to the charge-collecting electrode. Figure 4 illustrates a possible detector configuration and pulse-shapes, which allows for the determination of the depths of the interactions. The device shown consists of DSSD high-purity silicon (HPSi) or Lithium-drifted silicon (Si(Li)) detectors in the front and an HPGe detector in the back. The two detector types will be separated by a nominal distance of 2-10 cm that will be adjustable to allow some optimization in geometry to accommodate the imaging requirements of different source configurations.

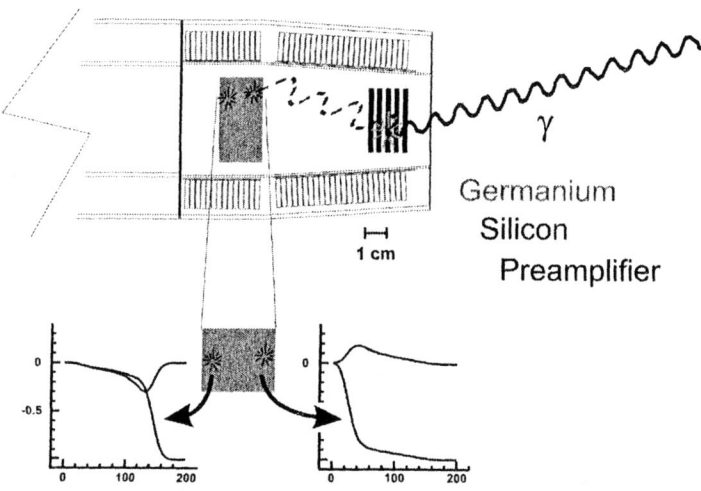

Figure 4: Schematic diagram of the front part of the imaging system in a configuration with 5 planar high-purity silicon detectors followed by one planar high-purity germanium detector. The compact preamplifiers are arranged outside of the cryostat with the semiconductor detectors. On the bottom, signals of charge collecting and adjacent segments are shown. They illustrate the sensitivity to obtain the depth of the interaction in the detector by pulse-shape analysis.

Photons are first incident on the HPSi detectors where, over the energy range of interest (60 keV $\leq E_\gamma \leq$ 1000 keV), the most probable interaction is incoherent Compton scattering. Photons scattering once in the HPSi at forward directions are then incident on the HPGe, detector where low-energy photons will interact most probably by photoelectric interactions and higher- energy photons will have a reasonable probability for either photoelectric interactions or total absorption by multiple interactions. A coincidence requirement will be imposed so that pairs or triples of valid events can be processed. The use of Si detectors in the front of the detector maximizes the probability that the first photon interaction will be due to incoherent scattering and thus, because of the low atomic number of silicon, should make possible tracking of photons with energies as low as about 60 keV. If HPGe were used throughout, the minimum photon energy for which a first Compton interaction is likely with reasonably large probability is at least 200-250 keV. In case of CdZnTe, the corresponding energy is about 400 keV. A second advantage of HPSi is the relatively low mean momentum of the atomic electrons, which minimizes the loss in spatial resolution due to scattering on electrons in motion as compared to the ideal of scattering on electrons at rest. A further reduction of the impact of the momenta of atomic electrons, the so-called Compton profile, is also achieved by the stack arrangement, which restrict scattering to forward angles. For gamma-ray energies above 500 keV, HPGe as well as CdZnTe detectors only provide sufficiently high probability for multiple interactions and the effect of the intrinsic momentum of the Compton electron is sufficiently reduced.

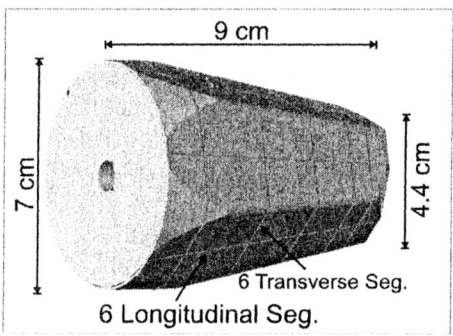

Figure 5: The 36-fold segmented, coaxial GRETA prototype detector.

Figure 5 shows a detector, which could be used for gamma-ray imaging; however, it was built for nuclear physics research. The tapered and hexagonal shape is designed to fit the detector in a closed shell of 120 of these detectors (*1*). This planned array is called Gamma-Ray Energy Tracking Array or GRETA and is currently under development at the Lawrence Berkeley National Laboratory (*17*). Here, the measurements of individual interactions are used not to primarily image gamma-ray sources but to improve the sensitivity in detecting weak gamma-ray transition by maximizing the full-energy efficiency and peak-to-background ratio, particularly for events in which up to 30 gamma rays are emitted simultaneously. Figure 6 shows measured and calculated signals in this detector to demonstrate the variety of pulse shapes that can be obtained in two-dimensionally segmented semiconductor detectors. The distance between the two interactions shown in Figure 6 is 8 mm. Particularly, the transient-charge signals of the non-charge collecting electrodes change from about –20 % to about +20 % relative to the net-charge signal which is measured in segment B4. The variations in the measured signals for each position is due to the finite opening angle of the collimation system. It was only possible to limit the single interactions to a voxel of about 1.5 mm^3. However, the observed spread of signals indicates already the sensitivity of better than 1 mm in all three dimensions, at least for the measured gamma-ray energy of 374 keV. More quantitative analysis has shown that position sensitivities of 0.5 mm can be achieved (*1*).

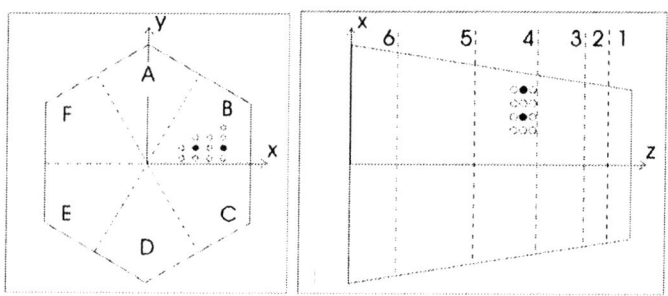

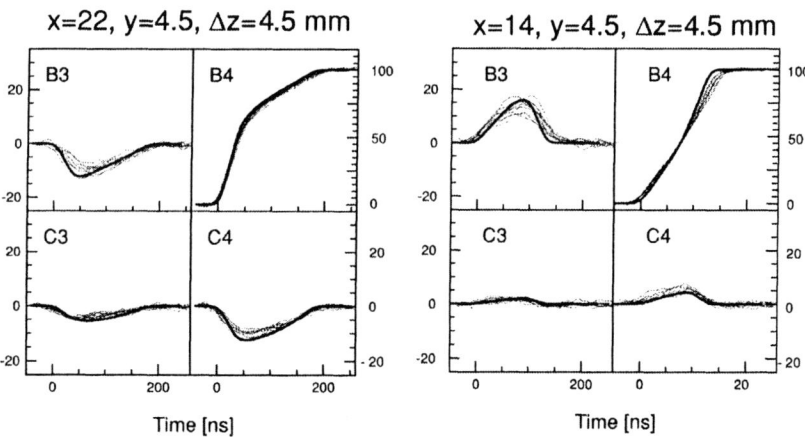

Figure 6: Positions of two interactions in segment B4 (top) and corresponding signals (bottom) measured in B4 (net-charge signals) and adjacent segments (transient-charge signals). Thin lines are measured, thick lines are calculated signals.

Algorithms have to be developed which are based on the underlying nature of the interactions such as Compton scattering or photoelectric effect in order to translate the obtained positions and energies of individual gamma-ray interactions into the path of the gamma ray (2). This is important since the incident angle is defined by the positions and energies of the first two interactions. Such gamma-ray tracking algorithms are necessary since the time resolution of this type of detectors is not sufficient to measure the time differences of the interactions. Figure 7 illustrates this problem. Assuming we measure three interactions, the tracking algorithm compares all 6 possible permutations of interactions with the Compton-scattering formula and extracts the best fit. This best fit provides the scattering sequence of all interactions (without actually measuring the time) and allows the first two interactions to be used in combination with the full gamma-ray energy to determine the incident angle of the gamma ray.

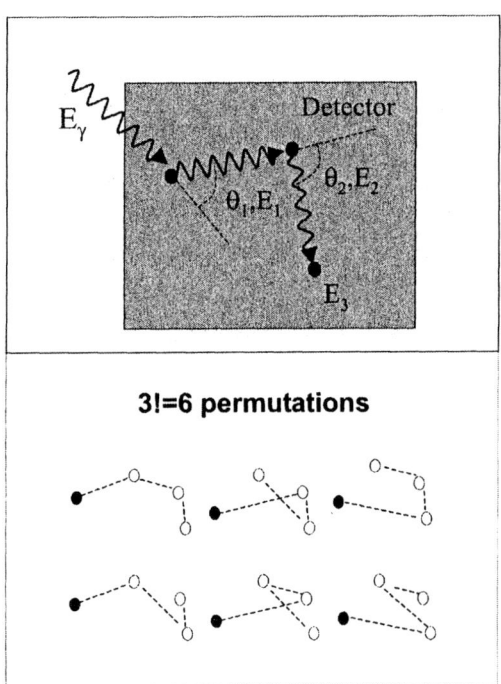

Figure 7: A potential scattering sequence of two Compton scatterings and one photoelectrical effect at the end. To obtain the correct scattering sequence, all 6 possible permutations of interactions are compared with the Compton scattering formula.

The advances in detector technology enable now the use of semi-conductor detectors as efficient gamma-ray tracking detectors with imaging capabilities. Not only do semiconductor detectors provide good position resolution, but also energy resolution, both of which are related due to the excellent and low noise characteristics of these type of detectors. The drawback of the low-noise characteristic, especially for HPGe, is the requirement to cool down the detector. HPGe detectors have to be cooled close to liquid nitrogen temperatures (about 100 K). Again, recent advances in cooling and detector technology allow to cool HPGe by mechanical means to avoid the use of liquid nitrogen. HPSi or Si(Li) detectors have only to be cooled down to –40 C which can be achieved by thermoelectrical cooling. CdZnTe detectors can be operated at room temperature without significant impact on the noise properties, and therefore position and energy resolution.

Applications of Gamma-Ray Imaging Technologies

Based on the previous discussions on the concept of collimator-less gamma-ray imaging in position-sensitive semi-conductor detectors, we will in this chapter briefly illustrate three examples which will benefit from these technologies. First, the GRETA project will be described as an instrument for basic nuclear physics experiments which uses the gamma-ray tracking capabilities to significantly increase the sensitivity in observing weak gamma-ray decays as compared with currently available instruments. Secondly, the gamma-ray tracking capabilities will be used for gamma-ray imaging in nuclear non-proliferation applications to increase the sensitivity in finding and localizing nuclear material, particularly special nuclear material. Finally, the same capabilities will be used to determine sensitivities of imaging gamma-ray sources in nuclear medicine applications.

Gamma-ray tracking for basic nuclear physics experiments

The capability to track gamma rays has profound consequences for nuclear physics experiments, which measure gamma rays in the range of 100 keV to 20 MeV. The current state-of-the-art detector arrays, such as Gammasphere, comprise approximately 100 and more Compton-suppressed Ge detectors. These systems, which were built in the 1990s, have an efficiency of about 10 % for detecting the full energy of a 1 MeV gamma ray and a peak-to-total ratio of about 55 % (*18*). Although these types of arrays represent a large increase in efficiency (about 10 times higher) and resolving power (about 100 times higher) compared to previous arrays, they are ultimately limited by the suppressor shields and the limited efficiency of individual Ge detectors. The removal of the Compton suppressor would allow to increase the solid angle coverage by roughly a factor of two and in addition, would enable the adding of energies of several detectors, e.g. increasing the efficiency for the detection of a 1.3 MeV gamma ray by another factor of two. Ultimately, the efficiency for detecting the full energy of 1.3 MeV for a closed shell purely built of Ge detectors can be estimated to be about 60%, limited only by the finite size of the Ge crystals, gaps and absorption in mounting structures. However, the inability to distinguish between two gamma rays emitted and detected simultaneously and one gamma ray interacting with two detectors prevented the introduction of pure Ge shells for nuclear physics experiments. To separate multiple and simultaneous gamma rays sufficiently, more than 1000 detectors would be required which is prohibitive due to the cost. An alternate approach, which circumvents the problem associated with a shell built purely of Ge detectors is to "track" the interactions of all gamma rays to identify and separate the emitted gamma rays. This is done by first clustering gamma-ray interactions according to their relative spatial positions and then tracking of the individual interactions in each cluster. In this way, gamma rays can be separated, classified as full-energy or escape events and their time sequence determined. A full-energy event is associated with one particular good fit of positions and energies with the Compton formula by probing all permutations of interactions in each cluster. In escape events only a partial energy is deposited due a gamma ray, which escapes the instrument after a scattering process.

The capability of gamma-ray tracking increases the sensitivity in the detection of gamma radiation due to several factors: 1.) The removal of anti-Compton shields and absorbers increases the solid angel coverage roughly by a factor of two; 2.) The removal also allows acceptance of cross-scattering between HPGe detectors, instead of being suppression, which increases the efficiency roughly by another factor of two; 3.) The determination of the position of the first interaction allows to correct for Doppler-shifted gamma-ray energies, particularly in experiments in which the gamma-ray emitting nucleus has high velocities; 4.) The determination of the positions of the first two interactions allows to accurately obtain the linear polarization.

Figure 8 shows a comparison between one the state-of-the-art instruments, Gammasphere (18) and the planned gamma-ray tracking array GRETA (17). It also compares the two different concepts of an anti-Compton and a tracking array. Figure 9 shows the evolution of the resolving power for gamma ray detectors and gamma-ray detector arrays. The resolving power is the inverse of the observational limit which indicates the weakest gamma-decay branch which can be observed in an experiment. For example, the current generation of gamma-ray arrays have a resolving power of 10^4 which implies that the weakest channel which can be observed is 10^{-4} weaker than the strongest reaction channel. A gamma-ray tracking arrays such as GRETA is expected to provide a resolving power of 10^7, which is three orders of magnitude larger than the current generation. Currently, the GRETA effort is acquiring the first module of three 36-fold segmented HPGe detectors, which are expected to be delivered by the end of 2003. With such a module, first physics experiments can be performed to show the performance of this new concept in "real" experiments.

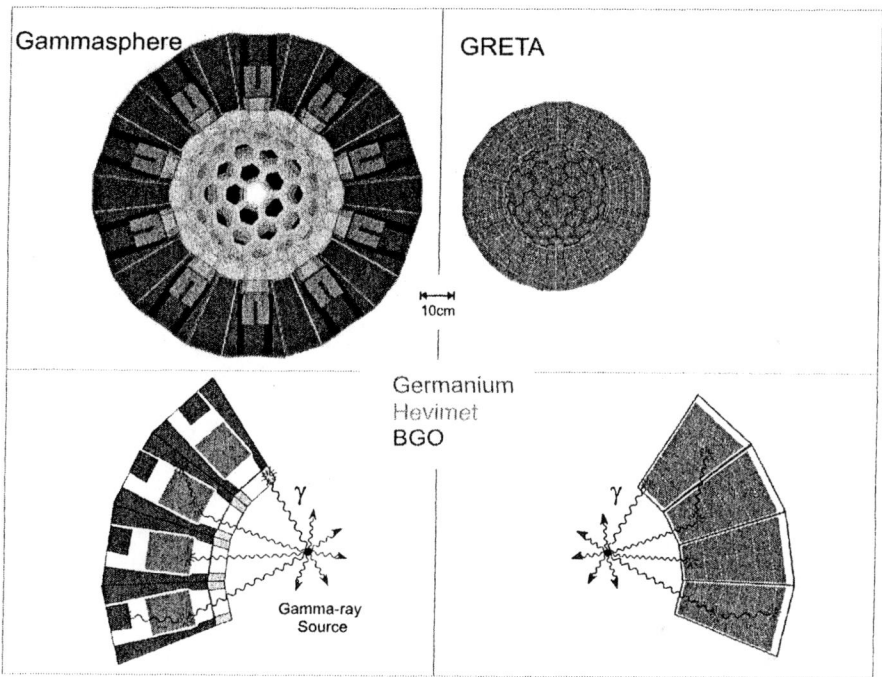

Figure 8: The state-of-the art gamma-ray spectrometer Gammasphere (left) in comparison with the planned gamma-ray tracking array GRETA (right). The top part is to scale, the lower part is not. It illustrates the different concept in the detection of multiple gamma rays. In Gammasphere, anti-Compton shields built of BGO are necessary to identify gamma-rays which escape the HPGe detectors and to suppress scattering between HPGe detectors and hevimet absorbers collimators in front of the BGO detectors prevent gamma rays from directly hitting these shields to reduce false suppression. In GRETA, anti-Compton shields and collimators are removed and all gamma rays are accepted.

Gamma-ray imaging based imaging for nun-proliferation applications

While in the previous chapter the concept of gamma-ray tracking was employed in nuclear physics experiments to increase the sensitivity in order to observe weak gamma-ray decay channels from a known source location, in this and the following paragraph, the imaging capability will be employed to increase the sensitivity in detecting and identifying gamma-ray sources of unknown location.

The aim for nuclear-nonproliferation is to prevent the illicit transport of nuclear material that consists, for example, of medical radioisotopes such as ^{60}Co or special nuclear material such as ^{235}U or ^{239}Pu that is related to nuclear weapons. Due to the increased use of radioisotopes in medicine, it is becoming more important to be able to distinguish between "legal" medical related and natural radioactivity and "illegal" proliferation of radioactive material. Currently, the nonproliferation of ^{235}U and ^{239}Pu is of very high interest due to the potential availability of this material and potential use by terrorist networks.

Collimator-less and gamma-ray tracking based gamma-ray imaging is expected to increase the sensitivity in detecting and localizing special nuclear material. This is due to the fact that the imaging capability is able to spread gamma-ray activity from the environment over the total FOV while maintaining all the activity of the source of interest in one location. In addition, the excellent energy resolution of the used semi-conductor detectors allows correlation of locations and energies to identify the detected sources. Figure 10 shows an example of calculated geometrical efficiencies for a coaxial HPGe detector, similar to the GRETA prototype detector shown in Figure 5. The plotted efficiencies are derived by Monte-Carlo simulations and take into account limiting factors such as finite position (1mm) and energy resolution (1 keV) and the Compton profile (for germanium). Shown are efficiencies and corresponding gains for a gamma-ray energy of 375 keV, which is one of the interesting gamma rays in ^{239}Pu.

To increase the obtainable angular resolution the distance between the first two interactions has to be increased which reduces the efficiency due to the attenuation of the scattered gamma ray. This drop in efficiency can be seen with the dashed line. An angular resolution of better than 2-3 degrees cannot be obtained at this energy due to the Compton profile and the finite energy resolution. The dotted line indicates the gain in terms of increasing the peak-to-background by increasing the angular resolution. The better the angular resolution, the more the unrelated background activity can be spread out over the total solid angle. The two dashed lines indicate the range of angular gains, which is expected with different image reconstruction procedures. The lower line indicates simple two-dimensional projections of the cones while the upper line indicates full maximum-likelihood reprocessing taking the exact point-spread functions into account. Finally, the solid lines show the gain due to the imaging capability, which is the product of the values of the dashed efficiency and the dotted angular gain lines, again for two types of image reconstruction. Both lines

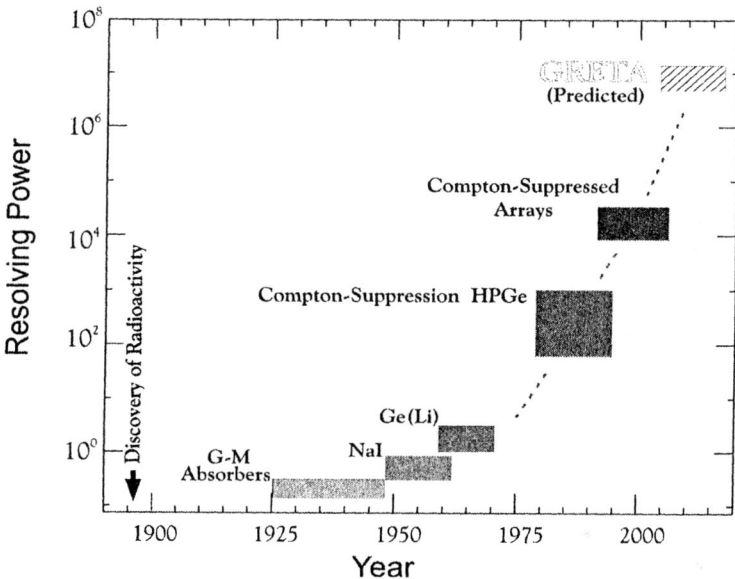

Figure 9: The evolution of the resolving power for gamma-ray detectors and gamma-ray detector arrays. Currently operating arrays such as Gammasphere (Compton-suppressed arrays) are characterized by a resolving power of about 10^4. The next generation of gamma-ray arrays, which are based on gamma-ray tracking, such as GRETA, are expected to have up to three orders of magnitude higher resolving power.

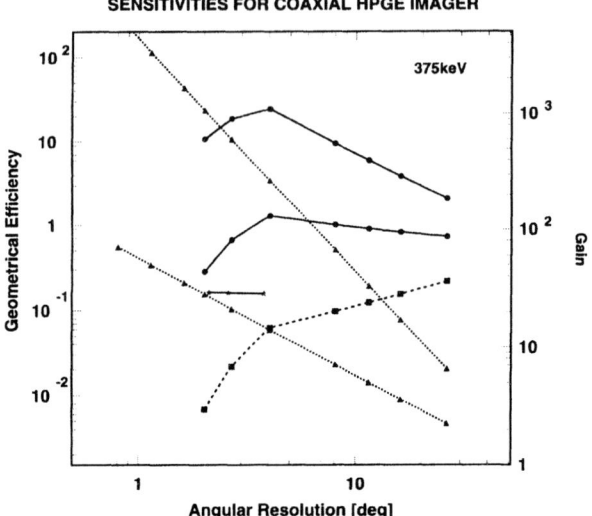

Figure 10: Geometrical efficiencies and gains as a function of angular resolution in a two-dimensionally segmented coaxial HPGe detector. The dashed line represents calculated efficiencies, which were derived by Monte-Carlo simulations and limits in the imaging process due to finite position and energy resolution as well as the Compton profile. Dotted lines indicate the gain in peak-to-background, which can be obtained for a given angular resolution. Solid lines show the overall gain, which is the product of the angular gain and the efficiency. The two lines indicate the range of gains, which are expected with different image reconstruction concepts.

converge to unity, which means no gain compared to a non-imaging operation for decreased angular resolution. For improved angular resolution the angular gain more than compensates for the loss in efficiency, in particular with full image reconstruction.

In Figure 11, gain factors are plotted for a collimator-less gamma-ray imaging system built of stack of HPSi and CdZnTe detectors, similar to figure 4, again assuming 1mm position resolution, achievable energy resolution values for HPSi and CdZnTe detectors and the Compton profile, now for silicon. Gain factors are plotted for 4 different energies of interest: 186 keV, 375 keV, 646 keV, and 1000 keV. Gains between 2 and 100 can be expected for all gamma-ray energies, which are remarkable, especially in comparison with collimator-based systems, which drop significantly for higher gamma-ray energies. Gain factors for parallel-hole systems are shown for 140 keV (in the 186 keV part) and 369 keV (in the 375 keV) for comparison for all three collimator arrangements (LEHR, LEGP, LEHS). In addition, the star on the upper left pert indicates the

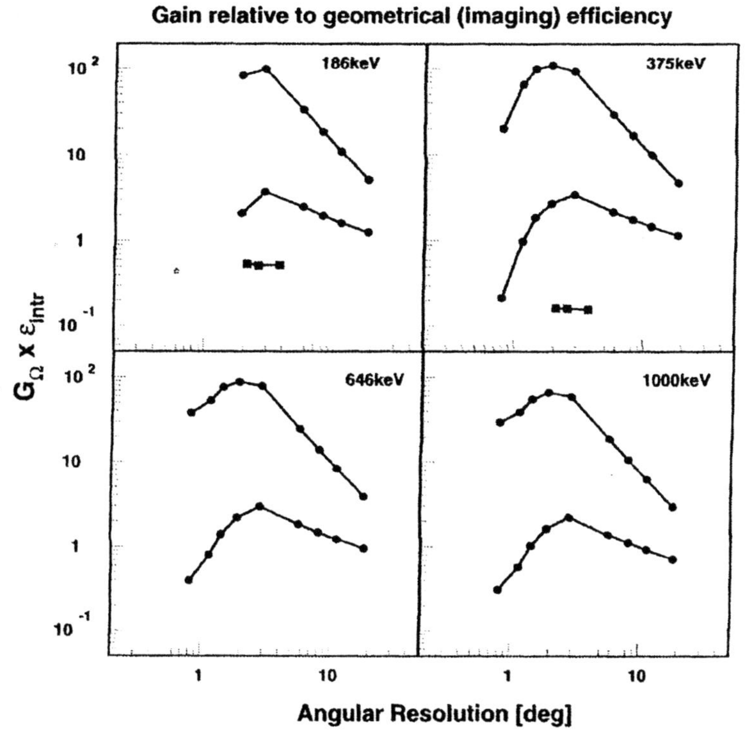

Gain relative to geometrical (imaging) efficiency

Figure 11: Relative gains of a hybrid collimator-less gamma-ray imaging system consisting of 4 5mm thick and position sensitive Si detectors and an array of position sensitive CdZnTe detectors. Circular markers show the values calculated for the hybrid imager, squares indicate the gain with the parallel-hole collimation systems from figure 2 (140 keV and 369keV for 186 keV and 375 keV, respectively). The star in the upper left figure at 186 keV indicates a pinhole collimator at 140 keV.

gain factor for a pinhole collimator. The gain of this collimator-less hybrid system in comparison with parallel-hole collimator-based systems is remarkable.

Collimator-less hybrid imager for medical radioisotopes

Due to the similarity of gamma-ray energies in nonproliferation (186 keV/375 keV) and the medical use of radioisotopes (e.g. 140keV/369keV), the same collimator-less gamma-ray imaging device can be employed to find, for example, cancer lesions with higher sensitivity than currently possible with collimator-based gamma-ray imaging systems. Here, we are not yet focusing on the measurements of the detailed activity distribution, but simply to find accumulated activity in cancer lesions quicker and more reliably, particularly for small lesion, which are smaller than 10 mm (which represents the current limit).

Figure 12 shows calculated geometrical efficiencies for the same system as above, now for slightly different energies reflecting the different radioisotopes of interest. As before, angular resolution of about 1 degree can be achieved with efficiencies of about 1 %. Efficiencies of 5-8 % are expected for angular resolution values of 3 degrees for energies between 140 keV and 511 keV. Again, the anticipated efficiencies are significantly higher than with parallel-hole collimator systems. In contrast to a collimator-based system, the sensitivities are very similar for all gamma-ray energies between 200 keV and 1000 keV.

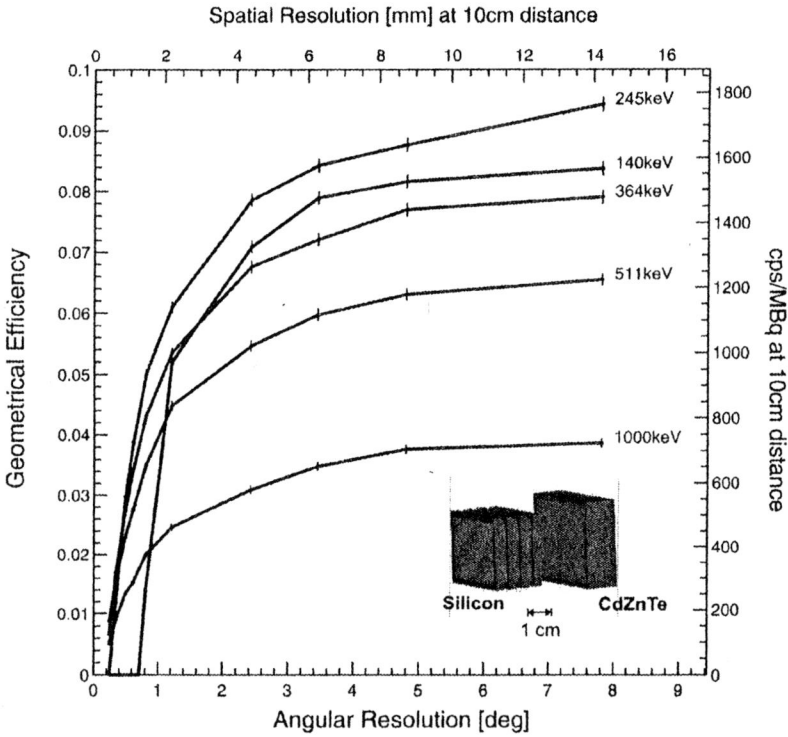

Figure 12: Calculated geometrical efficiencies and detector count rates at 10 cm distance of the indicated hybrid imager built of layers of Si detectors and CdZnTe detectors.

The improved sensitivity, reduced size and superior energy resolution as compared with state-of-the-art collimator-based gamma cameras should allow to significantly improve the possibility to detect cancer lesions in an earlier stage than currently possible and therefore increasing the chance of survival.

In addition, the improved sensitivity and the potentially improved spatial resolution can not only benefit the early detection of cancer, but also improve the treatment planning of cancer. The improved gamma-ray imaging capability is complementary to the development of pharmaceuticals with higher specificity to cancerous tissue (*19*). Currently, monoclonal antibodies are introduced for cancer diagnosis and treatment to increase the specificity for cancer as compared with other pharmaceuticals, which are primarily sensitive to increased metabolism (see for example (*20,21,22*)).

At the Glenn T. Seaborg Institute, new targeting agents are being developed to potentially increase the specificity for molecular targeted radiation therapy even beyond monoclonal antibodies. As can be seen in figure 12, efficiencies for higher gamma-ray energies than 140 keV are higher, especially for angular resolution of better than 1 degree. With the development of new targeting agents it might be possible to incorporate new radioisotopes with higher gamma-ray energies.

Conclusions

Due to recent advances in the manufacture of three-dimensional position sensitive semi-conductor detectors, compact and highly sensitive collimator-less gamma-ray imaging devices can be built. They are characterized by their superior sensitivity particularly for higher gamma-ray energies as compared with conventional collimator-based gamma-ray imaging devices. The underlying concept of gamma-ray tracking will allow building substantial more efficient gamma detector arrays for basic nuclear physics experiments. For imaging applications, this concept can significantly increase sensitivities in the detection of special nuclear material in non-proliferation applications as well as in the early detection and in the treatment planning of cancer.

Acknowledgments

The work was performed under the auspices of the U.S. Department of Energy by University of California Lawrence Livermore National Laboratory under contract No. W-7405-Eng-48.

References

(1) K. Vetter, A. Kuhn, M.A. Deleplanque, I.Y. Lee, F.S. Stephens, G.J. Schmid, D.A. Beckedahl, J.E. Kammeraad, J.J. Blair, R.M. Clark, M. Cromaz, R.M. Diamond, P. Fallon, G.J. Lane, A.O. Macchiavelli, C.E. Svensson, *Nuclear Instrumentation and Methods in Physics Research A* **2000**, 452, 223-238

(2) G.J. Schmid, M.A. Deleplanque, I.Y. Lee, F.S. Stephens, K. Vetter, R.M. Clark, M. Cromaz, R.M. Diamond, P. Fallon, A.O. Macchiavelli, *Nuclear Instrumentation and Methods in Physics Research A* **1999**, 430, 69-83

(3) K. Yamashita et al., *App. Opt.* **1998**, 37, 8067-8073

(4) M.D. Short, *Nuclear Instrumentation and Methods in Physics Research* **1984**, 221, 142-149

(5) *The physics of medical imaging*, S. Webb; Institute of Physics Publishing: Bristol, UK, 1998

(6) M.S. Gerber, D.W. Miller, P.A. Schlosser, J.W. Steidley, A.H. Deutschman, *IEEE Trans. Nucl. Sci.* **1977**, 24, 182-187

(7) M. Amman, P.N. Luke, *Nuclear Instrumentation and Methods in Physics Research A* **2000**, 452, 155-160

(8) Y.F. Du, Z. He, G.F. Knoll, D.K. Wehe, W. Li, *Nuclear Instrumentation and Methods in Physics Research A* **2001**, 457, 203-211

(9) *Physics in Nuclear Medicine*, 2nd ed., J.A. Sorenson and M. Phelps; W. B. Saunders Company: Philadelphia, PA 1987

(10) D. B. Everett, J.S. Fleming, R.W. Todds, J.M. Nightingale, *Proc. IEEE* **1977**, 124 , 995-1000

(11) M. Singh, *Medical Physics* **1983**, 10(4) , 421-427

(12) M. Singh and D. Doria, *Medical Physics* **1983**, 10(4) , 428-430

(13) M. Singh and R.R. Brechner, *Journal of Nuclear Medicine* **1990**, 31, 178-186

(14) G.W. Philips, *Nuclear Instrumentation and Methods in Physics Research B* **1995**, 99, 674-677

(15) B.F Phlips, S.E. Inderhees, R.A.Kroeger, W.N. Johnson, R.L. Kinzer, J.D.Kurfess, B.L.Graham, N. Gehrels, *IEEE Transactions on Nuclear Science*, 1996, 43(3), 1472-1475

(16) B. Hyams, U. Koetz, E. Belau, R. Klanner, G. Lutz, E. Neugebauer, A.Wylie, J. Kemmer, *Nuclear Instrumentation and Methods in Physics Research* **1983**, 205, 99-105

(17) M.A. Deleplanque, I.Y. Lee, K. Vetter, G.J. Schmid, F.S. Stephens, R.M. Clark, T.M. Diamond, P. Fallon, A.O. Macchiavelli, *Nuclear Instrumentation and Methods in Physics Research A* **1999**, 430, 292-310

(18) I.Y. Lee, *Nucl. Phys. A* **1990**, 520, 641c-655c

(19) C.L. Hartmann Siantar, K. Vetter, G.L. DeNardo, S.J. DeNardo, *Cancer Biotherapy and Radiopharmaceuticals* **2002**, 17(3), 267-280
(20) S-E Strand, B-A Jönsson, M. Ljungberg, J. Tennvall, *Acta Oncologica* **1993**, 32(7-8), 807-817
(21) P.B. Zanzonico, *J Nucl. Med.* **2000**, 41(2), 297-308
(22) G.L. DeNardo, M.E. Juweid, C.A. White, G.A. Wiseman, S.J. DeNardo, *Critical Reviews in Oncology and Hematology* **2001**, 39, 203-218

Chapter 6

Application of the Triple-to-Double Coincidence Ratio Method at National Institute of Standards and Technology for Absolute Standardization of Radionuclides by Liquid Scintillation Counting

B. E. Zimmerman[1], R. Collé[1], J. T. Cessna[1], R. Broda[2], and P. Cassette[3]

[1]Ionizing Radiation Division, Physics Laboratory, National Institute of Standard and Technology, Gaithersburg, MD 20899–8462
[2]Radioisotope Centre POLATOM, 05–400 Otwock-Swierk, Poland
[3]Labortoire National Henri Becquerel, 91191 Gif sur Yvette Cedex, France

A new liquid scintillation spectrometer that uses the Triple-to-Double Coincidence Ratio (TDCR) method to experimentally determine counting efficiencies has been constructed at the NIST for the primary standardization of β- emitting nuclides. This technique permits the efficiencies of the detector to be determined without the need for standard efficiency tracing and is thus considered a quasi-absolute method. This paper describes the new TDCR system at NIST and presents results of tests aimed at assessing the its operating characteristics. The results indicate that the measured activites for previously calibrated solutions of ^{3}H (tritiated water) and ^{63}Ni agree with certified activity values to within 0.04 % and 0.2 %, respectively.

Introduction

Because of the relatively high detection efficiency associated with liquid scintillation (LS) counting, it continues to be the preferred method used by most of the national metrology institutes (NMIs) around the world for the primary standardization and subsequent calibration of solutions for activity of radionuclides that undergo β- or α- decay. Acceptance of a particular technique by the metrology community requires that the methodology be accurate and reproducible, be based on a definable theoretical model, and have the ability to completely characterize the uncertainties associated with its application. Currently there are two such LS techniques in use by the various NMIs around the world: the CIEMAT/NIST* efficiency tracing method, and the Triple-to-Double Coincidence Ratio (TDCR) method.

The application of the CIEMAT/NIST efficiency tracing method (*1*), (*2*) requires that LS cocktails containing a tracing standard, such as ^{3}H or ^{14}C, be prepared so as to be chemically identical (in terms of amount of added water, ion and acid concentration, and amount of added imposed quenching agent) to a set of cocktails containing the radionuclide being investigated. The efficiency of the detection system for the radionuclide of interest is calculated from the experimentally determined efficiency for the standard through a calculational program, such as EFFY (*3*) or CIENIST 2001 (*4*), which calculates the counting efficiencies of both radionuclides as a function of a common, free variable. The degree to which the free variable has the same value for the matched cocktails of the standard and the nuclide of interest is monitored by a quenching variable (such as a quench indicating parameter determined by the spectrometer or composition variables such as total cocktail water fraction), through which corrections can be made to account for slight differences in composition that lead to differences in efficiency. The underlying assumption in the method is that the same mechanism for chemical quenching of the nuclide being studied is the same as that for the standard. While this assumption is often upheld, there are cases in which the chemistry of each cocktail is sufficiently different so as to result in incorrectable efficiency changes that result in erroneous values of the solution activity.

The TDCR method (*5, 6*) avoids this problem by being able to experimentally determine the detection efficiency of the system without the need for an external standard. As such, it is more of an "absolute" calibration

---

* The acronyms CIEMAT (Centro de Investigaciones Energéticas Medioambientales y Tecnológicas) and NIST (National Institute of Standards and Technology) stand for the names of the national metrology laboratories of Spain and the United States, respectively, who jointly developed this efficiency tracing technique.

technique. By observing the ratio of double and triple coincidences of photons that are produced in a liquid scintillator through the interaction with ionizing radiation in a three-photomultiplier tube (PMT) detector, the counting efficiency can be calculated from the assumed statistical distribution of photoelectrons and a theoretical quenching model. This paper describes the TDCR detection system currently in use at NIST and presents the first results of experiments aimed at characterizing its performance.

The TDCR Model

The theoretical model used to develop the TDCR method has been fully developed elsewhere (5, 6), but will be briefly summarized here to aid the reader. The counting efficiency, ε_x, in a particular counting channel (singles, doubles, triples, sum of doubles) for detecting pulses produced in an LS detection system can be given by

$$\varepsilon_x = \int_0^{E_{\beta,max}} S(E) P_x(E, \eta_0) dE$$

where $S(E)$ is the normalized β particle spectrum (with energy, E), and $P_x(E, \eta_0)$ is the energy-dependent detection probability for the particular counting channel given as as a function of E and the figure of merit, η_0 (in units of photoelectrons per keV). The expressions for P_x are (assuming a Poisson distribution of photons produced in the scintillator):

Double coincidence: $1-(2P_0-P_0^2)$,
Triple coincidence: $1-(3P_0-3P_0^2+P_0^3)$, and
Sum of double coincidences: $1-(3P_0^2-2P_0^3)$,

where $P_0 = \exp(-\eta_0 EQ(E))$. The quenching factor, $Q(E)$, is given by

$$Q(E) = \frac{1}{E} \int_0^E \frac{dE}{1+kB\frac{dE}{dx}}$$

where kB is the Birks value (in cm·keV^{-1}, and which is treated as an adjustable parameter) and dE/dx is the energy loss function for the scintillator medium.

Defining the ratio of double coincidences to triple coincidences (hereafter referred to as the TDCR) as a parameter, K, either as a ratio of experimental count rates in a doubles channel and triple coincidence channel or as a ratio of theoretical efficiencies in the doubles and triples coincidence channels, we seek in this application of the TDCR method a value of η_0 such that the difference between the theoretical K and experimental K is smaller than some arbitrary critical value, Δ:

$$\left|K_{th}(\eta_0) - K_{exp}\right| \leq \Delta$$

$$K_{th}(\eta_0) = \frac{\varphi_{T(\eta_0)}}{\varphi_{D(\eta_0)}}$$

$$K_{exp} = \frac{N_T}{N_D}$$

Once a value for η_0 is obtained, a system of equations is formed to allow for the calculation of the activity:

$$N_{ab} = N_0 \varphi_{ab}(\varepsilon_a, \varepsilon_b)$$
$$N_{bc} = N_0 \varphi_{bc}(\varepsilon_b, \varepsilon_c)$$
$$N_{ac} = N_0 \varphi_{ac}(\varepsilon_a, \varepsilon_c)$$
$$N_T = N_0 \varphi_T(\varepsilon_a, \varepsilon_b, \varepsilon_c)$$

where N_{xy} is the counting rate in the coincidence channel between PMTs x and y, N_T is the counting rate in the triples coincidence channel, N_0 is the activity, φ_{xy} is the efficiency for the appropriate coincidence channel, and ε_z is the detection efficiency for each individual PMT. The four variables (N_0, ε_a, ε_b, and ε_c) can be then solved for analytically using the system of equations.

The advantage of this particular approach is that unlike previous versions, it is no longer assumed that the efficiencies of the three phototubes are equal. Instead, the efficiencies of each PMT are directly determined. As will be discussed below, this improvement is very important for the system that has been installed at NIST.

The NIST TDCR System

The design of the NIST TDCR system is based on the systems currently in use in the laboratories at the Laboratoire National Henri Becquerel (LNHB) and POLATOM and is depicted schematically in Figure 1. The three PMTs were modified by reversing the high voltage polarity and applying it on the anode instead of the photocathode (thereby reducing thermal noise), by adding a high voltage input to the first dynode (permitting higher gain), and by adding an additional high voltage lead to the focusing electrode (allowing variable focusing voltage, thereby changing detection efficiency). The respective high voltage inputs for the anode, first dynode, and focusing electrode are connected at a common point for each of the three PMTs so that the voltage on each input can be simultaneously changed for each PMT using a single supply for each input. All of the data acquisition, as well as the high voltage on the focusing electrodes, is controlled through a LabView[1] program that was developed in-house specifically for this application.

The heart of the system is the MAC-3 unit (7), which was designed and built at LNHB and which contains all of the coincidence logic and extending deadtime correction circuitry in a single NIM module.

The three phototubes are arranged in a sample chamber at an angle of 120° apart and approximately 1 cm from the LS sample vial. All surfaces of the inside of the sample chamber have been painted with reflective paint to increase the light collection efficiency. The PMTs and sample chamber are enclosed in a light-tight aluminum box that has an upper chamber to allow for dark storage of counting sources and for manual changing of the sources. One important difference between the design of the instrument in use at LNHB and at NIST and the one at POLATOM is the ability of the systems at the former two laboratries to add a gamma-ray detector under the sample chamber that will allow the entire detector system to be used additionally as a β-γ coincidence or anticoincidence detector or as a traditional TDCR detector with gamma-ray gating capability. To date, the additions for the gamma-ray channel have been implemented only at LNHB.

Performance Tests

In order to test the behavior of the new TDCR system and determine the optimum operating characteristics, a series of LS samples was prepared with NIST standard solutions of ^{3}H (as tritiated water) and ^{63}Ni. Nominally 10 mL of OptiPhase HiSafe III LS scintillant (Wallac Oy, Turku, Finland) was added to glass vials, followed by the addition of nominally 100 mg of solution from either NIST Standard Reference Material 4226C (^{63}Ni) (8) or a gravimetric dilution of NIST Standard Reference Material 4927F (tritiated water) (9). Background blanks were prepared by the addition of 100 mg of distilled water to nominally 10 mL of HiSafe III scintillant in glass vials.

---

[1] Certain commercial equipment, instruments, or materials are identified in this paper to foster understanding. Such identification does not imply recommendation by the National Institute of Standards and Technology, nor does it imply that the materials or equipment identified are necessarily the best available for the purpose.

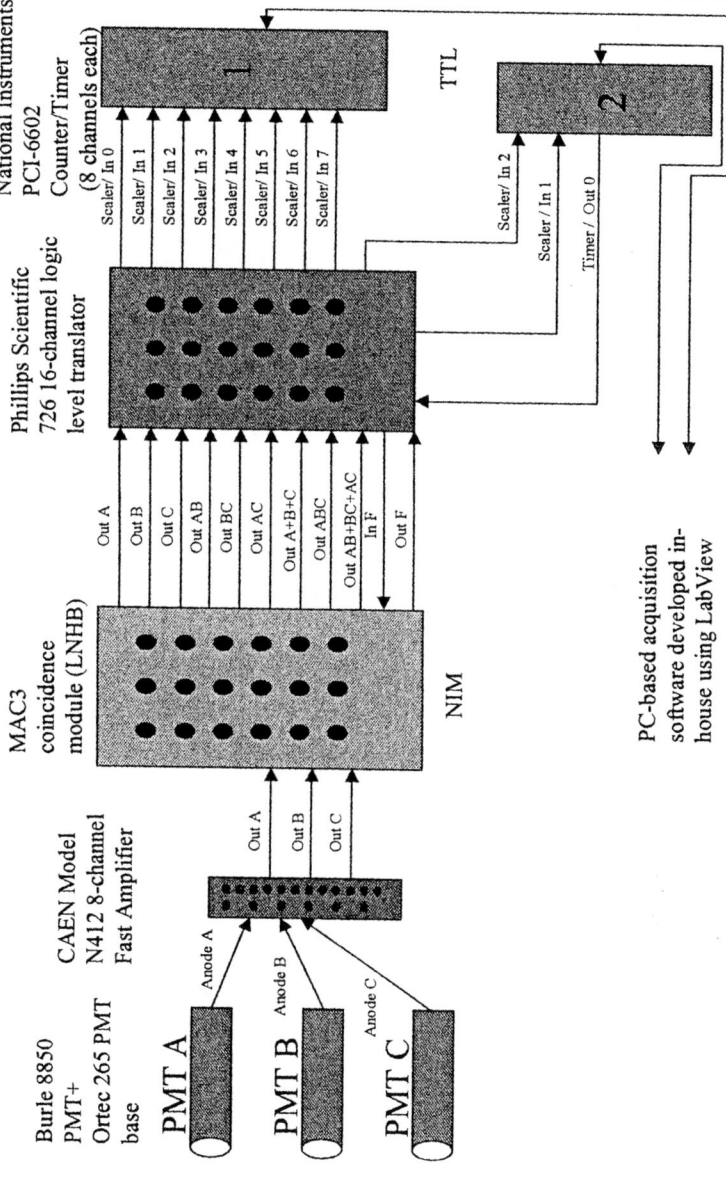

Figure 1. Schematic of acquisition hardware comprising the TDCR detection system installed at NIST.

Since the primary purpose of these experiments was to investigate the performance of the TDCR system and not to perform a full calibration of the solutions, there was no intentional experimental design. Instead, the intent was to merely count the LS sources several times under various conditions and check the resulting activities to see if changes to either the apparatus or analysis and acquisition codes were required. Therefore, the vials were counted in the NIST TDCR system at between 4 and 8 focusing voltages for between 1 and 3 minutes at each voltage. Each source was counted at least once, but the first source prepared for both the ^{3}H and ^{63}Ni (denoted "T-1" and "N-1", respectively) were counted at least 4 times each.

The data were analyzed using the computer codes TDCR-02B and TDCR-02P, which are modified versions of the TDCR-02 code (*10*). The TDCR-02B program contained only minor changes to the original, but the version TDCR-02P added the ability to calculate the detection efficiencies as a function of the TDCR using the Polya, or negative binomial, distribution in place of the Poisson distribution. This modification was made because of the fact that the Poisson distribution is not valid for extremely small numbers of photoelectrons, such as the case with low-energy β-emitters.

The results of such a counting experiment can be seen in Figure 2. The first striking characteristic of this instrument is the relatively large inequality of the counting efficiencies between the double coincidence channels despite modest attempts to match the PMTs. This inequality is most likely due to small differences remaining in nominal values of the resistors in the PMTs. Despite this, the program TDCR-02B is able to compensate for this effect.

Because of the low energies involved in the decay of both the ^{3}H and ^{63}Ni, the choice of statistical distribution used to describe the photoelectrons is critical. In most cases, the Poisson distribution is valid because the average number of photoelectrons produced is fairly large. In the case of very low energy decays, however, the average number of photoelectrons can very small, even less than 1. Under these conditions, the Poisson model is no longer valid, requiring another distribution such as the negative binomial, or Polya. For the ^{63}Ni data depicted in Figure 2, the average number of photoelectrons per PMT is calculated to be between 4 and 8. However, in the ^{3}H cocktails that were counted, the average number of photoelectrons per PMT is between 1 and 2, precluding the use of a Poisson distribution. Figure 3 shows the experimental efficiencies for the three double coincidence channels as a function of TDCR, along with the theoretical efficiencies calculated using the Poisson distribution and a kB value of 0.012 cm·keV^{-1}. It can be clearly seen that the calculated efficiencies are consistently higher than the experimental value. In comparison, Figure 4 shows the same experimental data with the theoretical efficiencies calculated using a Polya distribution and the same value for kB. In this case, much better agreement is obtained due to the fact that a much more appropriate distribution has been used to perform the calculations.

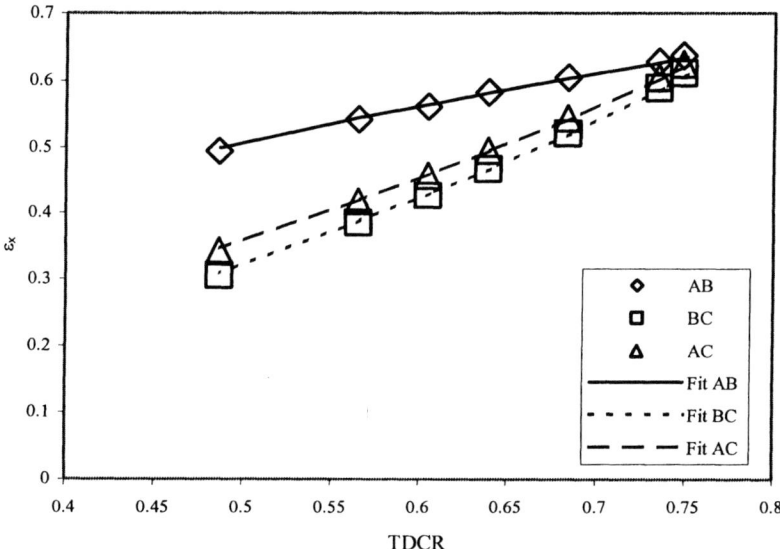

Figure 2. Experimental and theoretical counting efficiencies, ε_x, for double coincidence channels AB, BC, and AC as a function of the ratio of triple-photon to double-photon coincidences (TDCR) for ^{63}Ni LS source N-1. Each point represents two repeated counts at each focusing voltage (efficiency value), with the standard deviation of each measurement lying within the respective symbol for each point. The theoretical efficiencies were calculated with the program TDCR-02B using a Poisson distribution and a kB value of 0.012 cm·keV^{-1}.

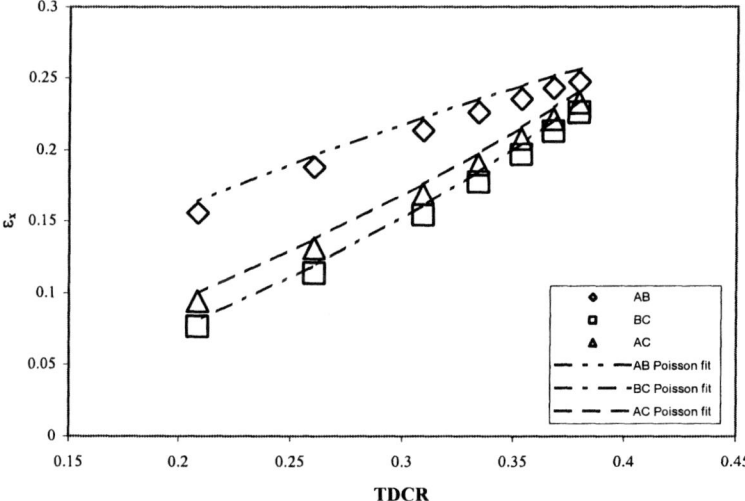

Figure 3. Experimental and theoretical counting efficiencies, ε_x, for double coincidence channels AB, BC, and AC as a function of TDCR for ^3H LS source T-1, which contained an aliquant of a gravimetrically diluted NIST standard solution of tritiated water. Each point represents three repeated counts at each focusing voltage (efficiency value), with the standard deviation of each measurement lying within the respective symbol for each point. The theoretical efficiencies were calculated with the program TDCR-02B using a Poisson distribution and a kB value of 0.012 $cm \cdot keV^{-1}$.

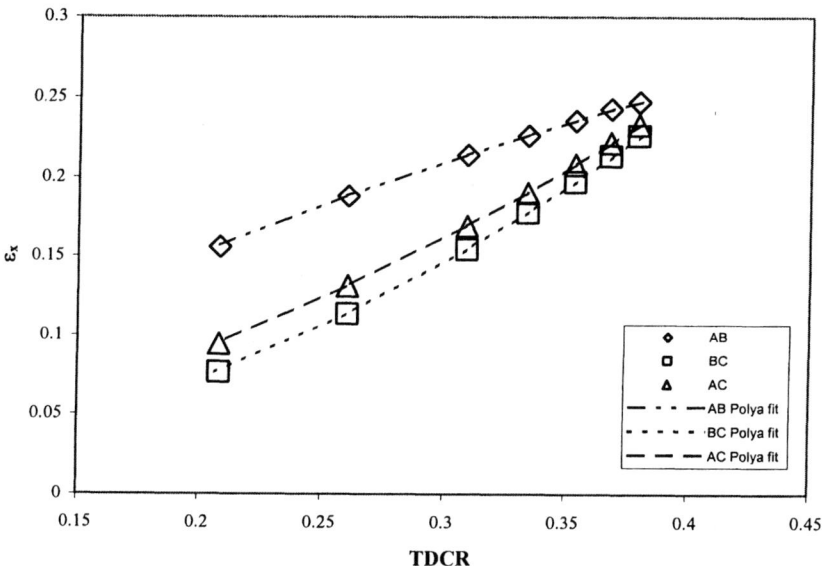

Figure 4. Experimental and theoretical counting efficiencies, ε_x, for double coincidence channels AB, BC, and AC as a function of TDCR for ^{3}H LS source T-1, which contained an aliquant of a gravimetrically diluted NIST standard solution of tritiated water. Each point represents three repeated counts at each focusing voltage (efficiency value), with the standard deviation of each measurement lying within the respective symbol for each point. The theoretical efficiencies were calculated with the program TDCR-02P using a Polya distribution and a kB value of 0.012 cm·keV^{-1}.

The ultimate test of the detection system and analysis software, of course, is the ability to correctly determine the activity of the solutions being measured. Through all of the testing and adjusting of the apparatus, a total of five activity measurements, with repeated measurements on two different sources, were able to be analyzed as independent determinations of the massic activity (in terms of Bq·g^{-1}) for the LS cocktails containing the ^{3}H standard. Likewise, four independent determinations of the massic activity of the ^{63}Ni standard solution were able to be analyzed, with three measurements on one source and a single measurement on a second. The results of the measurements are presented in Table I. For the ^{3}H using the Polya-fitted results, the average massic activity, C_A, was 28.10±0.29 kBq·g^{-1}, where the uncertainty is an expanded (k=2) uncertainty calculated from the quadratic addition of the standard deviation on the activity determinations for typically 8 points in a counting series having different efficiency values (average of 0.09 %) and the standard deviation on the 5 independent determinations of C_A with three sources (0.5 %). This result is in excellent agreement with the previously certified value of 28.11 ±0.20 kBq·g^{-1}, where the uncertainty is also an expanded (k=2) uncertainty.

Table I. TDCR counting results of NIST standard reference material solutions of ^{3}H and ^{63}Ni. Experimental massic activities C_A (in units of kBq·g^{-1}) are given, along with the NIST-certified value. All uncertainties are expanded (k=2) uncertainties.

Nuclide	Exp. C_A / kBq·g^{-1}	Certified C_A / kBq·g^{-1}
^{3}H	28.10 ±0.29	28.11 ±0.20
^{63}Ni	48.18 ±0.32	48.26 ±0.44

The average value of C_A for the ^{63}Ni solution was measured to be 48.18 ±0.32 kBq·g^{-1}, where, as in the case of the ^{3}H solution, the uncertainty is an expanded (k=2) uncertainty calculated from the quadratic addition of the standard deviation on the activity determinations for typically 8 points in a counting series having different efficiency values (average of 0.11 %) and the standard deviation on the 4 independent determinations of C_A with two sources (0.33 %). Again, this result is in excellent agreement with the previously certified massic activity value of 48.26±0.44 kBq·g^{-1}, where the uncertainty is an expanded (k=2) uncertainty.

Conclusion

A new TDCR spectrometry system, based on the design of systems currently in use at the French and Polish standards laboratories (LNHB and POLATOM, respectively), has been constructed at NIST. Tests of the apparatus indicate that despite attempts to match the PMTs, an asymmetry exists in the efficiencies of the three phototubes that make up the detector, meaning that older TDCR models that assume equality among the PMTs are invalid in this case.

Activity determinations on previously calibrated solutions of ^{3}H (as tritiated water) and ^{63}Ni gave excellent results when compared to the certified values. For the ^{3}H solution, the agreement was 0.04 % and for the ^{63}Ni, the agreement was 0.2 %, and were within the measurement uncertainties. In the case of the ^{3}H measurements, the best agreement was obtained when the Polya distribution was used to describe the photoelectrons in place of the Poisson. This is not entirely surprising, since the Poisson distribution is invalid when the number of photoelectrons approaches unity, as was found for most of the measurements of the ^{3}H sources in these experiments.

With the addition of this new spectrometry system, NIST adds another technique to those already available for the standardization of β-emitting radionuclides. The ease of sample preparation and the ability to internally determine the detection efficiency give the TDCR method a distinct advantage over other techniques such as efficiency tracing. Future plans include its application not only to β- emitters, but also electron capture emitters.

References

1. Coursey, B. M. et al. *Nucl. Instrum. Methods Phys. Res., Sect. A* **1986,** 279, 603-610.
2. Zimmerman, B. E.Collé, R. *J. Res. Natl. Inst. Stand. Technol.* **1997,** 102, 455-477.
3. Garcìa-Toraño, E.Grau Malonda, A. *Comput. Phys. Commun.* **1985,** 36, 307-312.
4. E. Günther, CIENIST 2001, private communication, 2001.
5. Pochwalski, K. et al. *Appl. Radiat. Isot.* **1988,** 39, 165-172.
6. Broda, R; Pochwalski, K. *Nucl. Instrum. Methods Phys. Res., Sect. A* **1992,** A312, 85-89.
7. Bouchard, J.; Cassette, P. *Appl. Radiat. Isot.* **2000,** 52, 669-672.
8. National Institute of Standards and Technology Standard Reference Material 4226C, Radioactivity Standard, Nickel-63, 1995.
9. National Institute of Standards and Technology Standard Reference Material 4927F, Radioactivity Standard, Hydrogen-3, 2000.
10. Broda, R. et al. *Appl. Radiat. Isot.* **2000,** 52, 673-678.

Chapter 7

Applications of Californium-252 Neutron Irradiations and Other Nondestructive Examination Methods at Oak Ridge National Laboratory

R. C. Martin[1], D. C. Glasgow[2], and M. Z. Martin[3]

[1]Nuclear Science and Technology Division, [2]Chemical Sciences Division, and [3]Environmental Sciences Division, Oak Ridge National Laboratory, Oak Ridge, TN 37831

The U.S. Department of Energy (DOE) program for the production and distribution of heavy (transplutonium) elements is described, along with the variety of applications for californium-252 (^{252}Cf) neutron sources and neutron irradiation experiments at the Californium User Facility for Neutron Science (CUF) at Oak Ridge National Laboratory (ORNL). ORNL's Radiochemical Engineering Development Center (REDC) is the center for the Transuranium Element Processing Program (TEPP), which produces curium, berkelium, californium, einsteinium, and fermium for basic research, and the distribution of ^{252}Cf neutron sources for commercial and research applications. Capabilities for neutron and laser-based in situ nondestructive examination techniques for elemental analysis in the REDC and collaborative ORNL programs are also described.

© 2004 American Chemical Society

Transuranium Element Processing Program

The TEPP was established in the late 1950s to produce transplutonium heavy elements and distribute them to interested researchers (1). The production is based at ORNL's REDC. The REDC has two large hot cell buildings. Building 7920 has nine heavily shielded hot cells for fabrication of target rods, dissolution of the irradiated rods, and subsequent radiochemical processing and purifications. Building 7930 has three hot cells for the preparation, handling, and distribution of ^{252}Cf neutron sources and recovery of the ^{248}Cm daughter. Building 7930 also has two unused hot cells, which are reserved for future radiochemical recovery of ^{238}Pu from irradiated ^{237}Np target rods for use as radioisotopic power sources in space and elsewhere (2).

Heavy element (HE) production begins with long, thin target rods containing 35 pellets made from a blend of curium oxide and aluminum. Each target contains up to 10 g of actinide. After a year-long irradiation within the High Flux Isotope Reactor (HFIR) at ORNL, the targets are returned to the REDC and dissolved. Radiochemical processing removes impurities and fission products, HEs are separated and purified using ion-exchange chromatography, and unreacted curium is recycled into the next batch of target rods (3). Existing actinide oxide inventory can maintain HE production at current rates for two decades. The remaining HFIR lifetime is projected at >30 years.

The HE production path of neutron capture and β decay is shown in Figure 1. HE Campaign 73 will begin in early 2003 with the dissolution of nine curium target rods and the subsequent recovery of the following estimated masses: ≥275 mg of ^{252}Cf, ≤10 mg of ^{249}Cf, ≤35 mg of ^{249}Bk, ≤1 mg of ^{253}Es (isotopically mixed), ≤0.1 mg of "milked" (isotopically pure) ^{253}Es, ≤4 μg of pure ^{254}Es, ≤5 ng per run of milked ^{255}Fm (decayed from the Es product), and ~0.5 pg of ^{257}Fm. This campaign should provide industrial customers with an uninterrupted supply of ^{252}Cf into mid-decade. The short-lived isotopes are available only at the conclusion of each HE campaign, currently conducted biennially. A Transuranium Research Isotope Users' Committee with representatives from the primary end-user organizations determines the distribution of the rare isotopes to researchers, without charge for the isotope. More than 1000 HE shipments have been made as part of the TEPP. Campaign schedules and other REDC information are posted at URL http://www.ornl.gov/nstd/redc.

Other actinides available from the REDC include ^{249}Cf and ^{251}Cf; ^{244}Cm, ^{245}Cm, and ^{248}Cm; ^{243}Am; plutonium isotopes ^{238}Pu through ^{242}Pu and ^{244}Pu; and ^{234}U. Other radioisotopes available from the REDC include ^{63}Ni, ^{99}Tc, and ^{209}Po. ORNL also produces additional medical isotopes (e.g., ^{188}W and ^{213}Bi).

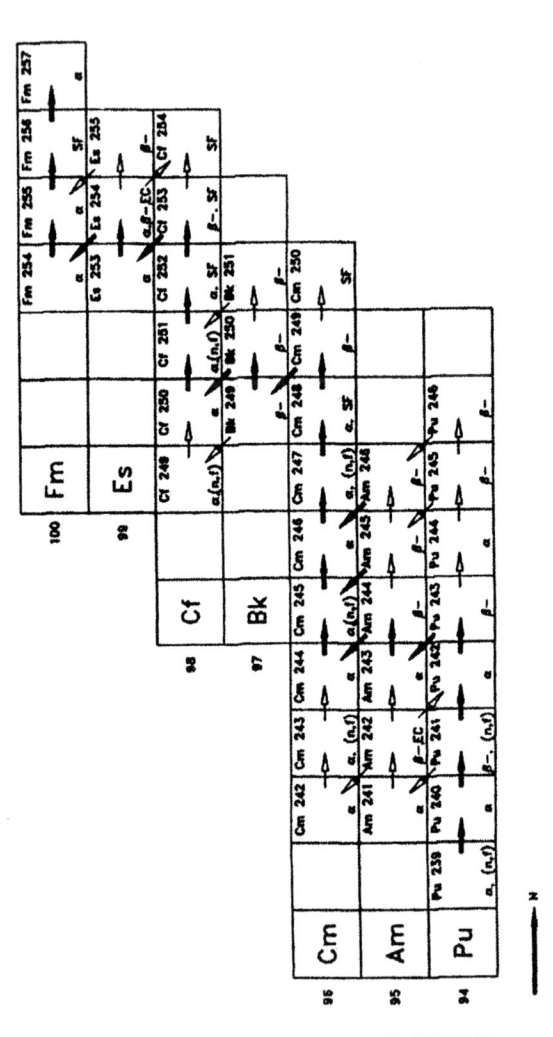

Figure 1. Transmutation of target material for heavy element production.

Californium-252 Properties and Source Forms

Californium metal melts at 900°C and boils at 1472°C, with a room-temperature density of 15.1 g/cm^3 (*4,5*). The pure metal is very reactive and apparently bright silver in color as prepared in an inert atmosphere. The ^{252}Cf radioisotope is an intense neutron emitter that is readily encapsulated in compact sealed sources. For ease of encapsulation, Cf is converted into either the oxide or oxysulfate form (*6*). Commercial and medical sources typically incorporate a cermet wire or pellet containing Cf$_2$O$_3$ in a palladium matrix. Industrial and research sources contain either a Cf$_2$O$_3$ powder (obtained by filtering and heating a Cf oxalate slurry inside a source capsule) or a pressed pellet containing aluminum powder and Cf oxysulfate microspheres (obtained by firing an ion-exchange resin containing Cf). The inherent safety of ^{252}Cf source encapsulations is well demonstrated by 30 years of experience, even under explosive impact (*6*).

Electroplates of ^{252}Cf can be encapsulated inside small ionization chambers (*7*) for use as timed neutron sources (i.e., the fission pulse acts as a trigger to time the release of the neutrons). Electrodepositions of trivalent actinides (e.g., ^{252}Cf and ^{244}Cm) at the REDC range from nanograms of actinide to surface loadings as high as 300 µg/cm^2.

With 3.092% of ^{252}Cf decay occurring via spontaneous fission, 1 µg of ^{252}Cf emits 2.31434 × 10^6 fast neutrons/s with a 2.645-year half-life (0.5362 mCi/µg). The average of the Maxwellian neutron energy distribution is 2.14 MeV, with the most probable energy at ~0.7 MeV. A cylindrical source capsule the size of a person's little finger can contain >50 mg of ^{252}Cf, emitting up to ~10^{11} neutrons/s. Radiological exposure in air, at 1 m from a 1-µg ^{252}Cf source, totals 2.21 mrem/h (22.1 µSv/h) from fast neutrons and 0.19 mR/h (1.9 µGy/h) from gamma rays.

Californium-252 sources contain several Cf isotopes. The isotopic distributions from a recent campaign (early 1999) are typical: 4.3 atom % ^{249}Cf; 10.8% ^{250}Cf; 3.3% ^{251}Cf; 81.5% ^{252}Cf; 0.04% ^{253}Cf; and 0.01% ^{254}Cf. Neutron emission from ^{250}Cf decay (13-year half-life) as well as from ^{252}Cf must be considered for sources >15 years old. Neutron emission from ^{254}Cf must be considered in assays of recently produced ^{252}Cf. With 99.7% of decay by spontaneous fission and a 60-day half-life, neutron emission from 2 µg of ^{254}Cf equals that from 1 mg of ^{252}Cf. Pure ^{249}Cf is obtained from decay of ^{249}Bk. The REDC recently supplied fairly pure ^{251}Cf (with acceptably low neutron emission for routine handling) from the reprocessing of 30-year-old ^{252}Cf sources.

Applications of ^{252}Cf Neutron Sources

Californium-252 neutron sources are used whenever compact, portable, and reliable neutron sources are required. Recent American Nuclear Society conference sessions surveyed the range of applications (2,8). A recent estimate indicates that ~70 publications per year involve either use of ^{252}Cf sources throughout the world or calculation of ^{252}Cf properties. Common industrial and research applications include prompt gamma neutron activation analysis (PGNAA) of coal, cement, and minerals and PGNAA-based detection and identification of explosives, land mines, and unexploded military ordnance. Other uses include neutron radiography, materials characterization and nuclear assay, fuel rod scanning for uniformity of fissile material, reactor start-up sources, cancer therapy, calibration standards, and standard neutron fields.

Some applications use the high-energy (60- to 80-MeV) fission-fragment (FF) emission from unsealed ^{252}Cf (typically an electroplate). Etching of thin plastic samples exposed to FFs produces tiny holes along the FF tracks, for use in specialized applications. A benchtop system uses a low-activity ^{252}Cf electroplate for FF simulation of cosmic ion damage to electronic devices.

Subcritical multiplier assemblies driven by ^{252}Cf were used in the past for enhanced flux neutron irradiations. When a ^{252}Cf source is placed within a carefully designed array of fissile fuel rods or plates, the total neutron emission can be 1 to 2 orders of magnitude greater than that of the ^{252}Cf source but remains safely below criticality. This approach has been suggested for generation of non-reactor-based external beams for cancer therapy.

Cancer Therapy

Brachytherapy is a method of cancer treatment that places a small radioactive source inside the patient's body, at the location of the tumor, to kill the cancerous cells. Intense sources are typically attached to a cable and inserted through a catheter to the tumor site. Californium-252 neutron brachytherapy (NBT) is more effective than photon and gamma brachytherapy in treating radioresistant tumors such as bulky and late-stage tumors, melanomas, and glioblastomas (brain tumors) and causes rapid regression of bulky, localized tumors (2). Over 6000 patients have been treated using NBT, with the number of treatments accelerating. The Linden Neutron Knife Company installed the first NBT treatment unit in China in 1999 and now has 10 treatment centers in operation (>350 patients treated to date) and up to 30 centers planned.

Iridium-192 photon brachytherapy is the current standard for high-dose-rate (HDR) brachytherapy. Small (~1-mm outer diameter) ^{192}Ir HDR sources can provide treatments in ≤10 minutes. Russian ^{252}Cf HDR sources with an outer diameter of ~3 mm are appropriate for intracavitary brachytherapy (e.g., gynecological, rectal, head, neck, and oral cavity treatments) but are too large for interstitial brachytherapy (inside organs, such as the brain and prostate). To date, ^{252}Cf interstitial brachytherapy has been limited to low-dose-rate treatments with lengthy treatment times because of the difficulty in miniaturizing high-activity ^{252}Cf sources. In 1999, ORNL and Isotron, Inc., entered into a Cooperative Research and Development Agreement to design a new family of medical sources suitable for interstitial and intracavitary HDR NBT. The first prototype sources have been fabricated for use with existing gamma-source applicators, achieving the desired miniaturization. Isotron's NBT system will provide optimized treatments of adult and pediatric cancer in 18 sites throughout the body.

Neutron Radiography

Unlike X-radiography, neutron radiography provides nondestructive examination and visual contrast for low-Z elements (such as hydrogen) and moisture. A ^{252}Cf-based facility for neutron radiography was used at McClelland Air Force Base in California in the 1990s to detect debonding of composites and moisture, fuel leakage, and corrosion within the aluminum honeycomb structure of F-111 and F-15 aircraft, without requiring disassembly (8). A major ^{252}Cf radiography facility (up to 150 mg of ^{252}Cf) is used for the nondestructive examination of components at the Pantex Plant in Amarillo, Texas. With collimated neutron fluxes of $\sim 2 \times 10^4$ cm^{-2} s^{-1}, up to nine radiographs per day can be obtained with image quality comparable to that of reactor-based radiography (9).

Fissile Material and Nuclear Waste Characterization

Passive-active neutron shufflers employing ^{252}Cf have been used to monitor vehicles' contents and to determine the fissile content of waste drums, spent fuel assemblies, and process materials. The source is placed near the sample, and the delayed neutrons from induced fissions are counted after source removal. Detection of milligram masses of fissile isotopes has been demonstrated.

ORNL researchers have developed a cart-portable Nuclear Materials Identification System (NMIS) containing a low-intensity ^{252}Cf ionization chamber (7) to probe and characterize fissile material (10). Neutrons from the ^{252}Cf source induce fission of fissile atoms present within the sample, releasing additional neutrons. These neutrons are detected by the detector array, providing a sample response to the incident neutrons that can be analyzed using standard time-correlation and/or frequency-analysis techniques. The NMIS uses the ^{252}Cf ionization chamber to time the ^{252}Cf neutrons so as to exclude their signal from the response of the detector array. Spontaneous fission of a ^{252}Cf atom is accompanied by production of energetic fission fragments as well as several neutrons. The fission fragments create an electronic spike within the ionization chamber that is used to time the detector array to reject the prompt ^{252}Cf neutron signal, but accept the subsequent fission-induced neutron signal from the sample.

This technique has been used to determine the spatial distribution, mass, and hydration of deposits of $UO_2F_2 \cdot nH_2O$ in process piping in order to estimate subcriticality and plan for deposit removal (8). Other applications include identification and inventory of items for nuclear materials control and accountability, measurement of ^{235}U mass flow rate in a UF_6 gas stream for downblending Russian weapons-grade uranium, and the determination of the effective neutron multiplication factor (k_{eff}) for optimized storage and shipping of fuel elements. A value of k_{eff} less than one represents a safe, subcritical system while a value greater than one represents a potential criticality accident.

Neutron Backscattering for Land Mine and Contraband Detection

Californium-252 ionization chambers have been incorporated into a handheld land mine detector. Modern land mines use a nonmetallic casing such as plastic to elude detection, but this detector measures the time delay between neutron emission and backscattering from the plastic (correlated detection) to differentiate readings from background (11). Land mine detection using uncorrelated neutron backscattering has also been demonstrated using ~3.5 μg of ^{252}Cf in a prototype handheld detector weighing ~8 kg. With this device, long-term radiological exposure to the operator is at an acceptable level (12).

Nova R&D, Inc., of Riverside, California, has developed a handheld instrument for the U.S. Coast Guard to detect contraband hidden in compartments behind metal and other structures (8). Fast neutrons from this Compact Integrated Narcotics Detection Instrument penetrate the barrier material, and unexpected backscatter of neutrons by any hydrogen-rich materials present suggests the potential presence of contraband.

Californium-252 neutron backscattering has also been used to measure the void fraction of boiling water in a nuclear fuel rod bundle testing facility (2).

Prompt Gamma Neutron Activation Analysis

On-line Process Monitoring and Borehole Logging

PGNAA involves capture of a neutron by an atom's nucleus. De-excitation of the nucleus occurs with instantaneous emission of a high-energy (multi-MeV) "prompt" gamma ray. The gamma spectrum peak energies identify the elements present. No significant radioactivity is induced by rapid sampling using lower-intensity neutron sources. PGNAA provides information concerning the principal components of a sample. Subpercent sensitivities are common.

PGNAA analysis of raw materials for process control or material management (blending, etc.) in the coal and cement industries is common. Conveyor belt and slurry analyzers quantify over 20 elements. Analysis of coal provides sulfur content (to meet smokestack release criteria), ash and moisture content, and information on its calorific content (Btu/lb). Over 200 analyzers are in service worldwide (8).

Subsurface PGNAA measurements of elemental chlorine in an old radiochemical processing waste pit were made using a high-purity germanium gamma detector and ~5 µg of ^{252}Cf in a cylindrical borehole probe (13). Although PGNAA measurements of H, Si, Ca, Fe, Al, and Cl were obtained within the boreholes, only chlorine was quantified, with concentrations ranging from 1,000 to 30,000 ppm and a minimum detection limit of 300 ppm.

Explosive and Land Mine Detection and Munitions Identification

Ancore Corporation uses ^{252}Cf sources in several explosives detection systems (e.g., Small Parcel Explosive Detection System, Vehicular Explosive and Drug Sensor, etc.) based on PGNAA detection of the nitrogen present in explosive compounds (14). A prototype system was demonstrated at airports for detection of explosives in luggage in the early 1990s. The Canadian Department of National Defence has developed a mobile PGNAA system to detect buried land mines; detection of nitrogen using <100 µg of ^{252}Cf can confirm the presence of most antitank mines and many antipersonnel mines within minutes (8).

The commercial Portable Isotopic Neutron-Spectroscopy (PINS) Chemical Assay System marketed by ORTEC is used for the nondestructive identification of the contents of munitions, chemical weapons, and general chemical-storage containers (8). PINS PGNAA systems use several micrograms of ^{252}Cf to differentiate between high explosives, nerve agents, poison gases, common military compounds, and inert contents.

Californium-based Instrumental Neutron Activation Analysis

Thermal neutron flux is maximized for optimum elemental sensitivities in instrumental neutron activation analysis (INAA) and gamma spectroscopy. Unlike radiochemical neutron activation analysis, INAA employs little or no sample treatment before or after activation. Pillay (15) evaluated the limits of detection for a low-flux (~2 × 10^6 cm^{-2} s^{-1}) INAA system based on a 50-µg ^{252}Cf source. He detected 19 elements with limits of detection (LODs) ranging from 29 to 8200 ppm by weight, with optimum sample masses of 100 mg. In order of increasing LOD, those elements with LODs <100 ppm are Au, As, Re, Mn, La, Br, and Sb. Those with LODs between 100 and 1000 ppm are Cu, Yb, W, Ga, Er, Dy, Mo, Pr, and Pd. The elements Cd, Hg, and Zn have LODs >1000 ppm.

Analysis of Environmental and Vitrified Samples

INAA systems typically employ a nuclear reactor, although several high-intensity ^{252}Cf-based INAA facilities have been demonstrated. A facility at the Savannah River Technology Center (SRTC) uses ~50 mg of ^{252}Cf (neutron flux ~5 × 10^8 cm^{-2} s^{-1}) to produce short-lived radioactive tracers and to provide INAA analyses of organic compounds, metal alloys, sediments, and site process solutions (8,16). Several dozen elements can be analyzed with parts-per-million sensitivity.

A major INAA system has been established by Fluor Hanford in Richland, Washington, for the high-precision analysis of sodium within vitrified fuel reprocessing wastes (8). INAA mitigates problems in analysis of highly radioactive samples (i.e., dilution of samples increases personnel radiological exposure and analytical uncertainties). This system also quantifies traces of fissionable isotopes (~0.1 ppm of U and Pu) via delayed neutron counting.

INAA at the Californium User Facility

The CUF, described below, represents an ideal ^{252}Cf-based INAA facility. Gamma spectroscopy capabilities are available, and a polyethylene irradiator has been fabricated with nearly cubic dimensions of ~26 cm on a side (2). Twelve ^{252}Cf sources containing 50 mg of ^{252}Cf can generate fast and thermal fluxes of ~10^9 cm^{-2} s^{-1} at a sample volume the size of a 30-mL plastic bottle. A cadmium cover is available to filter out thermal neutrons for epithermal and fast neutron (FN) activation. Future analysis of soil and environmental samples is planned.

INAA systems based on ^{252}Cf have lower flux and lower sensitivity than reactor-based systems, but they also have several advantages (17). Use of multiple point sources enables variable sample geometries (e.g., bulk samples and tailoring of fast-to-thermal flux ratios). FN fluxes can induce (n,α) and (n,p) reactions, improving sensitivity for certain elements (e.g., Fe, Ni, Si, P, and S), if interference free. Use of a cadmium sleeve is simple to implement for epithermal/fast INAA, and it also reduces total activation in sampling of short-lived isotopes. Low gamma and nuclear heating permit analysis of liquid and biological samples. Californium-252-based irradiators are not subject to reactor downtime, can be used for extended irradiations, and require less rigorous safety analyses than reactor facilities. Establishment of a ^{252}Cf-based INAA system is far more practical than a new reactor facility for a university or industry.

INAA at the High Flux Isotope Reactor

ORNL has perhaps the most flexible INAA capabilities of any nuclear establishment, with ultrasensitive thermal INAA at the HFIR reactor and epithermal and FN INAA capabilities 50 yards away at the CUF. The HFIR provides the most sensitive INAA capabilities in the world. Pneumatic transfer tube PT-1 terminates in the beryllium reflector and provides the highest thermal neutron flux available for neutron activation analysis (4×10^{14} cm^{-2} s^{-2}) at a thermal:epithermal flux ratio of ~35. A second pneumatic transfer tube (PT-2) terminates in the pressure vessel along the outside edge of the reflector and provides a highly thermalized (>99%) flux of 5×10^{13} cm^{-2} s^{-2}. PT-2 also has a delayed neutron counting station to measure picogram levels of fissile isotopes.

Trace elemental analyses at the HFIR include high-purity materials (silica, diamond films, pure metals, and plastics). Forensic analyses include determination of trace elements in hair and fingernails; characterization of evidentiary materials such as paint, plastic, bullets, wood, and metal; and analysis of rocks to determine their origin. Geological analyses include lunar and meteorite samples; baseline soil analyses for pollution assessment; analysis of marble to determine ancient trade routes; and analyses of pottery and bone (including trace levels of iridium in dinosaur bone). Nuclear security applications use delayed neutron counting to assess uranium enrichment and total fissile content in samples from International Atomic Energy Agency monitoring activities.

Comparison of NAA Detection Limits

Interference-free elemental detection limits for HFIR-based INAA include the following 64 elements, grouped by LOD:

0.1 to 1 µg:	Ca, Fe, Nb
10 to 100 ng:	Mg, K, Ti, Cr, Ge, Y, Zr, Mo, Ce, Tm, Th, Ta, Pt
1 to 10 ng:	Cl, Co, Cu, Zn, Se, Rb, Ru, Pd, Sn, Te, Ba, Pr, Nd, Tb, Yb
0.1 to 1 ng:	Na, Al, Ar, Sc, V, Ga, As, Br, Sr, Rh, Ag, Cd, Sb, I, Cs, La, Sm, Gd, Ho, Er, U, W, Re, Hg
<0.1 ng:	Mn, In, Hf, Eu, Dy, Lu, Hf, Ir, Au.

HFIR-based analyses are ~3 orders of magnitude more sensitive than those of the SRTC Cf-based facility (8) with ~50 mg of ^{252}Cf. (The SRTC data are based on 6-g samples, while HFIR typically uses ~0.5-g samples.) The CUF INAA capabilities should be comparable to those of SRTC. SRTC analyses are ~3 orders of magnitude more sensitive than the limits determined by Pillay (15) using a 50-µg source, consistent with the difference in neutron intensities.

Because an industrial PGNAA system (*2*) uses ^{252}Cf sources of comparable intensity (50 to 100 μg) to those in Pillay's INAA system, comparison of sensitivities is informative. Because PGNAA analyzes bulk samples, direct comparison of mass LODs is not possible. PGNAA system LODs (based on 10-minute analyses) range from <100 ppm by weight for elements Cl, Sc, Ti, Ni, Cd, Hg, Sm, Gd, Dy, and Ho; between 1 and 3% for C, Ge, Br, Sr, Zr, Ru, Pd, Sb, and Tl; and between 3 and 10% for Ne, Rb, Nb, Tc, and Sn. Elements He, B, O, F, Bi, and Pm are difficult to detect. Elements not listed above range between 100 ppm and 1% LOD. Direct comparison with Pillay's results shows that PGNAA is more sensitive by weight fraction for Dy, Cd, Hg, and Ba, while Pillay's INAA is more sensitive for Au, As, Re, Mn, La, Br, Sb, W, Ga, Mo, Pr, and Pd. Comparable sensitivities were indicated for Cu, Yb, Er, and Zn.

PGNAA systems are constrained by the neutron source-induced gamma background, making analysis of small sample masses more difficult. INAA has difficulty analyzing specific elements, most notably the light elements. These problem elements include H, He, Li, Be, B, C, N, O, Si, P, S, and Pb. Elements such as fluorine have very short half-lives for INAA. FN INAA can be advantageous for elements such as Si, P, S, Fe, and Ni.

Use of Laser-induced Breakdown Spectroscopy with INAA

Bulk elemental analysis usually requires sample preparation, typically dissolution. Disadvantages arise from excessive analytical uncertainty (e.g., dilution of radioactive samples), generation of waste, and destruction of irreplaceable objects. NAA avoids these problems, but induced radioactivity and inability to analyze light elements are disadvantages. Laser-induced breakdown spectroscopy (LIBS) complements INAA as a nondestructive in situ elemental technique. LIBS focuses a pulsed laser beam onto a sample surface, vaporizing a submicrogram sample mass into an ionized plasma. The ions de-excite in microseconds, and a spectrometer/detector system identifies the elements present based on the wavelength of the optical emissions. LODs typically approach the parts-per-million range under atmospheric conditions. LIBS can probe solid, liquid, gas, and aerosol samples and can provide elemental depth profiling of solids via repeated ablation.

LIBS has been used to sample contaminated soils and analyze mixed waste and radioactive contamination (Pu, Am, Sr, Pb, Cr, Be), including vitrified waste glass. Coupling to fiber optics permits remote sampling, which is ideal for radioactive materials and hazardous environments. Sampling has been demonstrated inside hot cells, glove boxes, and even within the pressure vessel of a nuclear reactor (in situ analysis of copper and chromium in tubes) (*18*).

As a microanalytical technique, LIBS analyses remain matrix dependent. As a complementary technique, simultaneous analysis by INAA can provide calibration of LIBS results. This has not been pursued to date, but preliminary work at the REDC evaluated the LODs of terbium solutions to measure the leaching of surrogate nonradioactive source capsules. Analysis via ^{252}Cf-based INAA indicated a LOD of ≤1 ppm in solution, while LIBS analysis of dried solutions demonstrated LODs below the parts-per-million level before optimization. HFIR INAA has a sensitivity for terbium approaching ppb levels, while inductively coupled plasma–mass spectrometry has a sensitivity on the order of several parts per trillion.

Unlike INAA, LIBS can detect hydrogen, oxygen, nitrogen, and helium, with sensitivity superior to that of PGNAA. Following are typical LIBS LODs for elements that are difficult to detect by INAA: Li (10 ppm in glass), Be (<10 ppm in soil), B (30 ppm in glass), Si (tens of ppm in alloys), P (≤15 ppm in air), Fe (tens of ppm in alloys), and Pb (tens of ppm in soil).

LIBS and PGNAA Detection of Carbon in Soils

Soil carbon content is important in evaluating carbon sequestration in soils (a mitigating factor in global warming from greenhouse gases). Conventional techniques require the field acquisition of soil core samples, followed by combustion in the laboratory to measure carbon dioxide generation, a long and expensive process. Recent LIBS in situ measurements of carbon in soils demonstrated an experimental LOD of ~0.17% by weight ([19]). Nitrogen has also been measured (LOD ≤100 ppm), to correlate with organic carbon content.

A field-portable LIBS instrument has previously been used to measure soil contamination. The only other technique proposed for practical in-field measurement of carbon (and nitrogen) is PGNAA. Detection of carbon using cold neutron beams has been reported, and extrapolation to a ^{252}Cf-based cold neutron irradiator ([20]) suggests subpercent carbon detection is feasible with a transportable system for in-field use. This concept has not been demonstrated in the laboratory.

Californium User Facility for Neutron Science

Facility Infrastructure and Experimental Capabilities

The CUF ([17]) is a unique facility for medium flux irradiations using a fission neutron spectrum. The CUF stores the DOE inventory of prefabricated ^{252}Cf sources and makes these sources available to researchers for experiments inside contamination-free hot cells. After hands-on experimental setup inside the hot cell, sources are pneumatically transferred into the cell for irradiation with up to ~50 mg of ^{252}Cf (fast and thermal neutron fluxes up to ~10^9 cm^{-2} s^{-1}). Extended irradiations are possible within the ^{252}Cf storage pool.

Two experiments are representative of CUF capabilities. The Canadian Department of National Defence's multisensor land mine detection system uses a moderate-intensity ^{252}Cf source for PGNAA detection of nitrogen in land mines. The high-count-rate gamma detection system was optimized at the CUF, with rates up to 1.5×10^6 counts/s using a 300-μg ^{252}Cf source (21). Another experiment relates to the development of boron-containing chemicals for potential use in boron neutron capture therapy (BNCT) of cancer. Living lung cancer cells were exposed to four experimental boron compounds developed at the University of Tennessee, Knoxville, and then irradiated in the CUF with 28 mg of ^{252}Cf (17). Comparison of cell killing with that of nonboronated cells indicates the relative effectiveness of the compounds.

Radiation Damage Testing of Semiconductor Detectors

A university consortium used up to 59 mg of ^{252}Cf at the CUF in a series of irradiations to test the neutron hardness of avalanche photodiode detectors. These photodetectors were candidates for use in a high-energy physics experiment at the CERN accelerator in Geneva, Switzerland (22). Experiments simulated the expected in-service FN flux of 2×10^5 cm^{-2} s^{-1} and the expected lifetime FN fluence of 1.2×10^{13} cm^{-2}. A specialized ^{252}Cf irradiator containing ~10 mg of ^{252}Cf was later designed and built at the University of Minnesota for quality control testing of these devices prior to use at CERN.

Radiation Damage Testing of Permanent Magnets

The Advanced Photon Source (APS) at Argonne National Laboratory produces high-intensity X-rays for scientific research. Neodymium-iron-boron (Nd-Fe-B) permanent magnets are used inside the insertion devices for beam tailoring but are exposed to high-energy radiation (e.g., bremsstrahlung-produced photoneutrons) from the 7-GeV APS storage ring Radiation-induced demagnetization would adversely impact operational lifetimes.

High photon radiation doses did not induce demagnetization (23). The APS Operations Division and the CUF collaborated to irradiate sample magnets with FNs. At a fluence of ~10^{13} cm^{-2}, the measured average degradation in magnetic field strength was 0.6%. At ~2×10^{13} cm^{-2}, the cumulative degradation averaged 1.3%. After 7-day irradiation to a fluence of ~10^{14} cm^{-2}, the magnetic field strength had degraded by 16%.

A thermalized neutron spectrum was then used to elucidate neutron damage mechanisms. A thermal fluence well above 10^{12} cm^{-2} produced no detectable magnet damage, despite production of energetic helium and lithium nuclei from neutron capture by ~8 atom % boron in the magnets. The data suggest that FNs are more damaging to the magnets than are thermal neutrons.

Fast-neutron-induced Mutation of Seeds

The development of modern tools of DNA sequencing has renewed interest in seed irradiations as a tool for plant breeding. FN irradiations are ideal because of their ability to induce more extensive genetic damage within plant chromosomes than other radiation sources (6). By noting the physical abnormalities in the germinated plant, followed by its DNA sequencing and comparison with normal DNA, one can conclude that missing DNA base sequences correspond to the plant's physical abnormalities. Subsequently, judicious cloning of genes that promote desirable plant properties could improve crop yields, nutritional value, or disease resistance.

To understand the functions of rice genes, a preliminary test was conducted at the CUF to generate rice deletion lines. Rice samples were irradiated to FN doses ranging from 12 to 37 Gy, based on a computer simulation for a standardized irradiation geometry consisting of a 2.5-cm-thick packet of rice sandwiched between two lead bricks and ten ^{252}Cf sources positioned on the outside of the bricks. The average neutron dose to the rice from 50 mg of ^{252}Cf was calculated to be 7.5 Gy/h. The lead bricks reduced the gamma dose to the sample to an estimated 5% that of neutron dose. A parametric study of dose effects was conducted for the U.S. Department of Agriculture, with CUF irradiations of rice samples from ~8 to ~163 Gy. Increasing dose increasingly stunted growth. Above 50 Gy, the embryos germinate but none survive as seedlings (24).

Californium-252 Source Loan Programs

Californium-252 sources for commercial applications are typically obtained from one of several commercial vendors. DOE's Californium Industrial Sales Program sells ^{252}Cf source material to these vendors for resale. DOE also has a loan program to provide sources to government agencies and subcontractors of the U.S. government and to universities (some foreign) for educational, research, and medical applications **without charge for the radioisotope**. To promote academic education and research, DOE also **waives loan fees** on lower-intensity sources under the Californium University Loan Program, with only transportation costs incurred by the academic user (6). Over 200 ^{252}Cf sources are on loan to ~70 U.S. government agencies, subcontractors, and educational and medical institutions. Several hundred ^{252}Cf sources have been fabricated at the REDC since the 1980s.

Low-intensity sources can often be purchased from a commercial vendor at lower cost than a loaned source. However, end-of-life disposal of a commercial source becomes the user's responsibility. Loan costs for high-intensity sources compare very favorably with procurement costs for electronic neutron generators and accelerators with comparable intensities. More moderation and shielding are required for 14-MeV neutrons from generators than for ^{252}Cf neutrons. For commercial analyzers, these added requirements increase the total system cost.

Summary

The CUF at ORNL is a unique facility for medium-flux fast neutron irradiations and facilitates research by alleviating the regulatory burdens on the experimenter. The DOE Californium Loan Program eliminates neutron source disposal concerns by providing compact, portable ^{252}Cf sources to governmental organizations, subcontractors, and universities at costs competitive with alternate sources of neutrons. DOE's University Loan Program is strongly supportive of academic research and training in experimentation with neutrons.

New miniature ^{252}Cf sources will provide unprecedented brachytherapy treatments of glioblastomas and a variety of other radioresistant tumors. INAA and LIBS provide complementary capabilities for nondestructive, in situ elemental analysis across the periodic table. Applications of ^{252}Cf neutron sources span many scientific disciplines, and continue to promote research at the frontiers of interdisciplinary science.

Acknowledgments

Development of miniature ^{252}Cf medical sources was performed under Cooperative Research and Development Agreement ORNL99-0558 between Isotron, Inc., and ORNL. The TEPP is administered by the Office of Basic Energy Sciences of the U.S. Department of Energy. The Californium Industrial/University Loan Program is administered by the Office of Nuclear Materials Production of the U.S. Department of Energy. Oak Ridge National Laboratory is managed by UT-Battelle, LLC, under contract DE-AC05-00OR22725 for the U.S. Department of Energy.

References

1. Bigelow, J. E.; Corbett, B. L.; King, L. J.; McGuire, S. C.; Sims, T. M. In *Transplutonium Elements—Production and Recovery;* Navratil, J. D.; Schulz, W. W., Eds.; ACS Symposium Series, No. 161; American Chemical Society: Washington, DC, 1981; pp 3–18.

2. *Trans. Am. Nucl. Soc.* **2002,** *86,* 378–383 (Production and applications of californium-252 and other transplutonium elements, session summaries).
3. King, L. J. *Proceedings of 27th Conference on Remote Systems Technology;* American Nuclear Society: La Grange Park, IL, 1979; pp 96–102.
4. Haire, R. G. In *The Chemistry of the Actinide Elements,* 2nd ed.; Katz, J. J.; Seaborg, G. T.; Morss, L. R., Eds.; Chapman and Hall: London, 1986; Vol. 2, pp 1025–1070.
5. Hettwer, U. *Np, Pu ... Transuranium Elements;* Gmelin Handbuch der Anorganischen Chemie; Springer-Verlag: Berlin, Germany, 1979.
6. Martin, R. C.; Knauer, J. B.; Balo, P. A. *Appl. Radiat. Isot.* **2000,** *53,* 785–792.
7. Chiles, M. M.; Mihalczo, J. T.; Fowler, C. E. *IEEE Trans. Nucl. Sci.* **1993,** *40,* 816–818.
8. *Trans. Am. Nucl. Soc.* **2000,** *82,* 94–110 (Applications of ^{252}Cf, session summaries).
9. Barton, J. P.; Sievers, W. L.; Rogers, A. H. In *Fifth World Conference on Neutron Radiography;* DGZfB: Berlin, Germany, 1997; pp 496–503.
10. Mihalczo, J. T., et al. *Field Use of NMIS at Oak Ridge;* Oak Ridge Y-12 Plant Report Y/LB-16,019: Oak Ridge, TN, 1999.
11. Craig, R. A.; Peurrung, A. J.; Stromswold, D. C. *Trans. Am. Nucl. Soc.* **2000,** *83,* 267–268.
12. Datema, C. P.; Bom, V. R.; van Eijk, C. W. E. *Nucl. Instr. Meth. B* **2002,** *488,* 441–450.
13. Kos, S. E.; Price, R. K.; Randall, R. R. *Geophysical Logging at Pit 9, INEEL;* Waste Management Federal Services, Inc.: Richland, WA, 2000.
14. Ancore Corporation, Santa Clara, CA, URL http://www.ancore.com.
15. Pillay, A. E. *Appl. Radiat. Isot.* **2002,** *56,* 577–580.
16. DiPrete, D. P.; Peterson, S. F.; Sigg, R. A. *J. Radioanal. Nucl. Chem.* **2000,** *244,* 343–347.
17. Martin, R. C.; Byrne, T. E.; Miller, L. F. *J. Radioanal. Nucl. Chem.* **1998,** *236,* 5–10. See also URL http://www.ornl.gov/nstd/cuf.
18. Whitehouse, A. I.; Young, J.; Evans, C. P. In *Laser Induced Plasma Spectroscopy and Applications;* OSA Trends in Optics and Photonics (TOPS), Vol. 81; Optical Society of America: Washington, DC, 2002; pp 2–4.
19. Martin, M.; Wullschleger, S.; Garten, C., Jr. *Proceedings of SPIE Conference on Advanced Environmental Sensing Technology II;* International Society for Optical Engineering: Bellingham, WA, 2002; Vol. 4576, pp 188–195.
20. Clark, D. D. *Trans. Am. Nucl. Soc.* **1994,** *71,* 162–163.

21. Clifford, E.; Ing, H.; McFee, J.; Cousins, T. *Proceedings of SPIE Conference on Penetrating Radiation Systems and Applications;* International Society for Optical Engineering: Bellingham, WA, 1999; Vol. 3769, pp 155–166.
22. Musienko, Y.; Reucroft, S.; Ruuska, D.; Swain, J. *Nucl. Instr. Meth. A* **2000,** *447,* 437–458.
23. Alderman, J.; Job, P. K.; Martin, R. C.; Simmons, C. M.; Owen, G. D. *Nucl. Instr. Meth. B* **2002,** *481,* 9–28.
24. Pinson, S. R. M. United States Department of Agriculture, unpublished results, 2002.

Chapter 8

Sequential and Simultaneous Radionuclide Separation–Measurement with Flow-Cell Radiation Detection

R. A. Fjeld[1], J. E. Roane[1], J. D. Leyba[2], A. Paulenova[3], and T. A. DeVol[1]

[1]Environmental Engineering and Science, L. G. Rich Environmental Research Laboratory, Clemson University, Clemson, SC 29634–0919 (fax: 864–656–0672, fjeld@clemson.edu)
[2]School of Science, Mathematics, and Allied Health, Newman University, Wichita, KS 67213
[3]Department of Nuclear Chemistry, Comenius University, Bratislava, Slovakia

Abstract Two techniques for measuring charged particle emitting radionuclides at low concentrations in aqueous solutions are presented. The first technique uses ion chromatography for elemental separation connected in series with a flow-cell scintillation detection for quantification of the eluted alpha- and beta-emitting radionuclides. The second technique uses scintillating extraction chromatographic resin contained within a flow-cell scintillation detection system for simultaneous separation and detection of actinides. The measured concentrations of Sr(II), Cs(I), Am(III), Cm(III) Pu(III, IV, V), and U(VI) in a high activity drain tank sample and in dissolved high-level waste tank sludge were compared with independently measured values and shown to be in good agreement. Advantages and disadvantages of these two rapid analytical methods are presented.

Introduction

The measurement of pure beta emitters, transuranics, and other radionuclides that emit no gamma rays, have low gamma emission probabilities, or emit gamma-rays of very low energy pose special problems in radiological characterization and monitoring. This is because their analysis typically requires labor and time intensive radiochemical procedures to separate them from a sample matrix. A class of new analysis techniques is being developed which combines chromatographic separation with either sequential or simultaneous radiation detection. Chromatography is used for either or both of two functions – to separate radionuclides from the matrix and to separate radionuclides from one another. The radionuclides are detected either following (sequential detection) or during (simultaneous detection) the chromatographic separation. Since these techniques combine separations with detection, they offer the potential for relatively rapid analysis of heretofore problematic radionuclides.

Approach

The sequential and simultaneous separation and detection configurations are illustrated schematically in Figure 1. Both configurations include a sample preparation step in which the radionuclides are transferred from the sample matrix to an aqueous matrix that facilitates separation. In the sequential separation/detection configuration, the radioactive analytes are loaded onto an ion exchange column and subsequently removed with chemical eluents (mobile phase). Elemental separation is achieved through the selection and sequencing of the eluents. The mobile phase is routed through a flow-cell detector where the separated radioactive fractions are counted as they pass through. In the simultaneous separation/detection configuration, the analytes are removed from the aqueous sample matrix by a specialized form of extraction chromatography in which the extractive resin is not only selective for a specific class of radionuclides (such as the actinides), it is also a scintillating material. By positioning a photodetector near the scintillator-packed flow-cell, it is thus possible to simultaneously separate the analyte from the mobile phase and detect the presence of radioactivity. The signal from multiple radioactive elements can be deconvolved through the subsequent selective elution of specific oxidation states from the flow-cell. Stop-flow operation can be utilized to increase counting time and thus reduce counting uncertainty. Whereas both techniques separate analytes based on oxidation state, the sequential technique has the advantage of a degree of elemental separation as well. For both techniques it is possible to collect effluent fractions for subsequent isotopic analyses.

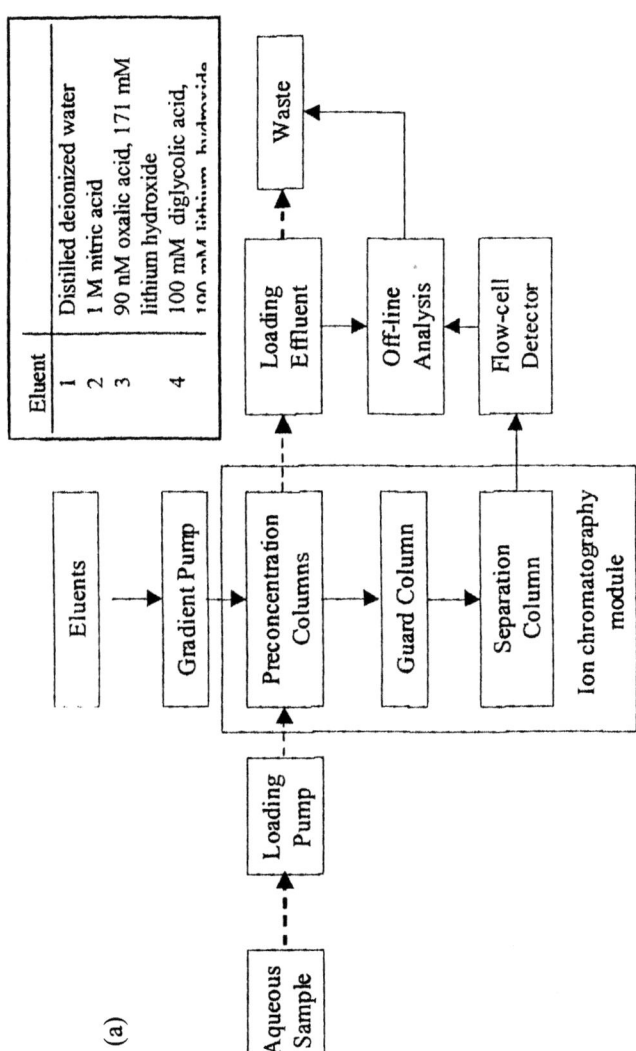

Figure 1. (a) *Schematic of the sequential separation and detection system.*
(b) *Schematic of the simultaneous separation and detection system. The dashed arrows connecting the aqueous sample to the waste collection represent the sample loading sequence. The solid arrows connect the analysis sequence*
Continued on next page.

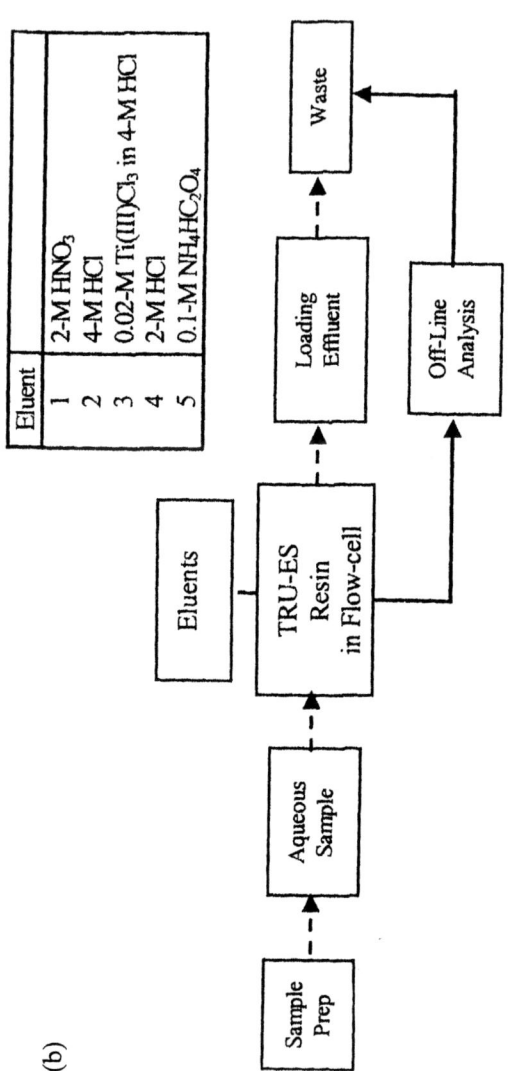

Figure 1. *Continued.*

Background

Flow-cell Scintillation Detector

Two general types of flow-cell scintillation detectors, heterogeneous and homogeneous, have been used to measure beta-emitting radionuclides in the biomedical and radiopharmaceutical industries for over 35 years *(1,2,3,4,5)*. Both types consist of a coiled transparent tube or an optical cell that is positioned adjacent to a single photomultiplier tube (PMT) or between a pair of PMTs operated in coincidence to lower the background signal. The heterogeneous flow-cell is packed with a granular solid scintillator. As a solution containing radioactivity flows through the interstitial space of the flow-cell, some of the emissions from the radionuclides interact with the scintillator granules and produce a signal. In the homogeneous flow-cell, a pre-mixed stream consisting of the aqueous sample and liquid scintillation cocktail produces visible photons that the photomultiplier tubes converts into the electronic signal. Today there are a variety of heterogeneous and homogeneous configurations. The principal solid scintillators are CaF_2:Eu, which is an inorganic crystal, and cerium doped yttrium silicate, which is a glass. Any of the commercial liquid scintillation cocktails can be used in the homogeneous flow-cells, but some lower viscosity solutions are specifically formulated for flow-cell applications. One of the major advantages of the homogeneous system is the miscibility of the sample-cocktail mixture. This allows nearly every charged particle to transfer some or all of its energy to the scintillator and yields high detection efficiencies.

The design of the flow-cell detector, whether homogeneous or heterogeneous, is a compromise among detector sensitivity, analysis time, and chromatographic resolution *(4,6)*. It is important for the volume of the flow-cell to be as small as possible to minimize peak broadening due to the detector volume; yet, any decrease in cell volume is accompanied by a decrease in the sensitivity. Since the accuracy of any radioactivity measurement depends on the number of events that are recorded, the longer the measurement period the more accurate the result. However, the counting time is linked to the transit time, which is governed by the flow rate and the flow-cell volume. Lowering the flow rate to increase counting time adversely affects chromatographic separation by causing peak broadening. Studies indicate that chromatographic resolution is preserved if the flow-cell volume is less than 1/10 the volume of the average chromatographic peak, resulting in relatively high sensitivities with only a 10% increase in peak width *(6)*. With pulse shape discrimination (PSD), it is possible to distinguish between scintillation events from alpha particles and beta particles. Heterogeneous PSD flow-cell detection has been demonstrated with polymer coated CsI:Tl granules using both analog *(7)* and digital electronics *(8)*. These systems offer good pulse shape discrimination of alpha/beta radioactivity in water with relatively high detection efficiencies. An alternative heterogeneous PSD flow-cell detection system utilizing a spiral channel positioned between two mylar covered CsI:Tl crystals was developed by Usada et al. (9). This configuration gave excellent pulse shape discrimination, but at the expense of detection efficiency. Hastie et al. (10) developed a homogeneous PSD flow-cell detection system which has both good alpha/beta PSD capability and good detection efficiency of the alpha and beta radiation.

Sequential Separation and Detection

The sequential separation and detection configuration illustrated in Figure 1a is also given the name ion exchange radiochromatography. Bradbury et al. (11) demonstrated the usefulness of ion exchange chromatography in separating radioactive alkali metals, alkaline earth, lanthanides and halides that emit no gamma radiation or emit very weak gamma radiation. Reboul and Fjeld (12) extended the technique by focusing on system performance, i.e. elution effectiveness, peak characteristics, response linearity, detection efficiencies, and sensitivities for those similar radionuclides. Additional work by Reboul (13) concentrated on the development of an elution program capable of separating actinides. Applications of the radiochromatography technique were tested with synthetic reactor coolant (14), and actual coolant from a pressurized water reactor (15). In most of these applications, the sample introduced into the coupled system was relatively "clean" with few or no interfering constituents. However, routine applications of the technique to environmental samples are limited by interferences from the sample matrix, which can affect the separation or detection of radionuclides. Fjeld et al. (14) investigated the effects of potential interferents (boric acid, lithium hydroxide, hydrazine, and ^{137}Cs) in coolant from a pressurized water reactor. Reboul and Fjeld (16) investigated the interfering effects from dissolved ions (Na^+, K^+, Ca^{+2}, Cl^-, and SO_4^{-2}), humic acid, and radium on the separation of actinides in solution. Roane et al., (17) investigated the effect of radiological and chemical interferences on the chromatographic separation of a mixture of activation and fission products. Fjeld et al. (18) subsequently applied the technique to waste stream and environmental samples from a Department of Energy facility.

Chromatography of radioisotopes has also been performed using additional analytical detection techniques. Isolation of ^{90}Sr and ^{137}Cs while separating and measuring the non-radioactive constituents of nuclear waste was conducted using a conductivity cell for detection of the non-radioactive constituents and liquid scintillation counting for the strontium and cesium (19). Garcia-Alonso et al. (20) replaced the scintillating flow-cell detector with an inductively coupled plasma mass spectrometer (ICP-MS) to provide isotopic information about the separated radionuclides in spent nuclear fuel. Farmer et al. (21) added an ICP-

MS to the effluent end of an on-line scintillation flow-cell to utilize the capability of the flow-cell detector to measure the short-lived radionuclides and the ICP-MS to quantify components of co-eluted peaks as well as the long-lived and stable nuclides in a Hanford waste tank sample. In lieu of having access to an ICP-MS or similar type instrumentation, other radiochemical methods are required to concentrate the radioactive analytes and reduce the interfering effects of the sample matrix.

Simultaneous Separation and Detection

Extraction chromatography combines all of the favorable features of a liquid-liquid system with the convenience of a chromatographic system and the handling ease of an ion-exchange resin. Extraction chromatography resin, like liquid-liquid extraction, involves partitioning accompanied by chemical changes in converting a hydrated metal ion to a neutral, organophilic metal complex. Further, if no interactions occur between the support and the molecules of either the extractant used or the complex formed, the chromatographic behavior of a metal ion can be correlated with the behavior of the same metal ion in a liquid-liquid system containing the same extractant (22).

Although extraction chromatographic materials were proposed as early as 1959 (23), relatively few materials found widespread application due to inadequate selectivity, poor stability, and limited range of useable loading conditions. By the late 1980's, a variety of improvements had been made both in extractants and in methods for impregnating the support material with the extractant. Beginning in early 2000, simultaneous separation and detection was demonstrated with scintillation extraction chromatographic resins (24).

Simultaneous separation and detection has the potential of being more sensitive than the sequential technique due to longer counting times available with analyte sorption and stop-flow operation. Anionic and cationic scintillating ion-exchange resins for the separation and detection of ionic radionuclides were developed and tested successfully with a variety of environmental and experimental scenarios (25). Although discussed in that early work, simultaneous separation and detection was not demonstrated. Li and Schlenoff (26) sulfonated the surface of a cube of plastic scintillator for the concentration and detection of cations in solution. Extractive scintillating resins, which are analyte selective, are produced by coating a thin, transparent extractant onto the scintillator. The concept of the extractive scintillator resin was demonstrated using the extractants octyl(phenyl)-N,N-diisobutyl-carbamoylmethylphosphine oxide (CMPO) in tributyl phosphate (TBP) and bis(2-ethylhexyl)methane-diphosphonic acid ($H_2DEH[MDP]$), tricaprylylmethylammonium chloride (Aliquat 336), and bis-4,4'(5")-tert-butylcyclohexano-18-crown-6 in 1-octanol for the concentration and detection of actinides, pertechnetate, and strontium, respectively *(27,28,29,30,3)*. A similar approach was used by Egorov *et al.* *(32)* to develop selective scintillating microspheres for the determination of ^{99}Tc or ^{90}Sr/^{90}Y in aqueous solutions using a bead injection radionuclide sensor system. An alternative approach undertaken by Headrick *et al.*, utilized a cesium selective dual-mechanism bifunctional polymer (DBFP) resin epoxied to scintillating fiber *(33)* This configuration yielded rapid transfer kinetics to bring cesium into the polymer matrix and close to the recognition sites that retain the analyte.

Analytical Systems

Sequential Separation and Detection

The apparatus for sequential separation and detection, which at one time was marketed under the trade name ANABET, consists of an ion chromatography module (Model CHA-6, Dionex Corp.), a gradient pump (Model GPM-2, Dionex Corp.), and an on-line flow-cell scintillation detector (Model β-RAM 1A, IN/US Systems, Inc.). A schematic diagram of the apparatus is presented in Figure 1a; a more complete description of the system components can be found in Reboul and Fjeld *(12)*.

A cation-exchange column (HPIC CG2, Dionex Corp.) separates and concentrates the cationic species in the sample matrix prior initiating the chromatography. This pre-concentration column has an ion-exchange capacity of 12 microequivalents. Non-ionic radioactive species (e.g., tritiated water) and anionic radioactive species are separated from the radionuclide mixture during the loading of the solution onto the cationic pre-concentration column. (The system is not presently designed for the separation and detection of non-ionic radioactive species, although the system can be configured for anionic radioactive species analysis.) The analysis is initiated by routing eluent flow through the pre-concentration column with separation of the cationic species occurring on the mixed-bed ion exchange column (IonPac CS5, Dionex Corp).

The mobile phase passes through the on-line scintillation detector that detects the radioactive ions exclusively. The on-line scintillation detector consists of Teflon tubing packed with either 3% Parylene C coated 63-90 μm diameter CsI:Tl particles or 4% Parylene C coated 90-125 μm diameter CsI:Tl particles, which is positioned in the pulse shape discriminating flow-cell detection system *(7)*. Peaks in the resultant chromatogram occur as the separated radionuclides proceed through the detector. Identification of the ionic species can be inferred from the peak elution time. The active volume of the scintillation flow-cell is ~0.4 mL, and for a flow rate of 1 mL min^{-1} yields an effective counting time of 24 seconds. Analysis of the chromatogram (i.e., determination of elution times and peak areas) is performed with Aptec data acquisition software and Microsoft Excel.

Simultaneous Separation and Detection

Simultaneous separation and detection is achieved with a combination of TRU-ES extractive scintillating resin (Eichrom Technologies, Inc.) and a flow-cell scintillation detector, Figure 1b. The TRU-ES resin beads have a metal ion uptake performance similar to the commercially available TRU resin (Eichrom Technologies, Inc.) but with the added advantage of the resin core consisting of a plastic scintillator *(27)*. The resin is is packed into Teflon tubing and formed into a coil shape, yielding a flow-cell with a pore volume of approximately 0.2-mL. Data acquisition and analysis is accomplished with a custom-built flow-cell detection system operated in coincidence mode that is capable of simultaneous acquisition of pulse height and multichannel scaling data (count rate vs. time)*(27)*. Solutions are pumped through the flow-cell at a flow rate of 0.5 mL-min^{-1}. The loading and elution protocol for the sequential separation of actinides from the TRU-ES resin is an adaptation of the procedures developed by Horwitz and co-workers *(34)*. Following resin conditioning with 2 M HNO$_3$, the sample is acidified to 2 M HNO$_3$ and loaded onto the column followed by a 2 M HNO$_3$ wash. Using stop-flow, the gross alpha count rate is established. The trivalent actinides are then eluted with 4 M HCl and again the flow through the column is stopped to establish the count rate for the retained radionuclides. Plutonium is reduced and eluted with 0.02-M Ti(III)Cl$_3$ in 4-M HCl after Grate and Egorov *(35)*. The Horwitz procedure uses hydroquinone as a reductant; however, hydroquinone is a strong chemical quenching agent that prevents its use in a scintillation counting application. Following a period of stop flow to establish the count rate for the retained radionuclides, tetravalent actinides An(IV) are eluted with 2 M HCl which also decreases color quench produced by the TiCl$_3$. Following a period of stop flow, hexavalent actinides are eluted with 0.1 M NH$_4$HC$_2$O$_4$ (ammonium bioxalate) and again the count rate is established.

The on-line detection efficiencies for ^{241}Am, ^{239}Pu and ^{233}U are 96.5%, 77.5% and 96.6%, respectively *(27)*. The relatively low ^{239}Pu detection efficiency is attributed to the photon quenching effect of the titanium chloride in the elution solution. The gross alpha detection efficiency for the on-line method is 89.3% when analyzing a solution containing nearly equal activities of ^{241}Am, ^{239}Pu and ^{233}U.

Results of Sample Analyses

Results of analyses of two samples are presented here. The first is a liquid waste sample from radiochemistry laboratories, and the second is acid digested sludge from a high-level radioactive waste tank. The results of these analyses are compared with independent measurements performed by the Savannah River Technology Center (SRTC) using traditional radioanalytical procedures.

High Activity Drain Tank

The high activity drain tank at SRTC is a reservoir that receives liquid laboratory wastes of relatively high activity. A 250 µL sample from this tank was analyzed in the sequential system without any sample preparation. The chromatograms are presented in Figure 2. Based on elution times, plutonium, americium, and curium were detected in the alpha chromatogram; and cesium, strontium, and yttrium were detected in the beta chromatogram. There was some spillover of curium into the beta chromatogram.

Five 10-µL aliquots from the tank were analyzed using the simultaneous system, and the results were averaged. One of the five chromatograms is presented in Figure 3. The 90 cps plateau from 20 min. to 44 min. is the due to all of the actinides and represents the gross alpha response. The trivalent actinides were stripped from the column beginning at 44 min., and the resulting plateau from 50 to 125 min. is due to the remaining tetravalent and hexavalent actinides. Thus, the concentration of trivalent actinides is determined from the difference between the gross alpha response plateau and the tetravalent/hexavalent actinide plateau. The plutonium reduction/elution begins at 125 min. and the plateau from 175 min. to 225 min. is due to the hexavalent uranium which remains. Uranium elution begins at 230 min.

The results of the SRTC, sequential, and simultaneous analyses of the high activity drain tank sample are presented in Table I. The sequential measurements and the SRTC measurements are all within 20% of one another. The simultaneous measurements of gross alpha and An(III) were approximately 30% below the other two, and the plutonium measurement was much lower. Since the simultaneous analysis was conducted approximately six months after the other two, it is hypothesized that the low measurement was due to actinide sorption on the walls of the sample container between the analyses. This is supported by liquid scintillation analyses of two aliquots of the tank supernatant which yielded a gross alpha concentration of 10,180 Bq mL^{-1}. The simultaneous measurement of the gross alpha concentration is within 5% of the liquid scintillation analysis.

TABLE I. Summary of High Activity Drain Tank Sample. Sequential measurement data obtained using a 4% 90-125 μm Parylene C coated CsI:Tl flow-cell. Simultaneous measurements obtained with TRU-ES resin.[a]

Radionuclide	SRTC Measurement (Bq mL^{-1})	Sequential Measurement (Bq mL^{-1})	Simultaneous Measurement (Bq mL^{-1})
^{90}Sr	-	1,130	
^{137}Cs	1,557	1,880	
Gross Alpha (Pu + Am + Cm)	12,649	11,790	9,622
^{238}Pu	828	690	163
An(III)	11,821	11,100	8,779
^{241}Am	1,165	1,080	
^{244}Cm	10,656	10,020	

a. Sample standard deviation of simultaneous measurements was less than 30% based on 5 samples. Uncertainty not determined for other two methods.

High Level Waste Tank Sludge

Sludge was obtained from the archives for high-level waste tank #8 at the Department of Energy's Savannah River Site. A 0.125 g sub-sample was dissolved into 100 mL of 0.1 M HNO3. A 250-μL aliquot was analyzed in the sequential system without sample preparation. Cesium, strontium and yttrium were detected in the beta chromatogram; plutonium, americium and curium were detected in the alpha chromatogram. Three 25-μL aliquots were analyzed in the simultaneous system, and the results were averaged. Presented in Table II are the results of the SRTC, sequential, and simultaneousanalyses. The sequential measurements and the SRTC measurements were within 15% of one another for strontium, americium and curium and within 30% for cesium and plutonium. The simultaneous measurements of gross alpha concentration was 10% lower than the sequential measurement and 35% lower than the SRTC measurement, and the plutonium measurement was 45% lower than the sequential measurement and 60% lower than the SRTC measurement. The An(III) measurement, however, was over three times higher than the sequential and SRTC measurements. The lower plutonium and higher An(III) results from the simultaneous analysis are possibly due to the presence of trivalent plutonium in the sample.

TABLE II. Summary of the HLW sludge sample analysis. Sequential measurement data obtained using a 3% 63 - 90 μm Parylene C coated CsI:Tl flow-cell. Simultaneous measurements obtained with TRU-ES resin.[a]

Radionuclide	SRTC Measurement ($Bq\ mL^{-1}$)	Sequential Measurement ($Bq\ mL^{-1}$)	Simultaneous Measurement ($Bq\ mL^{-1}$)
^{90}Sr	33,300	28,410	
^{137}Cs	1,170	850	
Gross Alpha (Pu + Am + Cm)	1,025	735	662
^{238}Pu	920	640	361
An(III)	105	95	341
^{241}Am	45	41	
^{244}Cm	60	54	

a. Sample standard deviation of simultaneous measurements was less than 22% based on 3 samples. Uncertainty not determined for other two methods.

Here again, the sequential and simultaneous analyses could have been affected, in part, by sorption to walls of the container. A liquid scintillation analysis at the time of the simultaneous measurement yielded a gross alpha concentration of 827 Bq mL^{-1}, which is considerably lower than the 1027 Bq mL^{-1} measured by SRTC before the sample was sent to our laboratory.

Detector degradation is one problematic area of both the sequential and simultaneous systems. In the sequential method degradation of the Parylene C coated CsI:Tl scintillator with continual exposure to the chromatography mobile phase results in a lower detection efficiency over time. In the simultaneous system, the primary limitation is leaching of the fluor from the resin during the elution procedure, which results in a decline in detection efficiency of 4 to 5 % after three uses. Additional problematic areas include: chemical and color quenching, and the non-uniform sensitivity of the photomultiplier tubes.

Summary

There are advantages and disadvantages to the sequential and simultaneous separation and detection methods. The sequential separation and detection system can separate and quantify the trivalent actinides Am and Cm, which are co-eluted in the simultaneous system. Additionally, the sequential system can

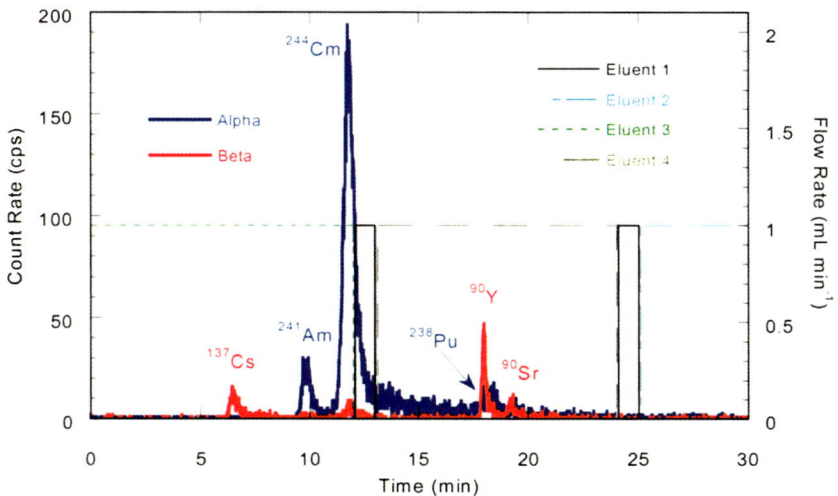

Figure 2. Alpha (blue) and beta (red) chromatograms for the supernatant from the high activity drain tank at SRTC.

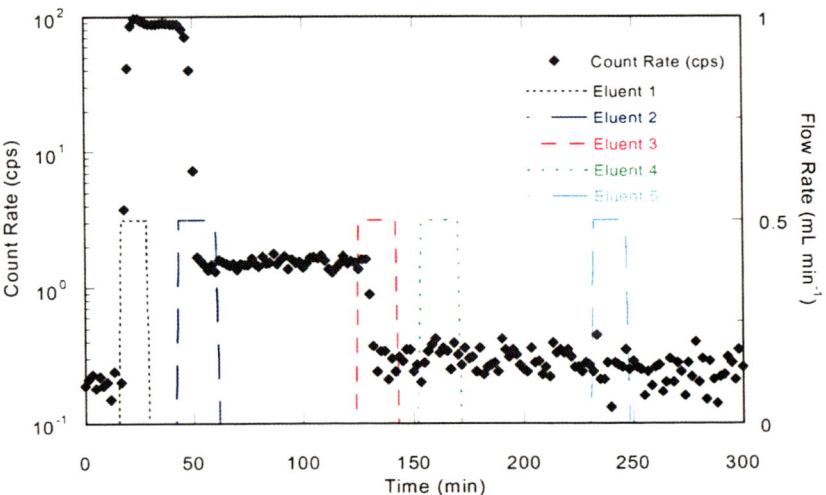

Figure 3. Chromatogram obtained with scintillating extraction chromatographic separation of the high activity drain tank supernatant.

quantify both alpha and beta emitting radionuclides at the same time, whereas the simultaneous system is selective to a group of radionuclides, e.g. actinides or individual radionuclides. The sequential separation was demonstrated as a semi-automated system, whereas the simultaneous system, although favorable for such automation, has not been demonstrated. The detection sensitivity of the simultaneous system is superior to that of the sequential system. This is due to a combination of a higher absolute detection efficiency, increased count time, and the fraction of the total activity in the flow-cell at any moment of time, which is evident when comparing the count rate and sample volumes used for the systems. The detection efficiency of the simultaneous system is high (85-95%) relative to the sequential system (55-65%) because the radioactive material adheres to the surface of the scintillator in the former system. The count time for the simultaneous system is inherently longer because the radionuclides are retained by the scintillator, and it can easily be extended by stop-flow operation. The sequential system, on the other hand, has a fixed count time of 24 seconds to assure good chromatographic resolution. The fixed and relatively small flow-cell pore volume results in only a fraction (~15%) of the total activity in a peak being present in the flow cell at any given time.

Acknowledgments

This research was sponsored in part through National Science Foundation SBIR, DOE/EMSP Grant 70179, DOE/SCUREF Cooperative Agreement Project 005, and DOE/NETL Contract DE-AR21-95MC32110. Acknowledgment is also extended to James Harvey and Joel Williamson at Eichrom Technologies, Inc. for preparation of the extractive scintillator resin

References

1. Schram, E.; Lombaert, R. *Anal. Biochem.* **1962**, *3*, 68-74.
2. Clifford, K. H.; Hewitt, A. J. W.; Popják, G. *J. Chromatogr.* **1969**, *40*, 377-385.
3. Scharpenseel, H. W.; Menke, K. H. *Tritium in the physical and biological sciences*; IAEA: Vienna, Austria, 1962; Vol. 1, p281.
4. Schutte, L. *J. Chromatogr.* **1972**, *72*, 303-309.
5. Rapkin, E.; Packard, L.E. University of New Mexico Conference on Organic Scintillator Detectors, Albuquerque 1960.
6. Potashner, S. J.; Lake, N.; Knowles, J. D. *Anal. Biochem.* **1981**, *112*, 82-89.

7. Chotoo, S. D.; DeVol, T.A.; Fjeld, R.A. *IEEE Trans. Nucl. Sci.* **1997**, *44 (4),* 1630-1634.
8. Tan, H.; Fjeld, R.A.; DeVol, T.A. *IEEE Trans. Nucl. Sci.* **2000**, *4 (4)* 1516-1521.
9. Usuda, S. *J. Radioanal. Nucl. Chem.* **1988**, *123(2),* 619-631.
10. Hastie, K.H.; DeVol, T.A.; Fjeld, R.A. *Nucl. Instrum. Methods Phys. Res. (Section A),* **1999**, *A422,* 133-138.
11. Bradbury, D.; Elder, G. R.; Dunn, M. J. *Waste Management '90* **1990**, 327-329.
12. Reboul, S.H.; Fjeld, R.A. *Radioact. Radiochem.* **1994**, *5,* 42-49.
13. Reboul, S. H. Ph.D. Thesis, Department of Environmental Systems Engineering, Clemson University, Clemson, SC, 1994.
14. Fjeld, R. A.; Guha, S.; DeVol, T. A.; Leyba, J. D. *J. Radioanal. Nucl. Chem.* **1995**, *194 (1),* 51-59.
15. Guha, S.; Roane, J. E.; DeVol; T. A.; Leyba, J. D.; Fjeld, R. A. *Waste Management '95, (Proceedings of the Symposium on Waste Management),* R. G. Post, Ed., Tucson, AZ, 1995.
16. Reboul, S.H.; Fjeld, R.A. *J. Health Phys.* **1995**, *68,* 585-589.
17. Roane, J.E.; DeVol, T.A.; Leyba, J.D.; Fjeld, R.A. *J. Radioanal. Nucl. Chem.* **1999**, *240 (1),* 197-208.
18. Fjeld, R. A.; DeVol, T. A.; Leyba, J. D.; Paulenova, A. (in review), *J. Radioanal. Nucl. Chem.*
19. Cadieux, Jr., J. R.; Salaymeh, S. R.; Griffin, H. C. *Radioact. Radiochem.* **1996**, *7 (2),* 39-49.
20. Garcia-Alonso, J. I.; Sena, F.; Arbore, P.; Betti, M.; Koch, L. *J. Anal. At. Spectrom.* **1995**, *10,* 381-393.
21. Farmer, III, O. T.; Reeves, J. H.; Wyse, E. J.; Clemetson, C. J.; Baringa, C. J.; Smith, M. R.; Koppenaal, D. W. In *Applications of Inductively Coupled Plasma-Mass Spectrometry to Radionuclide Determinations, ASTM STP 1291;* Morrow, R. W.; Crain, J. S.; Eds.; American Society of Testing and Materials, West Conshohocken, PA, 1995, pp 38-47.
22. Dietz, M.L.; Horwitz, E.P. *LC/GC,* **1993**, *11(6),* 424-436.
23. Siekierski, S.; Kotlinska, B. *Atomic Energiya* **1959**, *7,* 160.
24. DeVol, T. A.; Roane, J. E.; Williamson J. M.; Duffey, J. M.; Harvey, J. T. *Radioact. Radiochem.* **2000**, *11(1),* 34-46.
25. Heimbuch, A. M.; Gee, H. Y.; DeHaan, Jr., A.; Leventhall, L. *Proceedings of the International Atomic Energy Agency Symposium,* Vienna, Austria, **1965**, pp 505-519.
26. Li, M.; Schlenoff, J. B. *Anal. Chem.* **1994**, *66,* 824-29.
27. Roane, J.E.; DeVol, T.A. *Anal. Chem.* **2002**, *74,* 5629-5634.
28. Hughes, L.; DeVol, T. A. *in press Nucl. Instrum. Methods* **2002.**
29. DeVol, T.A.; Egorov, O.B.; Roane, J.E.; Paulenova, A.; Grate, J.W. *Radioanal. Nucl. Chem.* **2001**, *249 (1),* 181-189.

30. Ayaz, B.; DeVol, T. A. *in press Nucl. Instrum. Methods* **2002.**
31. DeVol, T.A.; Duffey, J.M.; Paulenova, A. *Radioanal. Nucl. Chem.* **2001,** *249 (2),* 295-301.
32. Egorov, O. B.; Fiskum, S. K.; O'Hara, M. J.; Grate, J. W. *Anal. Chem.* **1999,** *71,* 5420-5429.
33. Headrick, J.; Seaniak, M.; Alexandratos, S.; Datskos, P. *Anal. Chem.* **2000,** *72,* 1994-2000.
34. Horwitz, E. P.; Chiarizia, R.; Dietz, M. L.; Diamond, H. *Anal. Chim. Acta* **1993,** *281,* 361-372.
35. Grate, J. W.; Egorov, O. E. *Anal. Chem.* **1998,** *70 (18),* 3920-3929.

Chapter 9

α-Autoradiography: A Simple Method to Monitor the Migration of α-Emitters in the Environment

Carola A. Laue and David K. Smith

Analytical and Nuclear Chemistry Division, Lawrence Livermore National Laboratory, 7000 East Avenue, Livermore, CA 94550

Environments contaminated with α-emitters (mostly actinides) represent a serious hazards to humans and nature. Current technologies to monitor and remediate areas contaminated with α-emitters are often hindered by our limited understanding of the actinide fate and transport behavior in those environments. Here we demonstrate how the well-established simple technique of α-autoradiography can be deployed in a field setting to enhance our understanding of actinide transport, to determine the spatial distribution of actinides in soils, and to help identify the morphology of the association of actinides with soil constituents. Preliminary conclusions are drawn relative to controls that govern the fate and transport of actinides that were deposited more than 40 years ago in the semi-arid soils of the Nevada Test Site. Our experimental data clearly indicate that actinides move through desert soils. This observation contradicts previous assumptions that actinides are relatively immobile and remain within the very top soil layer. We were able to demonstrate that the designed tool can be applied for screening in monitoring and remediation efforts.

Introduction

The existence of α-emitters in our natural environment is a result of both natural and of anthropogenic factors. Of major concern to the public are the lower actinide elements such as uranium, plutonium and americium, all of which can be found as low-level background throughout the world today. This background distribution of actinide elements has several anthropogenic sources including global and local fallout from atmospheric weapons testing, reentry of space equipment powered by plutonium heat sources, accidental releases from fuel production and reprocessing, nuclear power generation and other military applications.

The fate and transport of actinide

The fate and transport of the actinide elements has become a major concern at contaminated U.S. Department of Energy nSite, the Hanford Reservation, and at the Nevada Test Site (NTS). Reasons for increased concern are motivated by path-to-closure remediation efforts of DOE complexes and originate in the toxicity and radioactivity of actinides. Although the toxic properties are comparable to other heavy metals including lead and mercury, the tissue damaging properties of their radioactivity, distinguish the actinides from heavy metals. Actinide-bearing aerosols pose serious health risks (*1*), as they can be directly inhaled or be deposited on food plants. Equally hazardous is the transport of actinides through the soil, creating a pathway to either enter water sources or to accumulate in food plants. Those concerns were already recognized in the mid-1970s by the Nevada Applied Ecology Group (NAEG). Extensive studies (*2-4*) have been performed investigating the resuspension of actinide and fission products and their potential pathways to human ingestion. Results from studies on the distribution of actinide in soils yielded the controversial conclusion that actinides are relatively immobile. The NAEG results showed that after an initial fast movement of the actinide within the top layer of the soil, the infiltration of the actinides stopped and did not change over the course of 20 years (*5, 6*). Recent studies (*7*) of an aquifer at the NTS, which is in contact with a test cavity indicate that plutonium can move in a saturated groundwater system. In a nuclear test cavity the actinides were thought to be incorporated in the melt glass formed after a nuclear detonation, and thereby, to be immobilized. Aside from the geology and weathering that occurs in the environment, the chemical nature of the contamination source strongly affects the fate and transport of the contaminant. One of our objectives was to investigate geochemical controls on actinide movement in desert soils. To accomplish this goal, a technique had to be devised that would allow us to study in-situ the range of actinide transport, the spatial distribution of the actinides in soils, and the morphology of the actinide association to soil constituents.

α-Autoradiography

α-Autoradiography, our technique of choice, relies on the actinide decay properties to visualize their spatial distribution. This simple and proven technique has been known for as long as radioactivity itself. The radiation creates latent tracks through its damaging interaction with the detector material, which is treated chemically to reveal the tracks after the exposure. In fact, Becquerel discovered simultaneously radioactivity and radiography by accidentally misplacing his photographic plates on a uranium mineral (8). Kinoshita (9) observed the first α-particle tracks induced in commercial photographic emulsion in 1910. Nuclear emulsions are now widely used in the nuclear chemistry and physics (10). After the discovery of tracks in dielectric solids, solid state nuclear track detectors (SSNTD) soon outcompeted emulsions for reasons including: ease of handling, transportability into remote and/or difficult to access areas, independence of any electronic equipment, speed of obtaining results, visualization of the results, possibility to automate track analysis, and their low cost. Common SSNTD materials are glass, mica, and several plastic materials. The most widely used plastic is CR-39, a polymer of polyallyl diglycol carbonate, which was also utilized in our study. CR-39 (11) responds uniform with less than 1% variation; a high sensitivity toward 1 MeV protons, 6 MeV alphas, and minimum-ionizing iron; and a superb optical quality making CR-39 ideal for identification of mass and charge of nuclear particles. Several studies (12, 13) have investigated the distribution of radioactive contaminants and transport mechanism of radioactive particulates in the environment using radiographic methods, but only one study (14) applied radiography for in-situ soil characterization.

Nevada Test Site and pertinent NAEG results

The NTS is a 3,561 km^2 vast desert area 105 kilometers north west of Las Vegas (see Figure 2). Atmospheric tests and safety experiment tests (SET) with non-critical or partial yield of nuclear weapons were performed at NTS. SETs involving a detonation of nuclear weapons by conventional explosives also have been done to assess the impacts of such terror attacts. As a result, extensive surficial actinide contamination occurred, which was thoroughly characterized immediately after testing; during the NAEG program in the 1970's (1-6, 15); and through ongoing efforts of the long-term stewardship program. In addition, the geological, hydrological, and weathering characteristics of the semi-arid environment at the NTS was well studied as part of U.S. nuclear testing efforts (16). Partial data from extensive NAEG studies pertaining to our investigation are summarized in Table I. The areas selected hosted SETs conducted at the surface resulting in dispersal of actinide material.

Table I: Summary NAEG investigation results documenting actinide occurrence in NTS surficial soils (15)

	Area 18 NS 201	Area 18 NS 219	Area 13	Area 11
Type of test	Safety w/ slight nuclear yield	Safety w/ slight nuclear yield	Safety	Safety
Documented type of actinide occurrence	Spherical glass or glass coated	Sponge-like highly porous, very fragile silicate glass	High density oxide particle	*
Particle size fraction w/ highest actinide conc.	0.1 to 2 mm	2-50 mm	20-50µm	*
Resuspendable Particle fraction bearing actinides (% of particles < 0.1 mm in Ø)	5	13	98	*
Depth with majority of actinide load in cm	top 5	top 5	top 2.5	top 2.5

* No information available

The NAEG investigations document that in locations where the safety tests had a slight nuclear yield, the particles bearing the majority of the actinides are a silicate based material and are so large as to not be susceptible to resuspension. In locations where the safety tests had no nuclear yield, the hot particles are extremely small, highly sintered actinide oxide particles. These differences result in variations in resuspension and migration behavior. The actinide contamination, which was measured directly after the safety tests, was found to be in the uppermost soil layer in the Areas 11 and 13. Twenty years later, NAEG studies determined that the actinides were mostly concentrated at a soil depth of 2.5 cm, but also found measurable amounts at depths of 25 cm. The vertical profile of the actinide contamination in Area 18, however, did not change with time indicating a strong correlation between source properties and transport behavior.

The objective of our investigation was to design and deploy a simple in-situ screening tool for α-emitting contaminants in soils. The spatial distribution of actinides in soils was of particular interest because actinide depth profiling allows us to evaluate the actinide fate and transport in desert environments. This in-situ technique was also developed due to its potential to support remedial actions that may be required in the future.

Experimental Work

Design of the in-situ soil exposure of CR-39 detectors to α-emitters

The design of the tool used to expose CR-39 track detectors to the soil column in the field was critical for our work. Features to minimize the soil disturbances, to avoid the cross contamination by transporting radioactive particulates downward while inserting the tool, and to protect the CR-39 from scratches that significantly decrease detector sensitivity, were incorporated into the design.

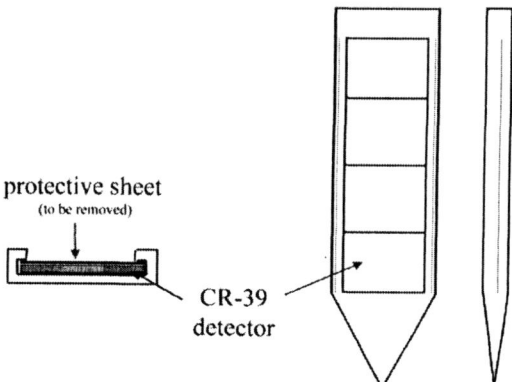

Figure 1: Schematic drawing of the exposure tool, designed and employed Drawing is not to scale.

A schematic of the in-situ exposure tool is provided in Figure 1. The actual dimensions of the tool are: 1 cm thick steel backing, 74 cm long, and 6 cm wide. As Figure 1 shows, the tool was constructed with a track to fix the detectors in place and to hold a removable shield in front of the detectors to avoid scratches during the emplacement in the ground. The CR-39 is a high quality SSNTD (TASTRAK™) obtained from Track Analysis System, Ltd. in Bristol, UK, 50x50x1 mm in dimension, each plate having a unique ID number engraved. The actual detector area exposed to the soil column is 46 mm wide and a maximum of 60 cm deep. The detectors were placed in sequential order and fixed in place by double sided tape.

CR-39 etching and evaluation procedure

Based on conventional etching procedures (17, 18), a 6 mol/L sodium hydroxide etch solution was heated and kept constant at 73°C in a water bath. During the etch process of 6 hours, the etch solution was stirred. The detectors, removed from exposure tool and transported from field site to our lab, were placed in the etch solution in a holder that spaced them vertically 1 cm apart. The etch process was stopped by rinsing the detectors several times with distilled water. An optical microscope equipped with a digital camera was used to examine the radiographic detectors.

Field site — Plutonium Valley, NTS

Plutonium Valley in Area 11 of the NTS was the location of four SETs in the mid-1950's, which contaminated the surface soils with uranium, plutonium and americium. Site D in Area 11 was selected in our study because the site had been previously studied by the NAEG. Area 11 offers the benefit of being secured and secluded (see its location in Figure 2) with disturbances to the desert soil limited to natural weathering processes and native animal activities subsequent to the last recorded soil suspension studies (1) and contamination surveys (19-21). Although actinide particulate partitioning had not been investigated here as for Area 13 and 18, substantial information regarding the contaminants exists (1-6, 15, 19-22). The references detail locations of those earlier studies, which enabled us to avoid areas with disturbed soil profiles. Shinn et al. (6, 22) published data based on NAEG studies on the vertical distribution of ^{239}Pu at Site D in 1993: in 1973, 99% of ^{239}Pu was located in the top 1.25 cm, while in 1981, 98.4% of the ^{239}Pu was in the top 7.5 cm of soil, and 1% was detected between 7.5 to 15 cm. However, the data from 1973 are based on chemical measurements, while the data from 1981 are based on γ-spectrometry, measuring ^{241}Am, and determining ^{239}Pu by a fixed ratio of ^{241}Am to ^{239}Pu of 1 to 7 (22, 23). Results described in (5) question the reliability of the fixed ^{241}Am to ^{239}Pu ratio to determine plutonium in soils as applied be measurements via the americium γ-activity. Americium, a trivalent cation, was observed to behave differently in soils than plutonium, which occurs as oxide, tetravalent cation or pentavalent plutonyl ion, and hence, obeys a different chemistry than americium. Earlier contamination surveys and later demarcation efforts (20) provided valuable information on the levels of contamination such as obtained by the 1994 aerial radiation survey measuring the ^{241}Am count rate per second for Plutonium Valley as shown in Figure 3, with Site D enlarged.

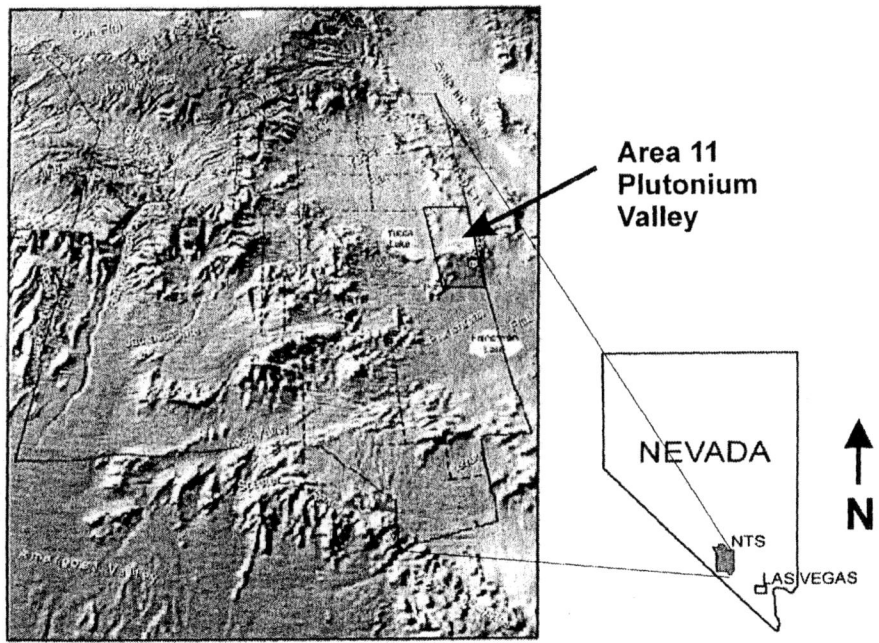

Figure 2: Topographic Map of NTS with superimposed area grid, Area 11 highlighted, showing NTSs location in Nevada. U.S. DOE Photograph. (Reproduced from reference 24).

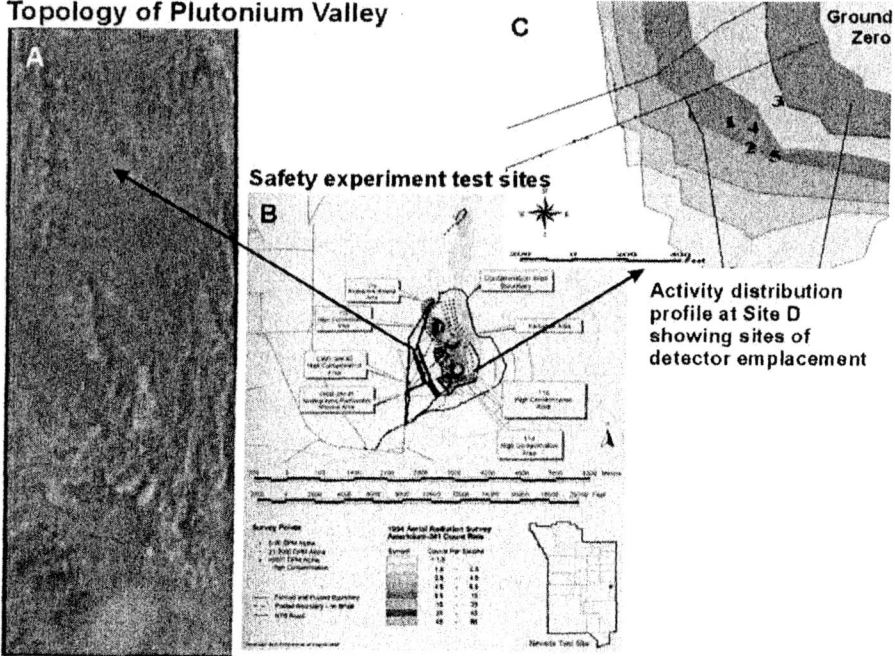

Figure 3: Plutonium Valley detailed: A) a topographical map identifying the SET sites;B) an activity level map of the SET sites as determined by 1994 arial gamma-ray survey (20); and C) our experimental site, Site D, indicating activity levels and detector deployment sites (1-5).

Field Exposure Experiments

Two in-situ exposure experiments were performed in October and December 2001. The exposure locations are labeled 1 through 5 in Figure 3C. After inserting the in-situ detector device into the ground using a field hammer, the detector protector sheet was carefully removed to avoid moving soil particulates up with the protector sheet. The in-situ detector device was covered to prevent rainwater from channeling down the surface of the detectors during the exposure. The exposure (St #1) in October lasted for 35 days. The analysis of the first set of detectors indicated that the ambient activity levels might result in overexposure. Hence, in December, the second set of in-situ experiments (St #2-5) were exposed to the soils for only two weeks. Table II summarizes the experimental parameters for each in-situ exposure. The distances of the exposure locations to ground zero of Site D range from 70 to 200 m. A manual survey with a fidler (NaI) gamma-ray probe provided an indication of the surface contamination at each location. "Characteristics" describes the susceptibility to wind erosion or to trapping of suspended particles, as observed in areas sheltered by rooted sage brush. The retrieval of the exposure tool was performed in a manner that avoided scratching of the detector material.

Table II: Experimental parameter of in-situ exposures of CR-39 detectors to soils in Plutonium Valley at the NTS, NV performed in 2001

ID	Duration	Exposure Depth	Experiment Location GPS coordinates	Fidler (NaI) reading dpm	Characteristics of Location
St #1	35 days	39 cm	N36°58'233" W115°57'409"	10^5	among sage brush, rain had wetted soil 15 cm deep
St #2	14 days	37.5 cm	N36°58'222" W115°57'395"	3×10^4	desert pavement
St #3	14 days	24 cm	N36°58'244" W115°57'376"	5×10^4	among sage brush
St #4	14 days	17.75 cm	N36°58'230" W115°57'393"	8×10^3	desert pavement water exposure
St #5	14 days	23 cm	N36°58'216" W115°57'380"	2×10^4	among sage brush

Observations and Discussion

Photographs of the etched CR-39 detectors as shown in Figures 4, 5, and 6 illustrate the distribution of the α-contamination in the soils studied. Points of concentrated α-activity (hereafter called 'hot spots') can be seen as white spots in the photographs of the detector plates in Figures 5 and 6. Evenly distributed α-tracks, random in their orientation, indicate evenly distributed α-emitters, which are observed but can only be seen using optical magnification. Hot spots are distinguishable by their generated α-tracks arrangement. In Figure 4, a schematic and several examples for the structure of hot spots as observed are shown. The center of a hot spots has perfect circular tracks but if the track density is too high the structure is erased. With increasing radial distance from the center, the tracks become more ellipsoidal with a larger diameter at the center side caused by the decreasing impact angle of the α–particle hitting the detector. The center size of a star-like structure is directly related to the size of the particle creating the hot spot, as snhown in the schematic in Figure 4. The track density observed is directly proportional to the activity and could be used as a measure for the in-situ radioactivity.

An observation, noted for all 5 exposures, is particularly conspicuous in experiment St #5: some very small hot spots have a high track density that erases the center structure and have nearly no surrounding ellipsoidal tracks (Fig.4c). Determining the size and nature of the particulates generating such small hot

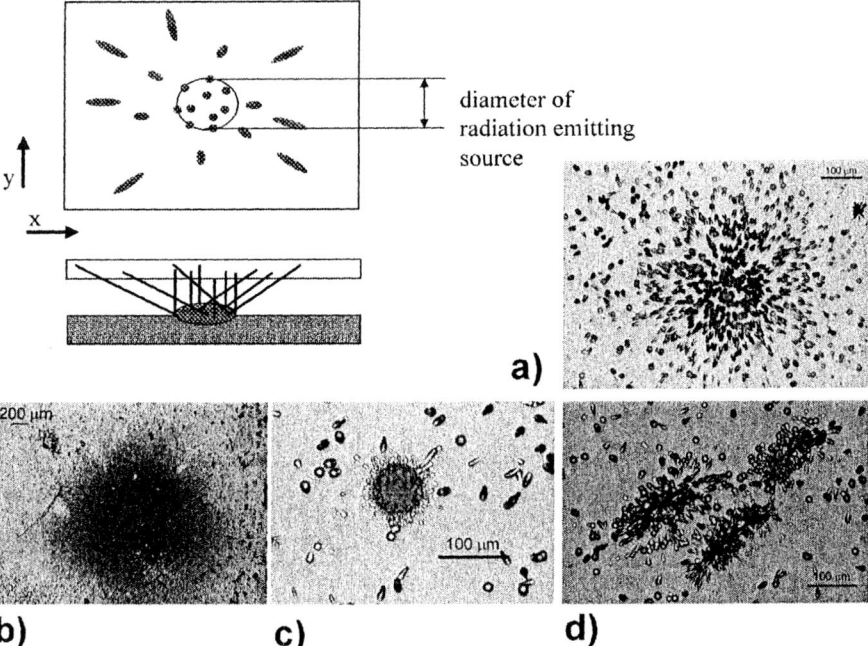

Figure 4: Schematic drawing of the radiographic process and photographic example images obtained in our experiments. a) Picture of a typical star-like structure generated by particles emitting α-radiation - circular tracks in the center that become more ellipsoidal with increasing distance to the center.
b) Same star-like appearance, but the particle emitting the α's is much larger and has a much higher activity level (its location is indicated by a (*) in Fig. 6).
c) Gun-shot like, very high track density that erases the center structure, no surrounding ellipsoidal tracks indicate a very small particle with exceedingly high activity. d) Rod shaped center surrounded by star-like ellipsoidal tracks is a newly observed accumulation of α-activity in soils (more detail in Fig. 6).

spots is speculative. However, such small hot spots with such high tracks density indicate the presense of particles that carry much higher activities than the larger particles generating the common star picture as shown in Figure 4a.

The composite picture of figure 5 shows clearly that α-emitting contamination has reached depths of at least 39 cm. The left and right column are photographs of the detector plates pasted together to resemble the actual arrangement during the in-situ exposure. Stake #1, left, reached a depth of 39 cm and Stake #2 a depth of 37.5 cm. White spots are particularily frequent and visible in the 2nd and 3rd detector plates. The white spots are the previously described hot spots that are large enough to be seen by the human eye without magnification. The observed band of hot particles on the 2nd and 3rd detector,

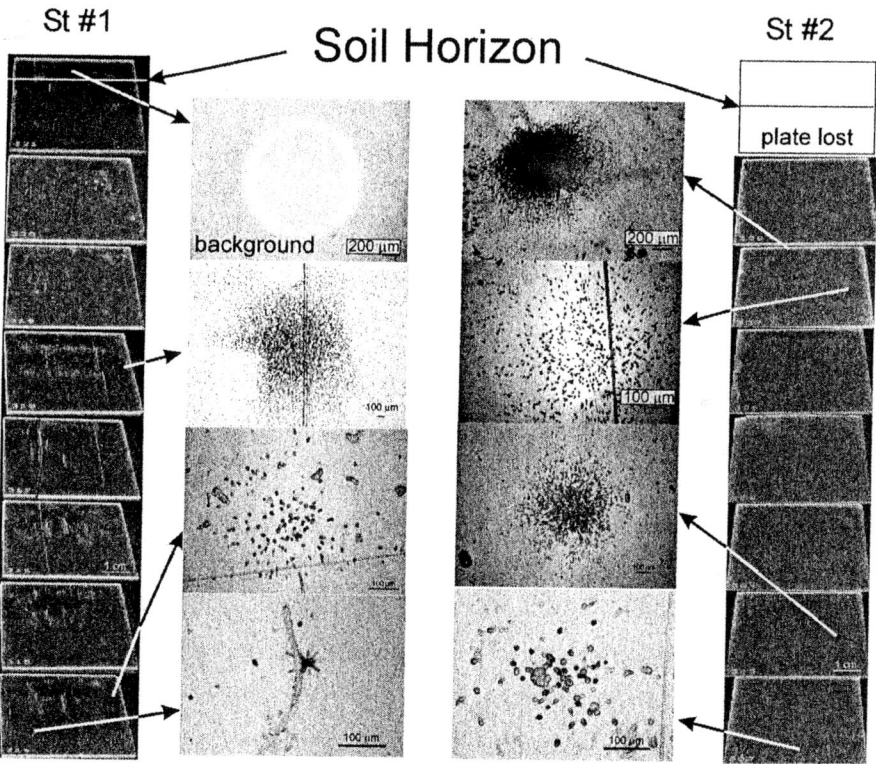

Figure 5: Left and right columns show photopgraphs of detector plates as emplaced in-situ. St #1 reached a depth of 39 cm and St #2 37.75 cm. The white circular spots on the plates are hot spots large enough to be visible to the eye, but most are too small so that optical magnification is necessary for identification. Examples for α-active particulates detected at greater depths are provided. (the perpenticular streaks are due to handling)

which can also be seen in the stakes shown in Figure 6, illustrates the region (4 to 20 cm) where currently most of the α-contamination exists. The photographs in the center section of Figure 5 show several examples of hot spots optically magnified and, for comparison, a section of the detector plate that was not exposed to the soil. Aside from the background photo, three photographs of hot spots from the major α-contamination zone are shown. The lower four photographs are magnifications of hot spots found at depth as low as 38 cm. The density of the randomly orientated tracks varies significantly for the different locations as well as for the different depths.

The detector plates of two other exposure experiments reveal a surprising activity distribution pattern. As in Figure 5, Figures 6 shows in the left and right columns the detector plates of the in-situ exposures for St #3 and St #5, as

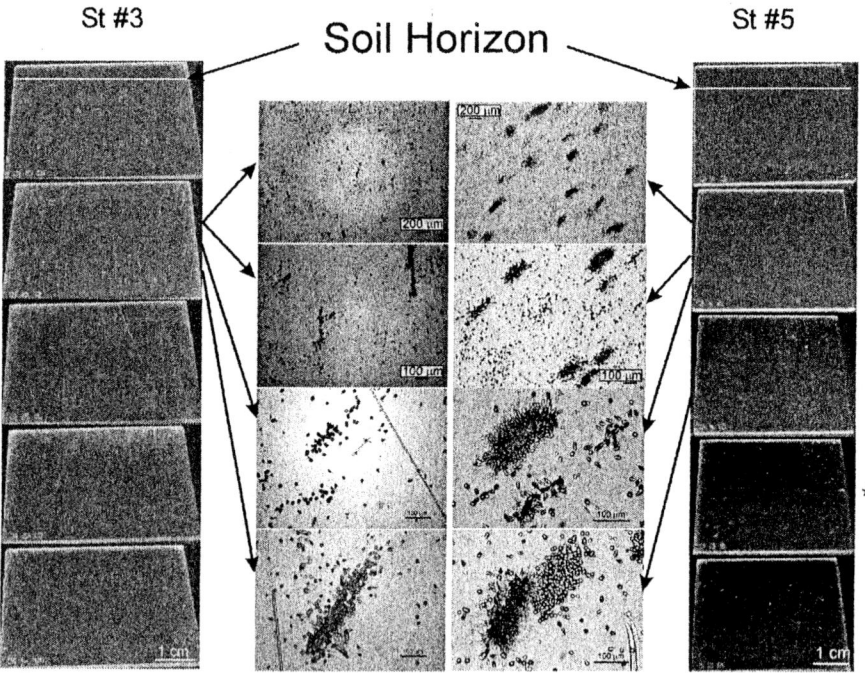

Figure 6: The left and right columns show photographs of St #3 & #5 detector plates, with maximum exposed depth of 24 and 23 cm, resp., both were in close proximity to sage brush. At depths below 5 cm an unsual but repetitive pattern of rod shape structures made of α-induced tracks is observed. Examples are pictured in appropriate magification in the center. The *indicates the particle that is magnified in Fig. 4.

reference for the magnified photographs in the center. Although, as illustrated by the photos, the pattern is much more dominant in Stake #5, the same pattern is still observed in Stake #3. The obvious pattern is characterized by rod shaped structures that all seemed to be aligned, and have a center that is not circular but rather an oval that is surrounded by star-like tracks beams. The lengths and thicknesses of those apparent ovals vary. The shape could resemble that of a microbe, however, the size exceedes microbial dimensions. Another, more reasonable explanation for such a rod shape with inherent orientation is offered by the presence of very fine root hairs of the sagebrush that is native to the field site. Fine root hairs were noticed during our experiments at the field site and were also found in soil samples taken after the exposure tools were removed. The detectors of the experiments (St #3 & #5) had been placed between sagebrush, approximated 40 cm from the sagebrush stem.

This observation poses the question, whether sagebrush roots adsorb or even take up α-active nuclides? Which nuclides are adsorbed, and furthermore, if those nuclides are accumulated by the plant? This question has to be posed since α-radiography can not easily distinguish between different α-emitters, such as ^{241}Am and ^{239}Pu mostly of concern here. Another possibility could be that the plant and its root system behave as a conduit for facilitating rapid transport of actinides and water. So far actinide uptake by plant material has been neglected due to the belief that the actinides ionic size, high valency, and oxy-ion structure make the actinides unfavorable for uptake in plant processes.

The 1994 aerial radiation survey, shown in Figure 3B, provides a valuable assessment tool of macroscopic scale, however, the survey does not and cannot reflect the microscopic vertical α-emitter profiling studied in the present investigation. The aerial contamination survey indicates a gradual decrease in contamination with increasing distance from the detonation center. In the enlarged section of Site D in Figure 3C our experimental locations are specified, demonstrating that in-situ exposures were performed in areas of different surface contamination according to the survey. The aerial survey accounts for the activity found on the surface only, while our technique can be used for determining the activity at the surface and also for depth profiling providing an additional dimension. Although we have not quantified our observations, we do observe that the macroscopic aerial survey is consistent with our microscopic screening assessment when we take only the top 5 mm of the exposed soils into account. At location #3 we find a much higher frequency of hotspots and overall track density (both measures of the activity) in the top 5mm of the soil than in any other location, as shown in the left and right columns of Figures 5 and 6, demonstrating the heterogeneity of the α-activity with depth. While for St #3 white spots are visible at the soil horizon, this is not the case for all other profiles. Noteworthy is the observation that in St #1 and St #5 a sudden apperance of hot spots with similar intensity as for St #3 is visible at ~4.5 cm and ~3 cm, respectively. Such observations were made by earlier NAEG studies, too, and where attributed to sand mound accumulation on top of the original soil surface among sagebrushes due to wind erosion. This shift in soil horizon could explain the sudden hot spot onset for the exposures St #1 and #5. This observation also illustrates the impact of the erosion processes on the depth profiling of the α-emitters.

Summary

By choosing a previously well-characterized field site, we were able to compare the vertical distribution of the actinides in the soil horizon with similar earlier investigations and draw conclusions on possible controls affecting the fate and transport of actinides in semi-arid desert environments. While large-scale surface survey maps are valuable assessment tools, the heterogeneities in the actinide distribution in the underlying surface profiles are not assessed by those surveys, and hence, yielding potentially incomplete contamination profiles. This techniques adds the third dimension to our understanding of actinide transport. In our field-deployed experiments, we found that α-emitting nuclides have moved to at least 39 cm in depth, far deeper than previously expected. We suggest that the α-active particulates found at that depth were transported by either, 1) the adsorption of the α-emitter to plant roots, which might be transported via translocation, 2) the plant-root systems that represents a conduit for facilitating the rapid transport of the α-active particulate, or 3) transport by successive sorption-desorption processes of the α-emitter on soil constitutents through rain water infiltration.

We observe different types of hot spots structures: star-like, rod-shaped, and gunshot-like hot spots without any structures. While the star-like structures are known, the other two were novel observations. The variety of observed hot spot structures can be explained through several formation modes of the particles that generate the hot spots. The larger, less active particles that generate the star-like structures are probably formed by adsorbing α-active nuclides, which can be mobilized by weathering processes, facilitating an actinide movement through soil by successive sorption-desorption steps. The rod-shaped structures are attributed to sorption of α-emitter to the surface of root hairs of the native sage brush. The very small and highly active particles are inducing gunshot-like structures, and are most likely finely dispersed plutonium metal or oxide particulates, that resulted from the weapon detonation initiated by conventional explosives. Possible pathways for moving those small highly active particles deeper in the soil are water induced channeling through cracks in the dry, hydrophobic desert soils, through translocation by the plant root systems, or animal activities. Those particle-size and adsorption specific processes as described explain the movement observed for the α-active contaminants to deeper layers as previously observed.

Our results not only demonstrate the utility of α-radiography to investigate the controls of the movement of actinides in desert soil, but also that this technique is simple, field-deployable, and can be applied for rapid, in-situ screening of α-contamination in upper soil layers. When comparing the efforts, the detector exposure time necessary and the subsequent etching procedure require less time than obtaining, processing, and analyzing numerous soil samples essential to achieve the same results by traditional wet chemical methods. Furthermore, this technique does not require transport and handling of radioactive material, and does not generate radioactive waste and minimizes human exposure due to soil disturbance. Field technicians can easily deploy and retrieve a network of detectors. Hence, we believe that this technique can be advantageously deployed, where α-activity contamination has to be rapidly assessed as part of an environmental remedial investigation or characterization.

Current and Future Work

Currently, we are analyzing the soil samples to establish a relationship between α-activity concentration, particle size, and composition of the α-activity bearing particles. The primary objective is to continue to identify the key factors that control the fate and transport of the actinides in semi-arid soils.

Acknowledgement

This work was performed under the auspices of the U. S. Department of Energy by the University of California, Lawrence Livermore National Laboratory under Contract No. W-7405-Eng-48, and funded by the Hydrologic Resources Management Resource Program of U.S. Department of Energy / National Nuclear Security Administration - Nevada Operations Office. We would have not been able to perform this work without the help of many colleagues at the Nevada Test Site: Jeff Haeberlin, Billy Hyatt, Don Felske, and CP Williams. Radiological control support in the field from Colleen Corlett and Ken Corville of Bechtel-Nevada was invaluable. Earl Kelly of LLNL's Engineering Directorate facilitated the design and timely production of the field-deployable exposure tool.

References

1. Shinn, J. H.; Homan, D.N. UCID-20514, Lawrence Livermore National Laboratory: Livermore, CA; 1985.
2. *The Radioecology of Plutonium and other Transuranics in Desert Environments*; White, M. G.; Dunaway, P. B., Eds.; NVO-153, U.S, Energy Research & Development Administration, Nevada Operations Office, Las Vegas, NV, 1975.
3. *Transuranics in Desert Ecosystems*; White, M. G.; Dunaway, P. B.; Wireman, D. L., Eds.; NVO-181, U.S. Department of Energy, Nevada Operations Office, Las Vegas, NV, 1977.
4. *Transuranic Elements in the Environment*; Hanson, W. C.; Ed.; TIC/U.S. Department of Energy: Springfield, VA, 1980.
5. Essington, E. H.; Fowler, E. B.; Gilbert, R. O.; Eberhardt, L. L. in *Transuranium Nuclides in the Environment*; Proceeding Series, IAEA, Vienna, 1976; pp 157-172.
6. Shinn, J. H.; Essington, E. H.; Gouveia, F. J.; Patton, S. E.; Romney, E. M. UCRL-JC-115992, Lawrence Livermore National Laboratory: Livermore, CA; 1993.
7. Kersting, A. B.; Efurd, D. W.; Finnegan, D. L.; Rokop, D. J.; Smith, D.K.; Thompson, J. L. *Nature* **1999**, *397* (N6714), 56-59.
8. Becquerel, H. *Cmpt. Rend.* **1896**, *122*, 501-503.
9. Kinoshita, S. *Proc. Roy. Soc.* **1910**, *83*, 432-453.
10. *Nuclear Tracks in Solids*; Fleischer R. L.; Price, P. B.; Walker, R. M., Eds.; University of California Press: Berkeley, CA, 1975.
11. Cartwright, B.G.; Shirk, E.K.; Price, P.B. *Nucl. Inst. Meth.* **1978**, *153*, 457-460.
12. Vapirev, E. I.; Kamenova, T.; Mandjoukov, I.G.; Mandjoukova, B. *Radiat. Prot. Dos.* **1990**, *30*, 121-124.
13. Akopova, A. B.; Viktorova, N. V.; Krishchian, V. M.; Magradze, N.V.; Ovnanian, K. M.;Tumanian, K. I.; Chalabian, T. S. *Nucl. Tracks Radiat. Meas.* **1993**, *21*, 323-328.
14. Espinosa, G; Silva, R., *J.Radioanal. Nucl. Chem. Art.* **1995**, *194*, 207-212.
15. Gilbert, R. O., Shinn, J. H.; Essington, E. H.; Tamura, T.; Romney, E. M.; Moor, K. S.; O'Farrell, T.P. *Health Phys.* **1988**, *55*, 869-887.
16. Borg, I. Y., Stone R., Levy H. B., Ramspott L. D. *Information Pertinent to the Migration of Radionuclides in Ground Water at the Nevada Test Site, Part I: Review and Analysis of Existing Information*; UCRL-52078; Lawrence Livermore National Laboratory: Livermore, CA; 1976, Part I.
17. Bondarenko, O.A.; Salmon, P.L.; Henshaw, D.L.; Fews, A.P.; Ross, A.N. *Nucl. Instr. and Meth. in Phys. Res. A* **1996**, *369*, 582-587.

18. Laue, C. A.; *α-Strahler und Korrosionsprodukte in den Kühlmedien von Leichtwasserreaktoren*; Verlag Görich & Weiershäuser GmbH: Marburg, Germany, 1996; Ph.D Thesis, Philips University Marburg.
19. Gilbert, R. O. in *Transuranics in Desert Ecosystems*; White, M. G.; Dunaway, P. B.; Wireman, D. L., Eds.; NVO-181, U.S, Department of Energy, Nevada Operations Office, Las Vegas, NV, 1977; pp 423-429.
20. Hendricks, T.J.; Riedhauer, S.R. *An Arial Radiological Survey of the Nevada Test Site*; DOE/NV/11718-324; Bechtel Nevada: Las Vegas, NV, 1999.
21. US Department of Energy Radionuclides in Surface Soil at the Nevada Test Site; DOE/NV 10845 – 02; Desert Research Institute: Las Vegas, NV, 1991.
22. Shinn, J.H.; Essington, E.H.; Miller, F.L.; O'Farrell, T.P.; Orcutt, J.A.; Romney, E.M.; Shugart, J.W.; Sorom, E.R. *Health Phys.* **1989**, *57*, 771-779.
23. Romney, E. M.; Wallace, A.; Gilbert, R. O.; Kinnear, J. J. in *Transuranium Nuclides in the Environment*; Proceeding Series, IAEA, Vienna, 1976; pp 479-491.
24. Townsend, Y. E.; Grossman, R. F., Eds.; Annual Site Environmental Report for Calendar Year 1999; DOE/NV/11718 – 463; Bechtel Nevada: Las Vegas, NV, 2000; pp 2-8.

Developments in Radioanalytical Methods

Separation Techniques

Chapter 10

Radioanalytical Techniques and the Characterization of New Separations Reagents

Kenneth L. Nash[1,*], Renato Chiarizia[1], Marian Borkowski[1], Paul G. Rickert[1], and Emmanuel Otu[2]

[1]Chemistry Division, Argonne National Laboratory, 9700 South Cass Avenue, Argonne, IL 60439–4831
[2]Division of Natural Sciences, Indiana University Southeast, 4201 Grant Line Road, New Albany, IN 47150–6405

Radioanalytical chemistry has proven historically to be a sensitive tool for the elaboration of reaction stoichiometries and equilibrium thermodynamics in solvent extraction. The rapid turn around of analytical results, the ability to use of low concentrations of metal ions, and the small volumes of solutions required for the experiments combine to make these methods well-suited to guiding the development of new reagents. In this chapter, the basic principles of the classical technique of "slope analysis" for determining the stoichiometry of extracted metal complexes in solvent extraction are demonstrated through a discussion of the features of the extraction of Am^{3+} by several new organophosphorus complexants. Both the advantages of the techniques and the precautions that must be taken in their application and interpretation of results are demonstrated through these examples.

Introduction

The development of new complexing agents for application in aqueous processing of metal ions by solvent extraction is a multi-step process involving conceptual design (often including modeling), synthesis of the ligand (or a series of ligands of systematically varied properties), and the characterization of the metal complexes they form. The process of ligand design must be guided by knowledge of the chemical properties of the target species (in the present context trivalent americium, Am^{3+}) and of the potential competing species in the medium of interest. Adjustments may include altering the arrangement and numbers of ligand donor groups, the affinity of the groups for the target species, and the solubility of the complexant in the medium. Modeling of conceptual species can provide guidance in the design of appropriate compounds, but ultimately new compounds must be synthesized in the laboratory.

Radioanalytical chemistry can be quite useful in the characterization of new compounds, particularly when synthesis procedures are complex and/or produce materials in small amounts. Among the several advantages of using radioanalytical chemistry for evaluation of the properties of new reagents is its ready adaptability to microchemical analysis. Depending on the isotope used for measurements, the properties of new species can be investigated at the nano-molar (and lower) concentration level of the metal complexes. The ability to apply such techniques is of exceptional value when the new reagents being synthesized are only available a few milligrams at a time. Exceeding solubility limits of complexants or their metal complexes is also readily avoided under most circumstances.

In the following discussion, the application of radioanalytical techniques to the task of developing new separations reagents for more efficient separations of actinide ions will be discussed. The primary emphasis will be on research we have conducted at Argonne National Laboratory on the development of new organophosphorus complexing agents, both water-soluble and lipophilic. We will discuss various aspects of the ligand design process and demonstrate the central role that radioanalytical chemistry plays in our approach to the creation of new reagents and the development of processes for their use.

Experimental

The radioanalytical techniques employed in the characterization of new complexants for separations are well known in chemistry. Several of the salient features and cautionary notes for the application of radioanalytical chemistry to solvent extraction are as described below:

1) Any radioactive isotope can in principle be employed in the characterization of new complexing agents. The most useful species are those whose half lives are between several months and about 10^4 years. The mostr straight-forward application is of isotopes that decay to produce stable (i.e., non-radioactive) daughters. If the daughters are radioactive, it is often necessary to apply a correction for the presence of a second isotope (or counting procedures must be adjusted to discriminate against the daughter). Species with half lives shorter than a few months must be replenished frequently, and may have to be corrected for decay losses during the radiometric analysis. It will be necessary to use higher concentrations of species with long half lives (e.g., ^{237}Np) to allow adequate precision in counting statistics. The use of higher concentrations of the radiotracer can generate conditions that challenge solubility limits for the new compounds or their metal complexes. Fortunately, mass balance can be used to monitor for problems generated by solubility limitations, as noted in item 5 below. Isotopes of intermediate half-lives producing radioactive daughter products can be used, but is often necessary to employ preliminary separations to remove those daughters prior to use of the parent isotope.

2) Alpha, beta, and gamma emitting isotopes are equally useful, assuming the appropriate counting equipment is available. Automatic counting of numerous samples is almost always desirable, as these experiments can require the generation of moderately large numbers of samples. Gamma-emitting isotopes require the least preparation to acquire counting samples, as they are readily detected using solid state devices (e.g., NaI(Tl) detectors). Energy discrimination is also easiest for such samples. Alpha and beta emitting isotopes require the preparation of either massless solid samples for gas-flow proportional counting, or liquid scintillation fluid for the counting of total activity using liquid scintillation counting systems. Alpha and beta spectroscopy (i.e., energy discrimination) is possible for solid samples. Alpha-beta discrimination can be done for liquid scintillation, however, liquid scintillation does not readily lend itself to good resolution of decays with closely spaced energies. Liquid scintillation cocktails are also susceptible to a variety of spectral interferences. Multiple channel discrimination (mainly between α and β particles) can allow the simultaneous determination of the properties of two element in the same sample. Neutron emitting isotopes offer no particular advantages and in fact present unique shielding challenges. Unless the element itself is of interest (e.g., ^{252}Cf), such an isotope would not ordinarily be selected for the characterization of new reagents. In the systems of our programmatic interest, ^{241}Am is almost always preferable for studies aimed at designing new actinide extractants/complexants because: 1) the trivalent actinides can be problematic in actinide separations, 2) ^{241}Am decays by α-emission with accompanying characteristic 60 kev γ-ray allowing a choice of detection trechniques, 3) it has a

moderately high specific activity, and 4) it produces long-lived ^{237}Np as the decay daughter.

3) Because most applications of radioanalytical chemistry to the characterization of new reagents is done without the aid of energy discrimination techniques (α- or γ-spectroscopy), it is important that the identity and purity of the isotopes be established prior to beginning the experiments. With few exceptions, radioactive decay daughters are chemically distinct, thus behaving differently from the parent. Incorrect assumptions concerning isotopic purity can lead to erroneous conclusions on the chemistry being evaluated.

4) In studies of the equilibria describing separations reactions, i.e., those involving the transfer of radioisotopes between immiscible liquid or liquid and solid phases, it is important to establish that the measurements being made represent a true chemical equilibrium condition. For this reason, it is important to allow adequate time for the establishment of chemical equilibrium. It is often preferable to repeat measurements using the radioactive material loaded into each phase separately (to verify that the same equilibria describe forward and reverse phase transfer reactions). Such precautions are particularly important when experimental observations are made as a function of temperature.

5) Demonstration of mass balance is essential. Because the radiotracer is usually present at very low concentrations, losses of very small amounts of radioactivity to precipitation or adsorption on "non-reactive" surfaces (e.g., vessel walls, cap liners) can compromise results. It is therefore important to verify that what went into the system stayed in the system.

Perhaps because of the often extremely low concentrations at which they are applied, careful attention to detail is necessary for the successful application of radioanalytical chemistry in the characterization of new reagents. Among the advantages of the use of radioanalytical chemistry is the comparative simplicity of the detection and analysis equipment, and the high throughput that is possible through its application. The principle disadvantage lies in the need to carefully manage the effluents and to properly dispose of wastes.

Results and Discussion

At more than 60 years after the discovery of the first transuranium actinide element (Np) and after almost 50 years of their large-scale production in fission reactors, there is a continuing need for new chelating agents for their analysis and processing. In reality, the discovery of the actinides and the investigation of their chemical properties has much to do with the design and evaluation of new chelating agents for their isolation and with the characterization of their interactions with complexing agents. The identification of new transplutonium elements was made possible only through the use of separations procedures involving water-soluble chelating agents and radioanalytical chemistry (*1*). Ultimately, the increasingly short half-lives and the limitation of creating only a few atoms at a time placed significant demands on the chelating agents employed to conduct the difficult separations of adjacent metal ions of nearly identical chemical properties.

For large scale production of transuranium elements, the PUREX process, based on the selective extraction of $UO_2(NO_3)_2$ and $Pu(NO_3)_4$ from the diverse mixture of highly-radioactive fission products by tributylphosphate (TBP), has been the standard method of actinide production around the world since the mid-1950's. The ability of TBP to reject the trivalent transplutonium actinides became a liability of the PUREX process in the late 1970's. The demand for options to recover and isolate the minor actinides (Np, Am, Cm) for waste volume minimization or for possible transmutation, led to the development over a period of years of the carbamoylmethylphosphine oxide chelating agents (CMPO's) and related organophosphorus compounds (*2*). Extractant development work of this sort has continued around the world during the past two decades leading to the development of new compounds sharing many characteristics of CMPO, some even containing no phosphorus (*3,4*) and a variety of different processes for their application (*5*).

The research supporting development of the CMPO-based TRUEX process for total actinide recovery from dissolved spent fuel or reprocessing wastes included a variety of reagent types including water soluble, lipophilic, and polymeric species. This work continues today with ongoing investigations of the design and characterization of new reagents for actinide separations. Though many different analytical methods are applied in the quest for new reagents, the ultimate evaluation of the utility of new reagents still relies on the analysis of distribution data using radioactive isotopes. The technique of slope analysis remains the best approach to learning the stoichiometry of complexes in each phase of a distribution experiment. The examples to follow illustrate both the power of the techniques of slope analysis and the potential for being led astray if careful control of experimental conditions is not practiced.

Basic Measurements and Methods

To apply radioanalytical chemistry for the characterization of new reagents, the fundamental measurement to be made is calculation of the distribution coefficient or ratio (D) as defined below:

$$D = \frac{cpm_{org}}{cpm_{aq}} = \frac{[M]_{org}}{[M]_{aq}} \qquad (1)$$

where cpm_{org} and cpm_{aq} are the numbers of detected radioactive decays per minute in each phase. As long as differences between the chemical/physical properties of the organic and aqueous media introduce no bias in the detection of radioactive decays, this D value represents the ratio of the molar concentrations of the metal ion in aqueous and organic phases. Furthermore, the reliance on distribution ratios eliminates the need to correct for the specific activity of the isotope or detection efficiency of the counting equipment. No simplifying assumptions are made in using this approach, hence the distribution ratio can be related to the analytical concentrations of reagents in the solutions, as the following discussion will illustrate.

If we consider as an example the distribution of a metal ion between immiscible liquids through the agency of a lipophilic reagent that exchanges H^+ ions for the metal ion in a manner to maintain electroneutrality in each phase, the following expression can be employed to describe the behavior of the species.

$$M^{n+}{}_{aq} + nHL_{org} \xrightleftharpoons{K_{ex}} ML_{n,org} + nH^+{}_{aq} \qquad (2)$$

The biphasic chemical equilibrium can be expressed mathematically as:

$$K_{ex} = \frac{[ML_{n,org}] \cdot [H^+{}_{aq}]^n}{[M^{n+}{}_{aq}] \cdot [HL_{org}]^n} \frac{\gamma_{ML_n} \cdot \gamma_H{}^n}{\gamma_M \cdot \gamma_{HL}{}^n} = D \cdot \frac{[H^+{}_{aq}]^n}{[HL_{org}]^n} \frac{\gamma_{ML_n} \cdot \gamma_H{}^n}{\gamma_M \cdot \gamma_{HL}{}^n} \qquad (3)$$

As long as the concentration of the radioisotope is below that of HL and H^+ (in this example), it can be assumed that the analytical concentrations of HL and H^+ are not measurably changed by their extraction of the metal ion and can be inserted directly into this expression. Activity coefficients in both aqueous and organic solutions often can be assumed constant, and so are often implicitly incorporated into a conditional extraction constant defining some appropriate "standard state":

$$K'_{ex} = K_{ex} \cdot \frac{\gamma_M \cdot \gamma_{HL}{}^n}{\gamma_{ML_n} \cdot \gamma_H{}^n} = D \cdot \frac{[H^+{}_{aq}]^n}{[HL_{org}]^n} \qquad (4)$$

Conversion of this equation to a logarithmic form allows simple graphical analysis to learn stoichiometries of the phase transfer reaction:

$$\log D = \log K'_{ex} + n \log [HL] - n \log [H^+] \qquad (5)$$

Least squares fitting of data sets after conversion to logarithmic presentation is the essence of "slope analysis". Data sets are typically created including demonstration of the effect of both [HL] and [H$^+$] on the equilibrium distribution ratio. When possible, data sets are obtained using an aqueous medium at constant ionic strength.

The so-called liquid cation exchanger extractants described in the above example represent only one class of the phase transfer reaction systems. With other classes of extractants, the assumption of constancy of activity coefficients over the full range of conditions can be more problematic. Several examples of other classes of extraction systems and their attendant equilibrium expressions are shown below (using a trivalent cation as an example and eliminating any consideration of activity coefficients):

Solvation:
$$M^{3+}{}_{aq} + 3X^-{}_{aq} + nS_{org} \rightleftharpoons (MX_3S_n)_{org} \qquad (6)$$

$$D = [MX_3S_n]_{org}/[M^{3+}]_{aq} = K_{ex}[X]_{aq}^3[S]_{org}^n \qquad (7)$$

Ion Pair Formation:
$$M^{3+}{}_{aq} + 3X^-{}_{aq} + (A^+X^-)_{org} \rightleftharpoons (MX_4A)_{org} \qquad (8)$$

$$D = [MX_4A]_{org}/[M^{3+}]_{aq} = K_{ex}[X]_{aq}^3[A^+X^-]_{org} \qquad (9)$$

Micellar Extractants:
$$M^{3+} + (HL)_{n,org} \rightleftharpoons M(H_{n-3}L_n)_{org} + 3H^+{}_{aq} \qquad (10)$$

$$D = [M(H_{n-3}L_n)]_{org}/[M^{3+}]_{aq} = K_{ex}[(HL)_n]_{org}/[H^+]^3_{aq} \qquad (11)$$

Synergism:
$$M^{3+}{}_{aq} + 3HL_{org} + nS_{org} \rightleftharpoons (ML_3S_n)_{org} + 3H^+{}_{aq} \qquad (12)$$

$$D = [ML_3S_n]_{org}/[M^{3+}]_{aq} = K_{ex}[HL]_{org}^3[S]_{org}^n/[H^+]^3_{aq} \qquad (13)$$

For solvating and ion pair extraction reactions in particular, wide variations in aqueous anion concentrations might be employed to accomplish extraction and back extraction. Because these anions participate in the phase transfer reaction, it is not possible to maintain a constant ionic strength and thus presume that the ionic medium in the aqueous solution represents a standard state. Substitution of other counter ions (for example perchlorate for nitrate) is likewise not possible, as the second ion can often participate in a parallel extraction reaction. Under these conditions, activity coefficients can vary substantially over the full range, and the changes in activity coefficients can impact the observed relationship between metal ion distribution and the concentration of reagents. Awareness of the effect is sometimes adequate to allow full interpretation of the results when learning the reaction stoichiometry is the primary objective. At other times, activity coefficient corrections become necessary.

Finally, one must always be aware of possible competing reactions that alter the conditions of extraction of the target species. For example, in solvating extractant systems the parallel extraction of mineral acid (e.g., HNO_3) by the extraction reagent S can compete with the metal ion for free extractant molecules. Water-soluble complexes that do not participate in the phase transfer reaction can also interfers. Usually, such complications occur in a predictable manner and can be readily accommodated in the analysis of the distribution data through the use of equilibrium stability constant data from the literature. Often a separate set of parallel measurements will be required to verify that such reactions are important. In the following examples from our recent research, the impact of these factors on the determination of extraction stoichiometry will be addressed.

Characterization of Diacidic Diphosphonate Extractant Systems

The application of radiometric slope analysis to elucidation of the stoichiometry of cation extraction by liquid cation exchangers usually lends itself to the most straight-forward interpretation of the results obtained. Many literature reports of the extraction of polyvalent cations by acidic chelating agents (6) demonstrate the clear exchange of n H^+ for M^{n+} in plots of log D as a function of pH. In such systems, control of the aqueous phase ionic strength is readily accomplished without complication, as the anion of the supporting electrolyte usually does not participate in the phase transfer reaction. Extractant dependencies in such systems are likewise well-behaved and indicative of the stoichiometry of the predominant extracted complex.

For acidic organophosphorus, carboxylate, or sulfonate extractants, the "self-assembly" of the free (i.e., non-complexed) extractant molecules in the nonpolar organic solution to form dimers or higher order aggregates is frequently observed. This phenomenon can result in a reduction in the apparent extractant dependence. Full interpretation of the slope analysis data in this case often requires the acquisition of supplementary data through the use of nonradioanalytical methods. For example, in our investigations of the extraction of selected metal ions by diacidic diphosphonic acid extractants, we have observed a complex pattern of extractant aggregation as a function of the length of the bridge between the two phosphonate groups, a condition initially indicated by slope analysis.

In Figure 1, the acid dependence for the extraction of Am^{3+} by bis P,P'(2-ethylhexyl) methylene, 1,2-ethylene, and 1,4-butylenediphosphonic acids into xylene reflects the expected 3:1 stoichiometry for exchange of H^+ for Am^{3+} in each system (7). The leveling effect that is observed at the highest nitric acid concentrations indicates a contribution to Am extraction from a solvating process in which the extractant does not exchange H^+ for Am^{3+}, but rather extracts an electroneutral Am species, likely $Am(NO_3)_3$. If we compare this

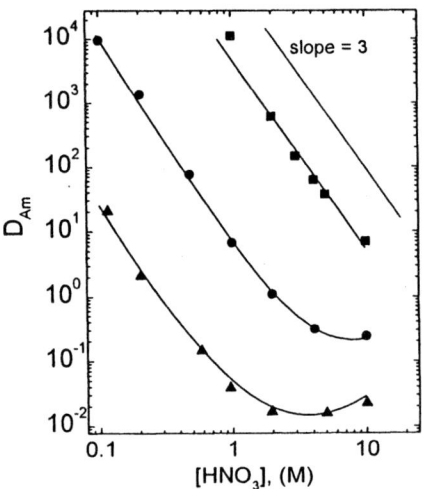

Figure 1. Acid dependence in the extraction of Am^{3+} by 0.1 (each) M $H_2DEH(MDP)$ (■), $H_2DEH(EDP)$ (●), and $H_2DEH(BuDP)$ (▲) in o-xylene

result with the results of slope analysis for extractant dependence in the same system (Figure 2), a very different picture emerges. For methylenediphosphonic acid, the extractant dependence is two; for 1,2-ethylenediphosphonic acid, the extractant dependence drops to 1; for the 1,4-butylenediphosphonic acid extraction system, extractant dependence lies somewhere between 1 and 2. Based on the structural features of the three extractant molecules, there is no obvious reason to predict such a wide variation in performance, nor is there any

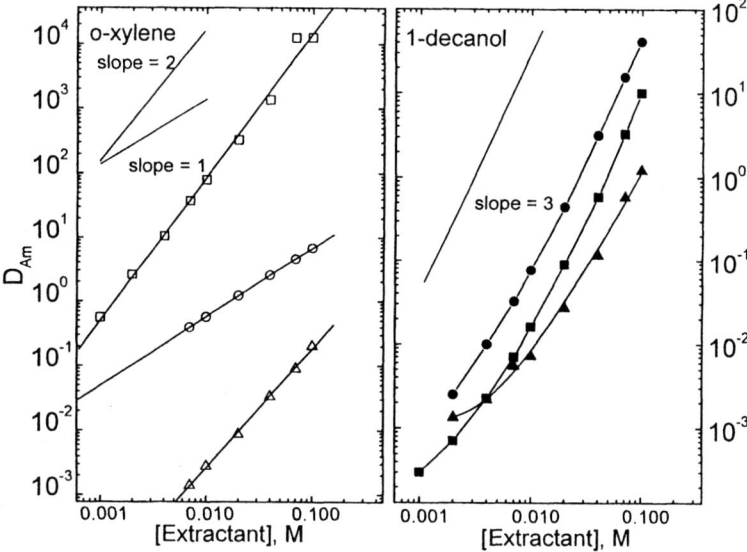

Figure 2. Extractant dependence for Am^{3+} extraction by $H_2DEH(MDP)$ (■, 1.0 M HNO_3), $H_2DEH(EDP)$ (●, 1.0 M HNO_3 in o-xylene, 0.02 M HNO_3 in decanol), and $H_2DEH(BuDP)$ (▲ 0.05 M HNO_3 in xylene, 0.01 M HNO_3 in decanol) in o-xylene (open symbols) and in 1-decanol (closed symbols).

rationale for the apparent minimum in extractant dependency for the ethylene derivative.

To gain the needed insight into this system, we conducted experiments on the colligative properties of these extractant solutions (vapor pressure osmometry in toluene) to determine that the respective aggregation numbers of the free methylene-, 1,2-ethylene-, and 1,4-butylene-diphosphonic acid extractants are 2, 6, and predominantly 3 in the three systems, respectively. The methylenediphosphonate dimer apparently behaves as a chelating agent binding the Am^{3+} ion with an unspecified denticity but closely associating the metal ion with two extractant molecule dimers. The ethylene diphosphonate hexamer gives an extractant dependency comparable to what is typically seen for micellar extraction systems, probably indicating zone bonding of the metal ion in the inner hydrophilic region of the extractant aggregate. The 1,4-butylene extractant appears to combine some features of each.

These apparent stoichiometries are constant with temperature between 25 and 60°C, thus allowing application of the van't Hoff relationship to calculate the enthalpies and entropies of the phase transfer reactions. The net thermodynamic equilibrium expressions for the respective extraction reactions are:

$$Am^{3+} + 2\ (H_2L)_2 \rightleftharpoons Am(H_5L_4) + 3\ H^+ \qquad (14)$$

$$Am^{3+} + (H_2L)_6 \rightleftharpoons Am(H_9L_6) + 3\ H^+ \qquad (15)$$

$$Am^{3+} + 1.5\ (H_2L)_3 \rightleftharpoons Am(H_6L_{4.5}) + 3\ H^+ \qquad (16)$$

We observe exothermic enthalpies for the methylenediphosphonate and athermic enthalpies for the ethylenediphosphonic acid systems. Based on comparison with literature reports, (8) these results confirm that the former engages in chelation while the latter extracts Am^{3+} into the hydrophilic region of a reverse micelle. The non-integral extractant dependency in the 1,4-butylene system indicates that the extraction reaction is proceeding by a more complex pathway, perhaps involving a combination of chelation and micellization reactions.

It is well known in solvent extraction that in organic solutions possessed of some intrinsic solvating ability, the aggregation of extractant molecules can be inhibited. Slope analysis of Am^{3+} extraction by these same three extractants into 1-decanol demonstrates the change in the nature of the predominant extracted complex (9). As was the case in the o-xylene study, we observed a third power acid dependence in decanol for all three extractants. In contrast to the o-xylene data, extractant dependencies in 1-decanol indicate a limiting tendency toward the formation of a 1:3 complex in this medium, as is shown in Figure 2. This result strongly suggests that the ability of the decanol to interact through hydrogen bonding with the phosphonates inhibits extractant self-assembly, thus allowing the formation of the AmL_3 as the predominant extracted complex.

Work by Herlinger and coworkers (10) indicated that the normal pattern of aggregation for alkyl-bridged diphosphphonic acid extractants was primarily dimeric for odd numbers of carbon atoms in the bridge, but higher ordered aggregates for even numbers of carbon atoms in the bridge. This was attributed to the staggered ordering of carbon atoms in the bridge, which tends to orient the phosphonate groups on the same "side" of the molecule for odd carbon numbers and opposite "sides" for the even number backbones. The eclipsed orientation will obviously favor intramolecular hydrogen bonding while the staggered orientation should promote intermolecular interactions. The orientation of the phosphonate groups is a less important determinant of extractant ordering in 1-decanol because the polarity of the diluent (combined with the inherent tendency of the diluent to engage in a specific hydrogen bonding interaction with the extractant) reduces the energetic driving force for the extractant aggregation reaction.

To test the limits of this organization of diacidic diphosphonate extractants, we determined that introduction of some ligand rigidity might produce interesting effects in the pattern of extractant aggregation. Introduction of benzene molecules as the bridge creates the possibility of creating 1,2-, 1,3-, and 1,4- variants in which the "bite angle" between the phosphonates is 60°, 120° and 180°. The 1,3 and 1,4 derivatives are far more likely to engage in cation bridging (chelating) interactions or intermolecular hydrogen bonding resulting in aggregation than the 1,2-derivatives. We synthesized P,P'-(2-ethylhexyl) benzene-1,2-diphosphonic acid (BzDPA) as the first example and examined its ability to extract Am^{3+} (11). Vapor pressure osmometry indicated that in freshly prepared solutions of BzDPA in o-xylene, the extractant was dimeric, as expected. Upon sitting at room temperature for about two weeks, partial decomposition of the extractant was indicated by the appearance of a white solid at the bottom of the solution. Subsequent analysis of the solid confirmed that it was water-soluble benzene-1,2-diphosphonic acid, the parent acid of the extractant. The evident de-esterification of about 30% of the initial BzDPA extractant introduced two equivalents of 2-ethylhexanol into the extractant phase, resulting in a nearly 1:1 ratio of the remaining BzDPA extractant to the lipophilic (and strongly solvating) alcohol. Follow up osmometric measurements now indicated that the BzDPA extractant was predominantly monomeric in the toluene (or xylene) solution. Refrigeration stabilized the remaining diacidic diester adequately to allow a complete investigation of the Am^{3+} extraction chemistry.

After correction of the experimentally measured D_{Am} values for the presence of the water-soluble complex $AmNO_3^{2+}$, slope analysis indicates the expected release of three H^+ and a similar extractant dependency, as shown in Figure 3. In these plots, the complete data sets are displayed, using a presumed stoichiometry of 1:3 for both extractant and acid dependence. The label for the ordinate in each plot demonstrates how this analysis is conducted. Though the data indicate a slight temperature-dependence in the apparent stoichiometry, the average values remain within ±5% of the mean in both the extractant dependence of +3 and the acid dependence of -3 over the entire temperature range.

It is tempting to try to analyze away variations in slopes by attributing them to changing activity coefficients or possibly as indicating a contribution from an additional extraction pathway. However, the analyst must keep in mind that two phase equilibria can be quite complex and reliable identification of a sole source for such variations is often unlikely. In the hands of a practiced experimentalist,

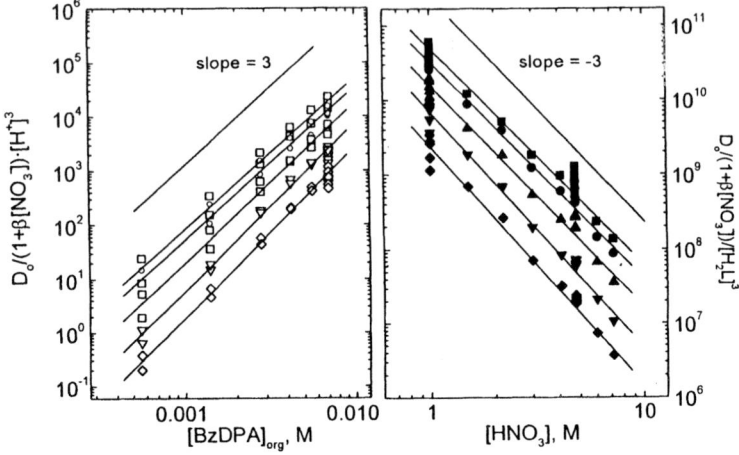

Figure 3. Ligand and acid dependencies for Am^{3+} extraction by $H_2DEH(1,2$-BzDPA) as a function of temperature (■ - 25.0°C, ● - 30.0°C, ▲ - 40.0°C, ▼ - 50.0°C, ♦ - 60.0°C).

two-phase distribution experiments should be considered to have a precision of not better than ±5%, and can be closer to ±10% in some cases. Within these limits, the slopes observed indicate a stoichiometry of 1:3 and a straight-forward cation exchange extraction reaction. The proposed predominant extraction reaction in this case is:

$$Am^{3+} + 3\ H_2L \rightleftharpoons Am(HL)_3 + 3\ H^+ \qquad (17)$$

The thermodynamic parameters developed from application of the van't Hoff relationship to these data result in exothermic heats of extraction, consistent with the simple chelation-cation exchange phase transfer reaction as was seen for the dimerized $H_2[DEH(MDPA)]$ extractants.

Characterization of Extraction by Solvating Ligands

For actinide processing by hydrometallurgical means, solvating extractants have been (and continue to be) the predominant technological reagents. The primary reasons for this preference are: 1) Spent nuclear fuels are typically dissolved in 3-6 M HNO$_3$, a medium from which cation exchangers of the type described above generally perform poorly; 2) Actinides are readily separated from most fission products in such systems; and 3) Dilute acid stripping of actinides from the loaded extractant phase yields a product that is readily manipulated for subsequent treatment options. A considerable amount of research continues around the world on the development of additional actinide-specific extractants of this general type (5). Identification of the species extracted in extractant systems of this type using the techniques of slope analysis can be complex, as these reagents can interact strongly with mineral acids and the range of acidities employed for extraction and stripping is often rather broad.

Research conducted at Argonne between the late 1970's and the early 1990's led to the development of CMPO and the TRUEX process for total actinide recycle (2), as we noted above. This process relied heavily on the application of radioanalytical chemistry and slope analysis. Continuing research in Russia has led to the development of another TRUEX process, based on the structurally symmetrical CMPO extractant diphenyl-N,N-dibutylcarbamoyl(methyl) phosphine oxide (12). Taking a clue from this development, we have begun an investigation of a series of substituted pyridine-2,6-bis(methylphosphine oxides) (NOPOPO) complexants prepared by Paine and coworkers (13-15). Crystal structures of lanthanide complexes with tetraphenyl(NOPOPO) indicate that the three oxygen donor atoms adopt a geometrically favorable arrangement for tridentate coordination of these metal ions in the solid state. At our prompting, the Paine group has synthesized tetraoctyl and tetra(2-ethylhexyl) derivatives of NOPOPO that we are evaluating as possible substitute extractants for CMPO in a next generation TRUEX process (16).

Our initial investigations indicate that the tetra(2-ethylhexyl)NOPOPO extractant can extract tri-, tetra- and hexavalent actinides with strength comparable to that reported for CMPO. As there are viable extractants (much cheaper and more fully investigated) for the isolation of tetravalent and hexavalent actinides, the extraction behavior of trivalent f-elements is considered as the benchmark for judging whether an actinide extractant is viable (assuming that total actinide recovery is the nuclear waste treatment objective). A profile of the extraction of Am^{3+} by tetra(2-ethylhexyl)NOPOPO in dodecane and in toluene is shown in Figure 4. The figure clearly indicates that maximum extraction efficiency is achieved from about 1 M HNO$_3$ and stripping is readily achieved at 0.01 M HNO$_3$.

The flattening of the extraction curve at [HNO$_3$] above 1 M reflects the ability of this extractant (as is true of all solvating organophosphorus extractants) to extract nitric acid from the aqueous to the organic phase. The contribution of the diluent to the apparent extraction reaction is seen from the different limiting slopes observed at low [HNO$_3$]. We expect that the limiting slope should reflect the need to extract electroneutral actinide cations, thus the dependence of D$_{Am}$ on nitrate concentration should approach 3. It is clear that determination of the stoichiometry of the extraction reaction through slope analysis will require that several corrections to the data must be made. The expected stoichiometry for the Am^{3+} extraction reaction is as defined in equation 6.

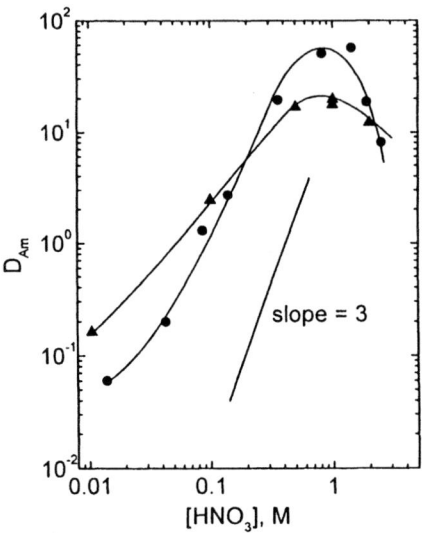

Figure 4. Extraction of $Am(NO_3)_3$ by 0.05 M TEH(NOPOPO) into toluene (▲) and n-dodecane (●).

Two approaches were used to facilitate the analysis of the stoichiometry of the extracted Am^{3+} complex in this system. First, the ligand dependence experiments were conducted from an aqueous medium comprised of 0.001 M HNO_3/0.5 M $NaNO_3$, from which nitric acid extraction has no impact on [NOPOPO] and the analytical concentration of the extractant represents that available for metal ion extraction. Secondly, the equilibrium constants for the extraction of HNO_3 by NOPOPO were determined as a function of temperature. Using these constants, the analytical concentrations of the extractant were corrected for that tied up by HNO_3 in the organic phase. An additional correction to the experimental D values must be made for the $AmNO_3^{2+}$ complex which remains in the aqueous phase (using equilibrium constant data from the literature). The appropriate equation describing the relationship between the observed distribution ratio, [NOPOPO], and [HNO$_3$] which corrects for these effects is:

$$K_{ex} = \frac{D_{exp}(1+\beta_1[NO_3^-]_{aq})(1+K_1[HNO_3]_{aq}+K_2[HNO_3]^2_{aq})^n}{[NO_3^-]^3_{aq}[TEHNOPOPO]^n_{org,initial}} \quad (18)$$

where the extraction of an electroneutral Am(NO$_3$)$_3$ complex is assumed (verified separately) and the extractant dependence is left undetermined as n. The constants β_1, K_1 and K_2 are either known from literature reports or were determined experimentally. Determination of the distribution of Am into NOPOPO solutions of different concentrations from 0.001 M HNO$_3$/0.5 M NaNO$_3$ established that the M:NOPOPO stoichiometry is 1:2, as demonstrated in Figure 5. Introducing the stoichiometric factor of 2 (for n in equation 18) allows for determination of the nitrate dependence, which was confirmed as 3.0 as shown in Figure 5. The overall stoichiometry of these reactions was thus determined to be:

$$Am^{3+}_{aq} + 3\ NO_3^-{}_{aq} + 2\ NOPOPO_{org} \rightleftharpoons Am(NO_3)_3(NOPOPO)_{2\ org} \quad (19)$$

Correction of the extractant dependence plot for the Am distribution from 0.5 M HNO$_3$ into NOPOPO for the extraction of HNO$_3$ gives the dashed line in the lower figure, indicating the ultimate agreement between the experiments and confirming the stoichiometry as written in 19.

These stoichiometric relationships were maintained over the range of temperatures between 10.0 and 40.0°C, hence it was legitimate to calculate enthalpies and entropies of extraction by applying the van't Hoff relationship. The observed enthalpies are exothermic and entropies slightly negative, consistent with most reported extraction reactions of this type (8). As nitrate anions tend to be bidentate in association with extracted actinide complexes, tridentate coordination of two NOPOPO molecules would require an inner sphere coordination number of 12 (6 nitrate oxygens, two NO oxygens, four PO oxygens). Typical trivalent actinide coordination numbers in systems of this type are 8-9, depending on the ligand system. We concluded that the net denticity of either nitrate or NOPOPO's must be lower than the maximum possible. Though slope analysis confirms the stoichiometries of the extracted metal complexes, it does not shed light on the question of total oxygen denticity or which oxygen atoms are bonded. We are planning additional experiments to gain further insight into this question.

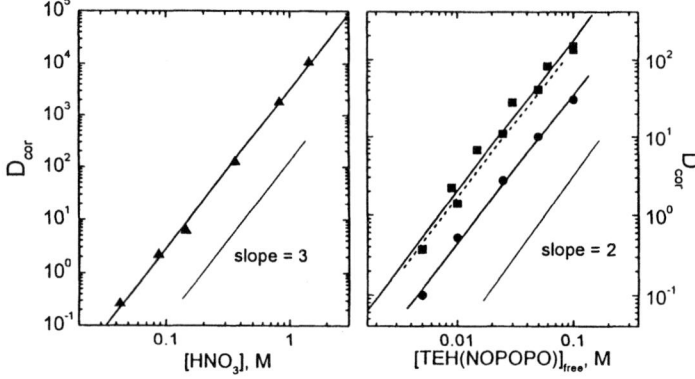

Figure 5. Dependencies of Am(NO₃)₃ extraction by TEH(NOPOPO) after application of corrections for HNO₃ extraction into the organic phase and AmNO₃²⁺ complex formation in the aqueous phase: nitrate dependence (▲), extractant dependence from 0.5 M HNO₃ (●) and from 0.001 M HNO₃/0.5 M NaNO₃ (■). Dashed line in lower plot represents D_{Am} 0.5 M HNO₃ extraction values after correction of [TEH(NOPOPO)] for the concentration of extractant associated with HNO₃ in the organic phase.

Characterization of Water-soluble Chelating Agents

In separations, water soluble chelating agents are employed as the basis of ion exchange-mediated separations, as holdback reagents to improve selectivity in solvent extraction, and as stripping agents to cleanse an organic extractant of contaminants prior to its recycle in a hydrometallurgical process. In each case, the thermodynamic data describing the complexes existing in the aqueous phase are helpful in demonstrating the effectiveness of (and predicting the outcome of) the contact between organic and aqueous phases. Equilibrium constants can be determined using a variety of conventional techniques (UV-visible spectrophotometry, NMR spectroscopy, potentiometry, fluorescence to name a few). When solubilities of the ligand or its complexes might be in question or only small amounts of material are available, radioanalytical techniques represent a viable alternative to the elucidation of such equilibria. Experiments of this type are most readily interpretable when the interactions between complexants in the aqueous phase and metal ion extractants are at a minimum. For this reason, liquid cation exchangers are most commonly employed as the phase separation medium. This approach also allows for more effective control of the ionic strength of the aqueous phase.

Slope analysis in this case can be employed to estimate the metal:ligand stoichiometry of complexes in the aqueous phase. For these experiments, it is not unusual to observe nonlinearity in log-log plots, as the metal complex species in water are not necessarily electroneutral (though electroneutrality must ultimately be maintained in the solution overall) and multiple species can be present simultaneously. The high dielectric constant of water can readily accommodate separation of cation and anion charges over some distance, hence some highly charged species can exist. One precautionary experiment that should be done in these experiments, but seldom is completed, is to verify that the phase transfer reaction is not altered by the presence of the water-soluble complexes. Slope analysis to confirm the constancy of extractant dependencies is always advisable. Under appropriate conditions, it is possible to establish the stoichiometry of the dominant aqueous complexes in such experiments by examining the limiting slopes of the log-log plots, as is shown for the case of americium complexes with phosphonoacetic acid in Figure 6 (*17*). Establishment of the average M:H:L stoichiometry in the aqueous phase is possible using slope analysis. However, the graphical techniques that were once the state-of-the-art for elucidating

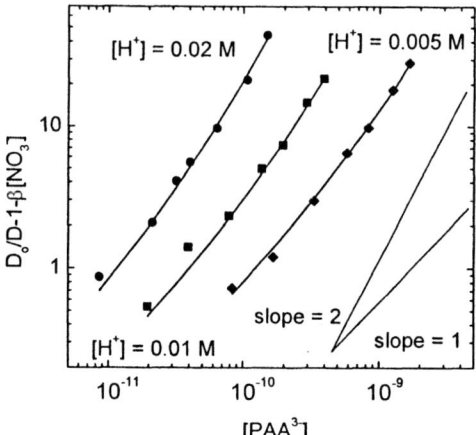

Figure 6. Distribution experiments to elucidate the stoichiometry and stability constants for aqueous complexes of Am^{3+} with phosphonoacetic acid (PAA^{3-}): $[H^+] = 0.02$ M - (●), $[H^+] = 0.01$ M - (■), $[H^+] = 0.005$ M - (♦).

aqueous complexation equilibrium constants have largely been displaced by non-linear least squares analysis.

Conclusions

The results described above indicate the versatility and sensitivity of the well-established procedures of "slope analysis" using radioanalytical techniques for the design and characterization of new chelating agents. Because these methods can be applied using extremely small amounts of materials, radioanalytical chemistry can be quite economical of precious samples. The ability to do real chemistry at micro- to nanomolar concentrations of metal complexes virtually ensures that solubility limits will not be exceeded. In our experience, this technique also allows for rapid turnaround on the evaluation of both the potential utility of new reagents, and the nature of their chemical interactions with metal ions. The technique is readily adaptable to a wide variety of possible different types of separations of systems.

However, because radioanalytical methods are microchemical in nature, their application to this task requires that the researcher be well aware of the limitations of the techniques. It is important to have well-characterized (both chemically and isotopically) radiotracer solutions and to select isotopes appropriate to the task. As with all investigations of the characteristics of new reagents, it is critically important to know the chemical purity of the reagents present at macroscopic concentrations as well. One must also be cognizant of possible side reactions that can complicate interpretations of experimental data. It will on occasion be necessary to supplement radioanalysis with the application of more conventional analytical methods, in the examples cited above, vapor pressure osmometry and potentiometry. Finally, these techniques can be applied to the task of elucidating the details of the thermodynamics of phase transfer reactions through application of the van't Hoff relationship to data taken as a function of temperature. It is critically important that such experiments be accompanied by sufficient experimental observations to verify that the stoichiometry of the phase transfer reactions do not change with the temperature.

Acknowledgments: Professor R. T. Paine, Dr. Xinmin Gan, University of New Mexico (NOPOPO synthesis), Professor D. D. Ensor, Tennessee Technological University (Phosphonoacetic acid studies), Ms. Claire Lavallette, ENSCP (NOPOPO distribution studies). Work performed under the auspices of the U.S. Department of Energy, Office of Science, Office of Basic Energy Sciences under contract number W-31-109-ENG-38.

The submitted manuscript has been created by the University of Chicago as Operator of Argonne National Laboratory. The U. S. Government retains for itself, and others acting on its behalf, a paid-up, non-exclusive, irrevocable worldwide license in said article to reproduce, prepare derivative works, distribute copies to the public, and perform publicly and display publicly, by or on behalf of the Government.

References

1. Choppin, G. R.; Nash, K. L. *Radiochim.Acta* **1996**, *70/71*, 225-236.
2. Horwitz, E. P.; Chiarizia, R. "Liquid Extraction, the TRUEX Process - Experimental Studies", in *Separation Techniques in Nuclear Waste Management*, Carlson, T. E.; Chipman, N. A.; Wai, C. M., Eds., CRC Press, Boca Raton, **1996**, Chapter 1. pp. 3-33.
3. Madic, C.; Hudson, M. J. "High-level Liquid Waste Partitioning by Means of Completely Incinerable Extractants", EUR 18038 EN, **(1998)**.
4. Madic, C.; Lecomte, M.; Testard, F.; Hudson, M. J.; Liljenzin, J.-O.; Sätmark, B.; Ferrando, M.; Facchini, A.; Geist, A.; Modolo, G.; Gonzalez-Espartero, A.; De Mendoza, J. "PARTNEW. An European Research Programme for the Partitioning of Minor Actinides from High Level Liquid Wastes" in *Proceedings of Global 2001, Back End of the Fuel Cycle: From Research to Solutions*, Paris, France, September 9-13, 2001, pp. 300-302.
5. Mathur, J. N.; Murali, M. S.; Nash, K. L. *Solvent Extr. Ion Exch.* **2001**, *19*, 357-390.
6. Poskanzer, A. M.; Foreman, B. M., Jr. *J. Inorg. Nucl. Chem.* **1961**, *16*, 323-336.
7. Otu, E. O.; Chiarizia, R. *Solvent Extr. Ion Exch.* **2001**, *19*, 885-904.
8. Nash, K. L. "Studies of the Thermodynamics of Extraction f-Elements", *Solvent Extraction for the 21st Century, Proceedings of the International Solvent Extraction Conference, 1999*, Cox, M.; Hidalgo, M.; Valiente, M., Eds., Society of Chemical Industry, London, **2001**, pp 555-559.
9. Otu, E. O.; Chiarizia, R. "Thermodynamics of the Extraction of Selected Metal Ions by Di(2-ethylhexyl)alkylenediphosphonic Acids", *Proceedings ISEC 2002*, Sole, K. C.; Cole, P. M.; Preston, J. S.; Robinson D. J., Eds., S. African Inst. of Mining and Metallurgy, Johannesburg, South Africa, **2002**, pp. 408-413.
10. McAlister, D. R.; Dietz, M. L.; Chiarizia, R.; Herlinger, A. W. "Aggregation and Metal Ion Extraction Properties of Novel Silicon-substituted Alkylenediphosphonic Acids" in *Proceedings ISEC 2002*, Sole, K. C.; Cole, P. M.; Preston, J. S.; Robinson, D. J., Eds., S. African Inst. of Mining and Metallurgy, Johannesburg, South Africa, **2002**, pp. 390-395.
11. Otu, E.O.; Chiarizia, R.; Rickert, P.G.; Nash, K. L. *Solvent Extr. Ion Exch.* **2002**, *20*, 607-632.

12. Myasoedov, B. F.; Chmutova, M. K; Smirnov, I. V.; Shadrin, A. U. in *Global '93: Future Nuclear Systems: Emerging Fuel Cycles and Waste Disposal Options*, American Nuclear Society, LaGrange Park, IL **1993**, p. 581.
13. Bond, E. M.; Engelhardt, U.; Deere, T. P.; Rapko, B. M.; Paine, R. T. *Solvent Extr. Ion Exch.* **1997**, *15*, 381-400.
14. Bond; E. M.; Engelhardt, U.; Deere, T. P.; Rapko, B. M.; Paine, R. T. *Solvent Extr. Ion Exch.* **1998**, *16*, 967-983.
15. Paine, R. T. "Design of Ligands for f Element Separations" in *Separations of f Elements: Proceedings of a Symposium at the 207th American Chemical Society Meeting, San Diego, CA, March, 1994*, Nash, K. L.; Choppin, G. R., Eds., Plenum Press, New York, **1995**, pp. 63-75.
16. Nash, K. L.; Lavallette, C.; Borkowski, M.; Paine, R. T.; Gan, X. *Inorg. Chem.* **2002**, *41*, 5849-5858.
17. Ensor, D. D.; Nash, K. L. "Separations of Americium from Europium by Solvent Extraction from Acidic Phosphonate Media" in *Separations of f Elements: Proceedings of a Symposium at the 207th American Chemical Society Meeting, San Diego, CA, March, 1994*, Nash, K. L.; Choppin, G. R., Eds., Plenum Press, New York, **1995**, pp. 143-152.

Chapter 11

Recent Progress in the Development of Extraction Chromatographic Methods for Radionuclide Separation and Preconcentration

Mark L. Dietz

Chemistry Division, Argonne National Laboratory, 9700 South Cass Avenue, Argonne, IL 60439-4831

Extraction chromatography (EXC), a form of liquid chromatography in which the stationary phase typically consists of an extractant solution sorbed on an inert support, provides a simple and effective means by which the separation and preconcentration of a variety of radionuclides and their isolation from major sample constituents can be achieved. Recent advances in extractant design, particularly the development of extractants capable of metal ion recognition or strong complex formation in acidic solution, have substantially improved the utility of the method. Advances in support design, particularly the introduction of functionalized supports capable of enhancing radionuclide retention or the stability of the chromatographic materials, promise to provide further improvements.

Background

Growing public health and safety concerns over the use of nuclear materials and technology, both in the production of power and the fabrication of nuclear weapons, have made rapid and reliable methods for the determination of various radionuclides in environmental and biological samples increasingly important. Such determinations, however, are rendered difficult by the complicated and variable composition of the samples encountered and by the low levels of the radionuclides typically present. Frequently, the radionuclide of interest must be both separated from the major constituents of the sample and/or possible interferents and preconcentrated prior to determination. In this chapter, we summarize recent progress in extraction chromatography, a separation and preconcentration technique that has attracted increasing attention over the last several years, and briefly describe several promising directions for future research efforts in this area.

Extraction chromatography (EXC) is a type of reversed-phase chromatography in which the stationary phase typically comprises an extractant or a solution of an extractant supported on an inert substrate (*e.g.*, Teflon™) (*1*). Unlike ordinary partition chromatography, in which a solute undergoes little if any chemical change upon sorption, the uptake of a metal ion in EXC involves the complex chemical changes associated with the conversion of a hydrated metal ion into a neutral, organophilic metal complex, just as in liquid-liquid extraction. This conversion process often involves a number of interactions and equilibria, many of which can be manipulated to provide systems capable of the efficient and selective separation of any of a number of metal ions (*2*).

Conventional EXC materials are prepared by the physical impregnation of a porous substrate with an extractant, which can be accomplished by any of several techniques (*1, 3, 4*). Most commonly, the support material is contacted with a solution of the extractant (or extractant-diluent mixture) in a volatile solvent, which after a period of equilibration, is slowly removed by evaporation under vacuum. For very hydrophobic extractants, better results (*i.e.*, more homogeneous distribution of the extractant on the support) can sometimes be obtained by contacting a solution of the extractant in a precalculated amount of diluent with a support until all of the liquid has been absorbed. In the alternative, the extractant can be incorporated into the support during its preparation. Macroporous styrene-divinylbenzene copolymers containing an extractant added to the mixture of monomers during polymerization, for example, have been prepared (*5, 6*). By proper choice of reaction conditions, the amount of extractant incorporated and other resin characteristics (*e.g.*, porosity and surface area) can be varied as desired (*7*). Although in principle, this approach could be applied for any extractant, in actual practice, only certain types of extractants (*e.g.*, neutral organophosphorus reagents) have appropriate physical and chemical properties (*e.g.*, solubility) for inclusion, since the polymerization process is affected by the extractant properties (*6*).

Regardless of the way in which an EXC material is prepared, the retention of the extractant by the support is solely the result of physical interactions, not covalent bond formation between the extractant and the support. Given this, it is not surprising that, with few exceptions (*8*), the properties of a supported extractant closely parallel those of the same compound in the corresponding liquid-liquid system. Liquid-liquid extraction data thus often represent a valuable guide to the design of EXC systems. Given the considerable body of data available on the extraction of various metal ions by any of a number of extractants, this clearly represents a major advantage of extraction chromatography over competing techniques (*e.g.*, ion exchange).

Certain aspects of the performance of an extraction chromatographic material are obviously not predictable from liquid-liquid extraction data, as there are many more factors involved in a dynamic chromatographic process than a batch (*i.e.*, static) liquid-liquid extraction (*9*). The behavior of EXC materials is normally defined in terms of seven parameters: retention, selectivity, efficiency, capacity, stability (both chemical and physical), ease of regeneration and reuse, and reproducibility. Like retention, the selectivity of an EXC material is governed primarily by the properties of the extractant and the mobile phase composition. Column efficiency, normally expressed in terms of HETP (height equivalent to a theoretical plate) or N (the number of theoretical plates), is a complex function of a number of system characteristics, including mobile phase velocity, diffusion coefficients of the metal ion in the mobile and stationary phases, particle diameter, temperature, kinetics of extraction and stationary phase thickness. (From an efficiency perspective, the ideal EXC resin is one consisting of uniform, small-particle size supports bearing a thin, homogeneous layer of a non-viscous extractant capable of rapid reaction with the metal ion(s) of interest. Few, if any, EXC materials approach this ideal.) The capacity of an EXC material is obviously dependent on the amount of extractant that can be loaded into the support. For analytical-scale applications, capacity *per se* is usually regarded as a secondary consideration. Maintaining a constant capacity (*i.e.*, stability), however, is very important; to be useful, an EXC material must exhibit satisfactory chemical and physical stability. Although chemical stability is not normally an issue (because prior experience with the extractant in liquid-liquid systems enables conditions under which it is unstable to be avoided), the physical stability of EXC materials is another matter. Because of weak support-extractant interactions (already noted), loss of extractant into the mobile phase is quite common. In fact, poor physical stability arising from the dissolution or the shearing off of the stationary phase from the support is regarded as the single biggest obstacle to the more widespread use of extraction chromatography (*10, 11*). Taken together, it is the physical and chemical stability of an EXC column that determines the ease with which it can be re-used and the reproducibility of results obtained with it. If one considers all of these parameters, their interrelationship, and their relative importance in the context of the way EXC is usually practiced, it becomes clear that improvements in EXC materials require extractants of improved selectivity and/or complexing power and supports or support/extractant/diluent combinations that provide higher efficiency and/or greater physical stability. Recently, there has been progress on several fronts.

Recent Developments

Extractants and diluents

Recent advances in molecular design and synthetic methodology have led to a wide variety of new extractants, among them such compounds as crown ethers (*12, 13*), cryptands (*13*), and calixarenes (*14*). These compounds are often capable of stronger or more selective binding of certain metal ions than has been possible with previously available reagents.

Nowhere are the improvements in chromatographic retention and selectivity made possible by these improved extractants more apparent than in the evolution of EXC methods for the separation and preconcentration of strontium ion. In the 1960's, Akaza (*15*) described the separation of Sr^{2+} from other alkaline earth cations using thenoyltrifluoroacetone (TTA) in methyl isobutyl ketone, supported on polytrifluorochloroethylene (Kel-F). Complete isolation of strontium from the variety of ions present in a typical environmental sample, however, required a complex, multi-column separation scheme. At about the same time, Lieser *et al.* (*16*) proposed the use of bis(2-ethylhexyl)phosphoric acid (HDEHP), also on Kel-F, for the separation of calcium and strontium, an important separation in environmental analysis. Although the quality of the separation was adequate at low calcium concentrations, it deteriorated markedly at higher (*i.e.*, "real world") concentrations. The problem with both of these reagents and other such "classical" extractants is that their strontium selectivity is simply not sufficient for applications as demanding as the analysis of environmental and biological samples. In addition, these reagents are not effective for highly acidic sample solutions, a significant drawback given that a number of procedures for the determination of radionuclides in urine (*17*), feces (*18*), soils (*19, 20*), and natural waters (*20*) involve either sample acidification or leaching and/or digestion of the sample with acid.

In the early 1990's, workers at Argonne National Laboratory developed a liquid-liquid extraction process for the removal of radiostrontium from acidic nuclear waste streams (*21-23*). This process, known as SREX (for strontium extraction) employs a macrocyclic polyether, di-(*tert*-butylcyclohexano)-18-crown-6 (DtBuCH18C6, Figure 1), in 1-octanol to extract strontium as a strontium-nitrato-crown ether complex. Shortly after the introduction of this process, it was demonstrated that impregnation of a polymeric support (*e.g.*, Amberlite XAD-7) with a 1 M solution of this crown ether in 1-octanol yields an EXC resin that exhibits both good retention of strontium from acidic nitrate media and excellent selectivity over most other cations (*24-26*). Especially noteworthy is the high selectivity of the material over calcium ion; in fact, this selectivity is such that the resin can tolerate levels of Ca^{2+} up to ca. 0.5 M with no significant decline in strontium retention. This is also the case for nearly all of the metallic constituents commonly encountered in environmental and biological samples, few of which have any impact on strontium retention. This behavior has its origins in the "peak selectivity" exhibited by crown ethers such as DtBuCH18C6, whereby those ions whose size corresponds most closely to that of the macrocycle cavity are capable of the strongest interaction with the crown (*27*). Thus, rather than employing a classical reagent that extracts many ions as well or better than strontium, this resin uses an extractant whose structure is "tuned" to minimize competition from other cations.

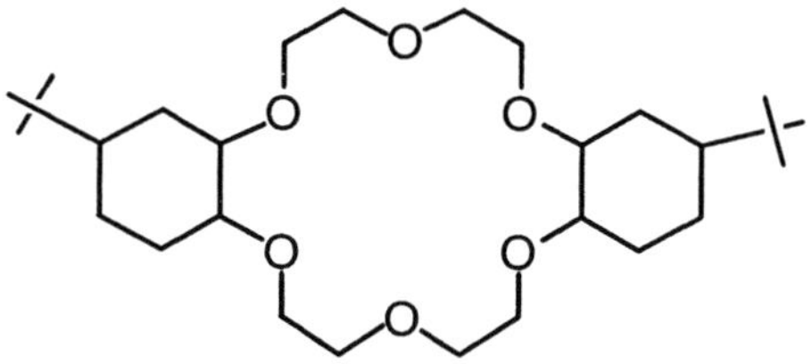

*Figure 1. Di-(*tert*-butylcyclohexano)-18-crown-6 (DtBuCH18C6).*

The application of this crown ether-based EXC resin in the separation and preconcentration of strontium is now well established, and to date, literally dozens of studies detailing its use in various analyses have appeared. (For representative examples, see refs. *28-32*). Ongoing work in this laboratory has demonstrated that despite its many advantages, the material does suffer from various limitations, the most important of which is inadequate physical stability. This is illustrated by the results presented in Figure 2, which depicts the effect of column washing on the elution behavior of ^{85}Sr on the crown ether resin. As can be seen, passage of a even a modest volume (250 free column volumes, FCV) of eluent (*i.e.*, 1 M HNO_3) through the column leads to a significant shift in the position of the elution band. Thus, attempts to reuse the material are complicated by changing column behavior induced by sample loading, column rinsing, and strontium recovery. That the elution band is shifted to higher

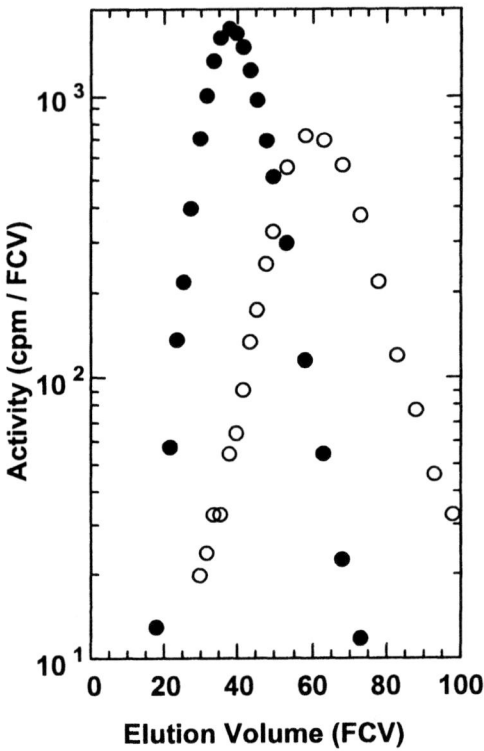

Figure 2. Effect of column washing on the elution behavior of Sr-85 on a conventional, strontium-selective extraction chromatographic material ("Sr Resin"). (Eluent: 1 M HNO_3; Flow rate: 1-2 $mL/cm^2/min$; Temperature: ca. 23°C; Particle size: 100-150 μm; Filled circles: unwashed resin; Open circles: washed (250 FCV) resin.)

volumes upon column washing is consistent with loss of diluent (here, 1-octanol) from the support, also noted by other workers (*33*), and suggests that the stability of the resin might be improved by changing or even eliminating the diluent.

Previous work, however, has shown that a change to a higher molecular weight (and thus, less water-soluble) alcohol would be expected to lead to a decrease in strontium retention by the resin (*21, 34*). Complete elimination of the diluent would also seem to be out of the question, as prior studies have clearly established the important role played by the alcohol (in particular, by the water dissolved therein) in promoting anion transfer and thus, strontium extraction, from acidic nitrate media (*21*).

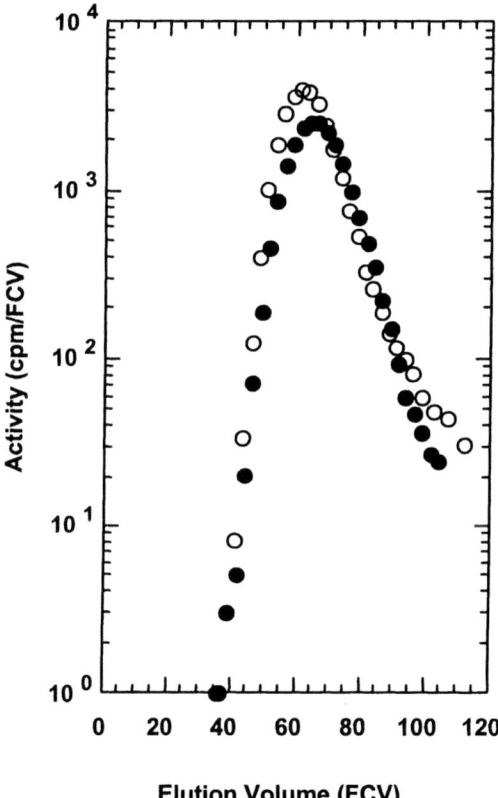

Figure 3. Comparison of the elution behavior of Sr-85 on a conventional (filled circles) and an octanol-free (open circles) strontium-selective extraction chromatographic material. (Conditions are as noted in Figure 2.)

Surprisingly, however, as shown in Figure 3, the elution behavior of ^{85}Sr on an EXC resin comprising only DtBuCH18C6 dispersed on an appropriate polymeric support is barely distinguishable from that observed on the conventional resin in which the stationary phase consists of an 1-octanol solution of the crown ether. The absence of an appreciable effect on elution behavior extends to other elements as well. Thus, the selectivity of the resin for Sr^{2+} over such cations as Na^+ and Ca^{2+} is preserved, despite the absence of a diluent (*35*). Figure 4, which relates the amount of water extracted into 1-octanol by an equimolar mixture of the *cis-syn-cis* (A) and *cis-anti-cis* (B) isomers of dicyclohexano-18-crown-6 (DCH18C6) to crown ether concentration provides a partial explanation for this unexpected result. That is, DCH18C6 (and by analogy, DtBuCH18C6), like aliphatic alcohols, is able to extract significant amounts (nearly 2 moles per mole of crown ether) of water. Thus, in the absence of a diluent, DtBuCH18C6 can apparently fulfill the role ordinarily played by the alcohol: bringing water into the stationary phase to facilitate transfer of the metal ion-nitrato-crown ether complex.

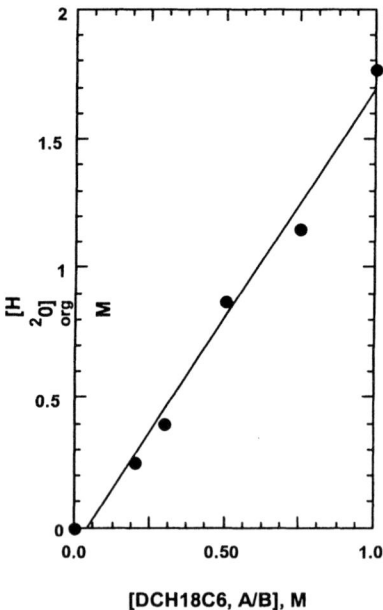

Figure 4. Crown ether concentration dependence of the extraction of water from 1 M nitric acid into 1-octanol by DCH18C6 (mixture of A and B isomers).

Figure 5 shows the effect of column washing on the elution behavior of ^{85}Sr on the octanol-free extraction chromatographic resin. As shown, a volume of mobile phase that had induced a significant change in the elution behavior of ^{85}Sr on the original resin (Figure 2) has no discernible effect on strontium elution on this material. Taken together, the enhanced physical stability resulting from the elimination of the diluent and the strong and selective strontium sorption provided by DtBuCH18C6 constitute an excellent illustration of the possibilities afforded by improved extractants and by judicious choice of diluent (or lack thereof) in the design of new EXC materials.

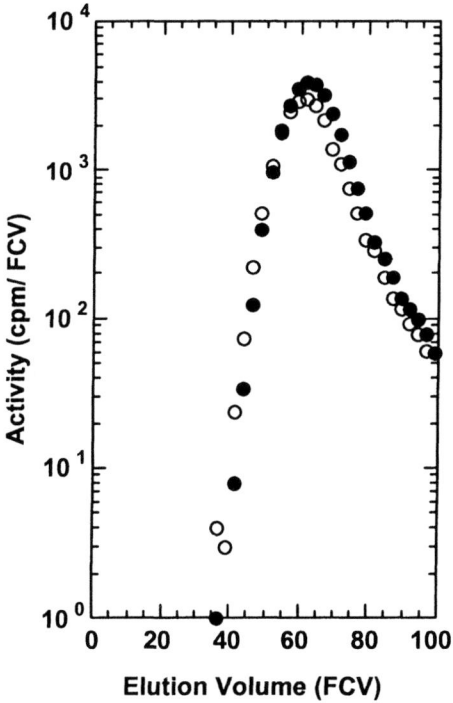

Figure 5. Effect of column washing on the elution behavior of Sr-85 on the octanol-free, strontium-selective extraction chromatographic material. (Eluent: 1 M HNO_3; Flow rate: 1-2 mL/cm² /min; Temperature: ca. 23°C; Particle size: 100-150 μm; Filled circles: unwashed resin; Open circles: washed (261 FCV) resin).

Supports

As already noted, in conventional extraction chromatography, supports are specifically chosen to be "inert" to the extractant, diluent (if any), and the constituents of the sample. In the last few years, however, there has been growing awareness of the possibility of employing support properties to enhance the performance of EXC materials. It has long been known, of course, that the *physical* properties of a support influence certain aspects of chromatographic separations. For example, the improved column efficiency resulting from the use of smaller particle size supports is well-established (*36*). Recently, however, interest has turned to the effect of support *chemistry* on the behavior of EXC materials. In these systems, hereafter denoted as "active-substrate" EXC materials, the support is chosen or designed specifically to either interact with the extractant through other than the weak adsorptive forces that typify extraction chromatography or to actually participate in metal ion uptake. In the first category are EXC resins prepared by the impregnation of a support capable of acid-base interactions or ion exchange with an extractant bearing an appropriate functional group (*e.g.*, an anionic functionality capable of interaction with an anion-exchanger). Table I summarizes the various studies that have been performed with such systems, extending from early investigations by Akaiwa (*37*), Tanaka (*38-41*), and Lee (*42*) to more recent work by Sarzanini (*43*), Warshawsky (*44, 45*), and Khalifa (*46*). Most frequently, a sulfonated extractant has been sorbed on a strong anion-exchanger and the metal ion sorption properties of the resultant materials determined. Not unexpectedly, these studies have shown that the retention of the extractant is likely due to a combination of adsorption and ion-exchange (with the latter dominating). Moreover, this retention is sometimes sufficiently strong for the resin to withstand contact with acids and bases. Two limitations in these materials have become evident, however. First, the capacity of certain of the resins is less than that expected on the basis of extractant loading, suggesting that not all of the extractant is available for complexation (*42*). In addition, it appears that the binding ability of certain immobilized extractants is less than that of the free extractant (*43*). In the absence of a detailed, systematic comparison of the performance of EXC materials based on functionalized supports to that of analogous "inert substrate" materials, it remains unclear if substrates bearing ion-exchange or acid-base functionalities offer a compelling advantage over conventional ones. For now then, this approach can only be regarded as a possible step toward improved (in particular, more stable) EXC materials, but one that clearly warrants additional investigation.

Table I. Extraction Chromatographic Systems Employing Functionalized Supports

Target Ion	Extractant	Support	Reference
Hg(II), Cu(II), Zn(II), Cd(II), Fe(II), Mn(II)	8-quinolinol-5-sulfonic acid	Diaion SA-100	37
Hg(II)	disodium(4-sulphophenyl)-1-[2-(4-sulphophenyl)hydrazide]diazenecarbothioate	Amberlite IRA-400	38
Hg(II), Cu(II), Cd(II)	disodium 4,4'-(4-diazenediyl-5-mercapto-3-methyl-1,2-diazacyclopenta-2,4-diene-1-yl)dibenzenesulphonate	Amberlite IRA-400	39, 41
Hg(II), Cu(II)	disodium(4-sulphophenyl)-1-[2-(4-sulphophenyl)hydrazide]diazenecarbothioate	Amberlite IRA-400	40
	tetraphenylporphinetrisulphonic acid		
Fe(III), Cu(II), Pb(II), Cr(VI)	7-iodo-8-hydroxyquinoline-5-sulfonic acid	Dowex 1-X8	42
Al(III)	Pyrocatechol Violet	BioRad AG MP-1	43
Pb(II), Zn(II), Ni(II)	di(2-ethylhexyl) dithiophosphoric acid	Reillex™ HP, Reillex™ 425	44, 45
U(VI)	Alizarin Red S	Duolite A 101	46

Recently, another approach to the stabilization of EXC materials has been proposed, one which also involves some novel support chemistry. In this approach (Figure 6), first described by Horwitz and Dietz (*47*) and subsequently elaborated by Alexandratos and co-workers (*48*), macroporous beads of a co-polymer of chloromethylstyrene and methylmethacrylate are prepared (using divinylbenzene as a cross-linker), then surface-functionalized to leave unreacted carbon-carbon double bonds. Following impregnation of the support with an extractant, a pair of water-soluble monomers (N,N' methylenebis(acrylamide /

glycidyl methacrylate) are polymerized in the presence of the beads, resulting in the formation of a polymer film anchored to the bead by what had been the surface C=C bonds. The effectiveness of this approach as a means of stabilizing an EXC resin is illustrated by results reported for the uptake of copper ion from a pH 8 buffer by a bis(2-ethylhexyl)phosphoric acid (HDEHP)-loaded resin, both in the presence and absence of a polymer coating. Even after 5 days of contact with the buffer solution, no diminution in the uptake of copper ion by the coated resin is observed. In contrast, prolonged contact of the uncoated resin with the same buffer leads to leaching off of the extractant and a pronounced decrease in copper uptake. One problem that has been identified with this approach is that the polymer film decomposes upon contact with acids to release formaldehyde. If this problem can be addressed and if it can be demonstrated that the encapsulation process does not lead to unacceptably long equilibration times (as has been the case for EXC materials encapsulated by a Nylon 6-10 film (*49*), for which 2-3 hours are required for the attainment of sorption equilibrium), the use of "anchored coatings" would seem to offer much promise as a means of producing improved EXC materials.

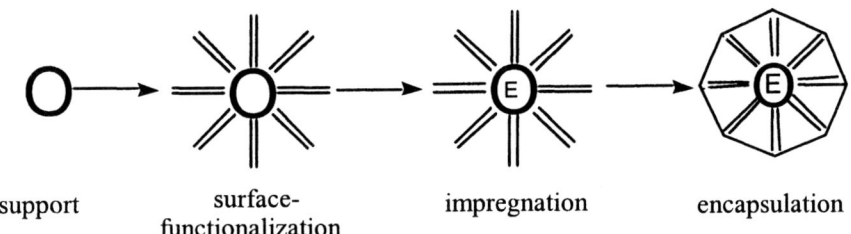

Figure 6. Schematic diagram showing the preparation of an extraction chromatographic resin stabilized via polymer encapsulation.

To date, little effort has been devoted to the study of the second type of active-substrate EXC material, in which the support actually participates in the metal ion extraction process. Moyer et al. (*50*) have examined the uptake of copper ion by EXC materials comprising a neutral extractant, tetrathia-14-crown-4, sorbed on a series of polystyrene-divinylbenzene-based strong-acid cation-exchange resins, among them the commercially available Dowex 50W-X8. While neither unfunctionalized polystyrene-divinylbenzene resin nor the same resin loaded with the macrocycle extracted any detectable copper, impregnation of the cation-exchanger produced a 10-100-fold enhancement in the observed copper distribution ratio *vs.* the cation-exchanger alone. This enhancement was attributed to a synergistic effect involving coordination of the copper by the macrocycle and cation-exchange by the polymer-bound sulfonic acid functional groups. Although such synergistic effects are common in liquid-

liquid extraction systems involving mixtures of crown ethers and liquid cation-exchangers (*51*), Moyer's results represent the first demonstration of synergism in an EXC material involving a functionalized support. Such supported synergistic systems appear to offer a wealth of opportunities for the development of new EXC materials exhibiting enhanced metal ion uptake and selectivity.

Conclusions

In recent years, the field of extraction chromatography has evolved in several important respects. Extractants of limited selectivity or complexing ability (particularly in highly acidic media), for example, are being supplanted by those exhibiting high selectivity and/or complexing power. At the same time, interest in purely "passive" supports, whose chemistry plays no significant role in either metal ion uptake or resin stability, is slowly giving way to recognition of the possibilities offered by functionalized (*i.e.*, "active") supports. Progress in support and extractant chemistry has already yielded notable improvements in the performance of EXC materials, but considerable room for improvement clearly remains, particularly in the area of physical stability. If this limitation can be overcome, the utility of extraction chromatography as a method for the separation and preconcentration of radionuclides is certain to increase.

Acknowledgements

This paper is based in part on publications with my colleagues, past and present, in the Chemistry Division at Argonne National Laboratory. Their efforts are gratefully acknowledged. This work was performed under the auspices of the Office of Basic Energy Sciences, Division of Chemical Sciences, U.S. Department of Energy, under contract number W-31-109-ENG-38.

References

1. *Extraction Chromatography*; Braun, T.; Ghersini, G., Eds.; Elsevier: New York, 1975.
2. Dietz, M. L.; Horwitz, E. P. *LC.GC* **1993**, *11*, 424-436.
3. Warshawsky, A. In *Ion Exchange and Solvent Extraction-A Series of Advances*; Marinsky, J. A., Marcus, Y., Eds.; Marcel Dekker: New York, 1981; Vol. 8, p 229.

4. Cortina, J. L.; Warshawsky, A. In *Ion Exchange and Solvent Extraction*; Marinsky, J. A., Marcus, Y., Eds.; Marcel Dekker: New York, 1997; Vol. 13, p 195.
5. Krochbel, R.; Meyer, A. In *Proceedings of the International Solvent Extraction Conference-1974*; Society of Chemical Industry, London, 1974; p 2095.
6. Kauczor, H. W.; Meyer, A. *Hydrometallurgy* **1978**, *3*, 65-73.
7. Poinescu, I.; Popescu, V.; Carpov, A. *Angew. Makromol. Chem.* **1985**, *135*, 21-32.
8. Cortina, J. L.; Miralles, N.; Aguilar, M.; Sastre, A. M. *Solvent Extr. Ion Exch.* **1994**, *12*, 371-391.
9. Ghersini, G. In *Extraction Chromatography*; Braun, T., Ghersini, G., Eds.; Elsevier: New York, 1975; p 68.
10. Alexandratos, S.; Ripperger, K. P. *Ind. Eng. Chem. Res.* **1998**, 37, 4756-4760.
11. Dietz, M. L.; Horwitz, E. P.; Bond, A. H. In *Metal-Ion Separation and Preconcentration: Progress and Opportunities*; Bond, A. H.; Dietz, M. L.; Rogers, R. D., Eds.; ACS Symposium Series 716; American Chemical Society, Washington, DC, 1999, pp 234-250.
12. McDowell, W. J. *Sep. Sci. Technol.* **1988**, 23, 1251-1268.
13. Gokel, G. *Crown Ethers and Cryptands*; Royal Society of Chemistry: Cambridge, England, 1991.
14. Lumetta, G. J.; Rogers, R. D.; Gopalan, A. S., Eds.; *Calixarenes for Separations*; ACS Symposium Series 757; American Chemical Society: Washington, DC, 2000.
15. Akaza, I. *Bull. Chem. Soc. Jpn.* **1966**, *39*, 980-989.
16. Lieser, K. H.; Bernhard, H. *Z. Anal. Chim.* **1966**, *219*, 401-408.
17. Dietz, M. L.; Horwitz, E. P.; Nelson, D. M.; Wahlgren, M. *Health Physics* **1991**, *61*, 871-877.
18. Veselsky, J. C. *Mikrochim. Acta* **1978**, *1*, 79-88.
19. Juznick, K.; Korun, M. *J. Radioanal. Nucl. Chem. Lett*. **1989**, *137*, 235-242.
20. Burnett, W. C.; Corbett, D. R.; Schultz, M.; Horwitz, E. P.; Chiarizia, R.; Dietz, M. L.; Thakkar, A.; Fern, M. *J. Radioanal. Nucl. Chem.* **1997**, *226*, 121-127.
21. Horwitz, E. P.; Dietz, M. L.; Fisher, D. E. *Solvent Extr. Ion Exch.* **1990**, *8*, 199-208.
22. Horwitz, E. P.; Dietz, M. L.; Fisher, D. E. *Solvent Extr. Ion Exch.* **1990**, *8*, 557-572.
23. Horwitz, E. P.; Dietz, M. L.; Fisher, D. E. *Solvent Extr. Ion Exch.* **1991**, *9*, 1-25.
24. Horwitz, E. P.; Dietz, M. L.; Fisher, D. E. *Anal. Chem.* **1991**, *63*, 522-525.

25. Horwitz, E. P.; Chiarizia, R.; Dietz, M. L. *Solvent Extr. Ion Exch.* **1992**, *10*, 313-336.
26. Chiarizia, R.; Horwitz, E. P.; Dietz, M. L. *Solvent Extr. Ion Exch.* **1992**, *10*, 337-361.
27. Izatt, R. M.; Bradshaw, J. S.; Nielsen, S. A.; Lamb, J. D.; Christensen, J. J. *Chem. Rev.* **1985**, *85*, 271-339.
28. Mellado, J.; Llaurado, M.; Rauret, G. *Anal. Chim. Acta* **2002**, *458*, 367-374.
29. DeVol, T. A.; Duffey, J. M.; Paulenova, A. *J. Radioanal. Nucl. Chem.* **2001**, *249*, 295-301.
30. Torres, J. M.; Llaurado, M.; Rauret, G.; Bickel, M.; Altzitzoglou, T.; Pilvio, R. *Anal. Chim. Acta* **2000**, *414*, 101-111.
31. Grahek, Z.; Zecevic, N.; Lulic, S. *Anal. Chim. Acta* **1999**, *399*, 237-247.
32. Filss, M.; Botsch, W.; Handl, J.; Michel, R.; Slavov, V. P.; Borschtschenko, V. V. *Radiochim. Acta* **1998**, *83*, 81-92.
33. Grate, J. W.; Fadeff, S. K.; Egorov, O. *Analyst* **1999**, *124*, 203-210.
34. Horwitz, E. P.; Dietz, M. L.; Rhoads, S.; Felinto, C.; Gale, N. H.; Houghton, J. *Anal. Chim. Acta* **1994**, *292*, 263-273.
35. Dietz, M. L.; Horwitz, E. P. U.S. Patent 6,511,603 B1, 2003.
36. Krstulocic, A. M.; Brown, P. R. *Reversed-Phase High-Performance Liquid Chromatography*, Wiley-Interscience, New York, NY, 1982, pp 16-25.
37. Akaiwa, H.; Kawamoto, H.; Nakata, N.; Ozeki, Y. *Chem. Lett.* **1975**, *10*, 1049-1050.
38. Tanaka, H.; Chikuma, M.; Harada, A.; Ueda, T.; Yube, S. *Talanta* **1976**, *23*, 489-491.
39. Chikuma, M.; Nakayama, M.; Tanaka, T.; Tanaka, H. *Talanta* **1979**, *26*, 911-912.
40. Chikuma, M.; Nakayama, M.; Itoh, T.; Tanaka, H.; Itoh, K. *Talanta* **1980**, *27*, 807-810.
41. Nakayama, M.; Chikuma, M.; Tanaka, H.; Tanaka, T. *Talanta* **1982**, *29*, 503-506.
42. Lee, K. S.; Lee, W.; Lee, D. W. *Anal. Chem.* **1978**, *50*, 255-258.
43. Sarzanini, C.; Mentasti, E.; Porta, V.; Gennaro, M. C. *Anal. Chem.* **1987**, *59*, 484-486.
44. Warshawsky, A.; Strikovsky, A. G.; Jerabek, K.; Cortina, J. L. *Solvent Extr. Ion Exch.* **1997**, *15*, 259-283.
45. Warshawsky, A.; Cortina, J. L.; Aguilar, M.; Jerabek, K. In *Solvent Extraction for the 21st Century (Proceedings of ISEC '99)*; Cox, M., Hidalgo, M., Valiente, M., Eds; Society of Chemical Industry, London, 2001; Vol. 1, pp 1267-1272.
46. Khalifi, M. E. *Sep. Sci. Technol.* **1998**, *33*, 2123-2141.

47. Horwitz, E. P.; Dietz, M. L. "A Novel Approach to Improving the Stability and Column Efficiency of Extraction Chromatographic Systems", Invention Report, Argonne National Laboratory, 1998.
48. Alexandratos, S. D.; Ripperger, K. P. *Ind. Eng. Chem. Res.* **1998**, 37, 4756-4760.
49. Nii, S.; Masutani, M.; Takeuchi, H. In *Solvent Extraction for the 21st Century (Proceedings of ISEC '99)*; Cox, M., Hidalgo, M., Valiente, M., Eds; Society of Chemical Industry, London, 2001; Vol. 1, pp 1279-1283.
50. Moyer, B. A.; Case, G. N.; Alexandratos, S. D.; Kriger, A. A. *Anal. Chem.* **1993**, *65*, 3389-3395.
51. Bond, A. H.; Dietz, M. L.; Chiarizia, R. *Ind. Eng. Chem. Res.* **2000**, *39*, 3442-3464.

Chapter 12

Radioanalytical Methods in the Discovery and Characterization of Non-Pertechnetate (^{99}Tc) Species in Hanford Tank Wastes

Rebecca M. Chamberlin[1,*], Kenneth R. Ashley[2], Jason R. Ball[1], Eve Bauer[1], Jonathan G. Bernard[1], Douglas E. Berning[1], Norman C. Schroeder[1], and Paul Sylvester[3,4]

[1]Chemistry Division, Los Alamos National Laboratory, Los Alamos, NM 87545
[2]Department of Chemistry, Texas A&M University at Commerce, Commerce, TX 75429
[3]Lynntech, Inc., 7610 Eastmark Drive, College Station, TX 77840
[4]Current address: Triton Systems Inc., 200 Turnpike Road, Chelmsford, MA 01824
[*]Corresponding author: rmchamberlin@lanl.gov

The existence of reduced technetium complexes in Hanford nuclear tank wastes was first proven using radioanalytical methods in anion exchange experiments. Subsequent efforts to understand the generation and stability of non-pertechnetate species in alkaline solution used radiochemistry in conjunction with spectroscopic methods. Collectively, the results indicate that aminocarboxylates such as EDTA do not form stable ^{99}Tc complexes in strong base, while sugar derivatives such as gluconate support Tc(VII) reduction under a wide range of conditions, including realistic tank waste simulants.

© 2004 American Chemical Society

As the U.S. Department of Energy prepares to dispose of defense nuclear wastes that have been generated and stored at the Hanford Site since the Manhattan project, increasing attention has centered on the highly complex alkaline wastes stored in underground tanks. Hanford's tank wastes consist of neutralized effluents from several chemical processes designed to recover plutonium from irradiated fuel rods, along with a host of minor effluents from processes such as equipment flushing, fission product recovery, and sugar denitration (1). While the reactivity of the initially acidic solutions was well-understood within the context of plutonium recovery processes, relatively little consideration was given until recently to the chemical dynamics of the wastes after neutralization to pH 11-14 for long-term storage in the tanks.

Of particular concern in this context is the chemical evolution of the long-lived fission product technetium-99. Removal of a significant fraction of the ^{99}Tc from the waste is currently required to meet NRC Class "A" limits, so the waste can be managed by vitrification and on-site storage. To support this plan, and earlier disposal concepts, significant effort has been directed at developing and characterizing methods to separate the pertechnetate (TcO_4^-) anion from high-nitrate alkaline solutions typical of Hanford wastes. This focus on pertechnetate separation was based on the knowledge that technetium emerged from fuel rod dissolution and plutonium recovery operations as the fully oxidized anion.

In recent years, however, it has been established (2-4) and generally accepted (5-7) that the ^{99}Tc in certain classes of Hanford waste has, over decades of storage, been converted to lower oxidation state complexes. These complexes resist conventional separations, and are difficult to selectively oxidize back to the more tractable TcO_4^- anion. These reduced complexes are characteristic of the Envelope C tanks, which contain large quantities of organic compounds that can serve as both reducing and complexing agents for ^{99}Tc. This discovery emerged from an experimental program that relied on radioanalytical chemistry as its primary methodology. Later studies using modern spectroscopic techniques have provided additional corroboration of these "non-pertechnetate" species.

This paper describes the role of radiochemistry in the discovery of non-pertechnetate species in Hanford wastes, and presents an overview of the radioanalytical methods that have subsequently been used—alone or in concert with other techniques—to study their potential origin and identity.

Discovery of the Non-Pertechnetate Problem

Samples of supernate from Hanford tanks 101-SY and 103-SY were provided to Los Alamos National Laboratory in 1995 for validation of the performance of commercial anion exchange resin Reillex-HPQ in removing $^{99}TcO_4^-$ from alkaline wastes. Later, a sample of 107-AN supernate was also provided to LANL by Pacific Northwest National Laboratory. Prior studies had established the sorption behavior of pertechnetate in the presence of strong nitrate and hydroxide solutions, and characterized ideal column load and stripping conditions (8).

The key challenge in the radiochemical analysis of ^{99}Tc in tank waste is the abundance of the ~30-year nuclides ^{137}Cs and ^{90}Sr, whose beta activity is several orders of magnitude greater than that of ^{99}Tc. Careful radiochemical purification allows the ^{99}Tc to be separated from these interfering nuclides, but chemical losses during the purification must be accounted for by concurrently processing a *thoroughly equilibrated* tracer nuclide such as ^{95m}Tc. The radiochemical method developed at LANL (3) for total ^{99}Tc analysis in waste samples, as well as in post-contact solutions for batch distribution coefficient (K_d) determinations, is summarized in Figure 1.

In this analysis, a measured volume of waste solution is spiked with the gamma-emitting tracer ^{95m}Tc, for subsequent chemical yield determination. To oxidize the technetium and organics, and fully equilibrate the ^{95m}Tc/^{99}Tc isotopes, Ce(IV) ion and concentrated nitric acid are added and the solution is evaporated to incipient dryness. Two additional cycles of nitric acid addition and evaporation follow, then the sample is dissolved in water and passed through a column of cation exchange resin to remove cationic radionuclides such as ^{137}Cs. The ^{99}TcO$_4^-$-containing effluent is neutralized, then contacted for two hours with Reillex-HPQ to selectively sorb the technetium as TcO$_4^-$. After thorough column washes, the ^{99}Tc and ^{95m}Tc are co-eluted by reductive complexation with alkaline Sn(II)/ethylenediamine solution. After measuring the final sample activity using both NaI gamma counting and liquid scintillation counting, the chemical yield (typically 65%) is calculated based on recovery of ^{95m}Tc. The LSC results then provide an accurate measure of the total ^{99}Tc present in the initial sample.

Determination of the total ^{99}Tc concentrations in the 101-SY and 103-SY samples by this method gave values of 5.9×10^{-5} and 7.2×10^{-5} mol/L, respectively (Table I). Significantly lower values of 1.7×10^{-5} and 4.3×10^{-5} mol/L, also determined radiochemically but using a milder equilibration with hydrogen peroxide in 4 mol/L HNO$_3$, had been reported by PNNL. Because of this disparity in the radiochemistry results, the ^{99}Tc concentration was determined by two other methods. First, total ^{99}Tc was measured by ICP-MS using a standard addition technique, and assuming that all signal at mass 99 could be attributed to technetium. The species-independent ICP-MS determination gave values that agreed within uncertainty with the Ce(IV) radiochemistry results. Finally, the TcO$_4^-$ concentration in the unoxidized waste samples and Ce(IV)-oxidized samples was measured by ^{99}Tc NMR using a signal-to-noise ratio method (4). The vigorously oxidized NMR sample indicated a pertechnetate concentration consistent with the ICP-MS and Ce(IV)/HNO$_3$ radiochemistry results.

Collectively, these results indicate that pertechnetate-selective methods such as anion exchange and ^{99}Tc NMR greatly underestimate the technetium concentration in the wastes, unless extreme steps are taken to oxidize the samples. Similar results were obtained for the 107-AN samples. While strongly suggestive of the presence of reduced ^{99}Tc species, further evidence was

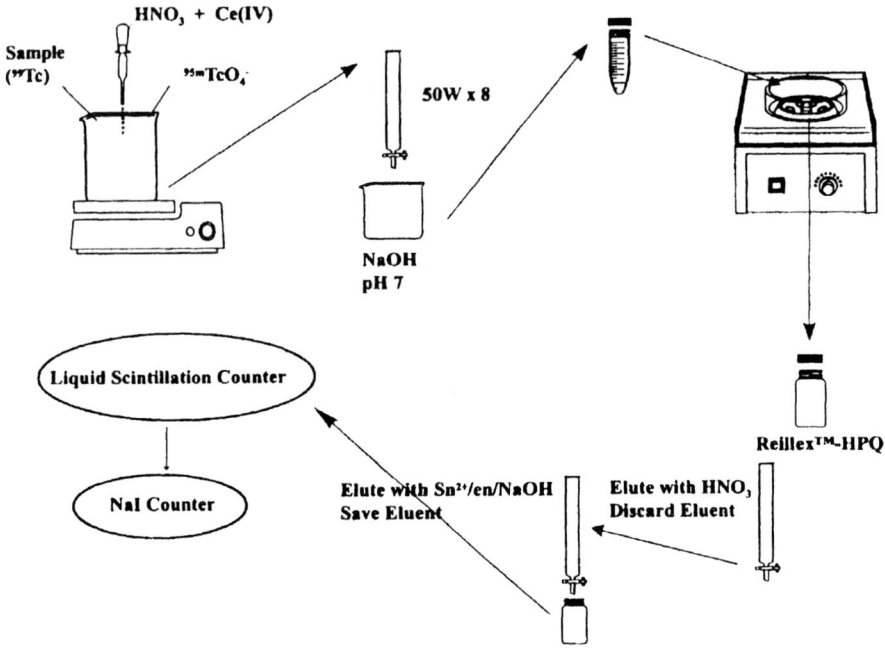

Figure 1. Radiochemical method for total ^{99}Tc analysis in tank wastes.

Table I. ^{99}Tc Concentrations in 101-SY and 103-SY Samples, by Method.

Analytical Method	Oxidation Method	101-SY (10^{-5} mol/L)	103-SY (10^{-5} mol/L)
Radiochemistry	Ce(IV) / conc HNO$_3$	5.92 ± 0.88	7.22 ± 0.35
NMR, oxidized	Ce(IV) / conc HNO$_3$	4.19 ± 1.22	7.70 ± 1.96
ICP-MS	--	5.18 ± 0.30	7.76 ± 0.44
Radiochemistry	H$_2$O$_2$ / 4 mol/L HNO$_3$	1.72	4.25
NMR, unoxidized	--	1.57 ± 0.46	2.30 ± 0.58

provided by measuring the partitioning behavior of ^{99}Tc from the waste samples onto Reillex-HPQ resin.

Batch distribution coefficients (K_d) were measured under experimental conditions developed in prior simulant studies (8). The "apparent" ^{99}Tc K_d for the waste samples is defined as the ratio of the post-contact ^{99}Tc activity sorbed onto the anion exchange resin (per g of resin), to the ^{99}Tc activity remaining in solution (per mL of solution), without regard to chemical speciation. In practice, the K_d is determined by accurately measuring the ^{99}Tc activity in the pre- and post-contact solutions using the method outlined above. The activity on the resin is then calculated by difference, assuming that any decrement is due to sorption.

The apparent ^{99}Tc K_d values for the three tank waste samples ranged from 4.0 to 5.2 mL/g (Table II). A "true" K_d value for pertechnetate ion was concurrently measured by adding gamma-emitter ^{95m}Tc (204 keV) as ^{95m}TcO$_4^-$; its detection was facilitated by removing most of the ^{137}Cs in a batch contact with a proprietary Cs$^+$-selective cation exchange resin. The true K_d accurately reflects any competition or inhibition of sorption caused by the waste matrix, and thus reflects the ^{99}Tc partitioning behavior expected if pertechnetate is the only species present. The predicted K_d values in Table 2 are calculated from the empirical behavior of TcO$_4^-$ in simple NaOH/NaNO$_3$ solutions.

Table II. Apparent, True and Predicted K_d Values for Tank Waste Samples

Sample	Apparent ^{99}Tc K_d	True ^{95m}TcO$_4^-$ K_d	Predicted TcO$_4^-$ K_d	% non-TcO$_4^-$
101-SY	3.99	545	525	70 ± 11
103-SY	5.16	543	492	64 ± 19
107-AN	5.20	321	318	54 ± 7

NOTE: K_d values in mL/g. Samples were diluted (101-SY, 1:3.94; 103-SY, 1:2.87; 107-AN, 1:3.76) and/or their OH$^-$ concentration adjusted before measurement (2-4).

Remarkably, the true K_d values, which range from 321 to 545 mL/g, are two orders of magnitude greater than their corresponding apparent K_d values. This difference implies that there are ^{99}Tc species present that are not being sorbed efficiently, and are therefore in a chemical form other than TcO$_4^-$. Calculation of the % non-pertechnetate species in each supernate is possible if the non-pertechnetate fraction is assumed to have no sorption ($K_d = 0$) and slow or no equilibration with pertechnetate (3). Since normal oxidants for simple reduced forms of technetium did not oxidize the 54-70% non-pertechnetate fraction back to pertechnetate, it seemed likely that a reduced complex was present.

To corroborate this conclusion, supernate samples from the three tanks were analyzed by PNNL using x-ray absorption near edge spectroscopy. Coincident near-edge energies were reported for the three samples (5), indicating that the same chemical species is present. Although a small shoulder corresponding to TcO$_4^-$ was observed, the main signal had an energy intermediate between Tc(0) and Tc(VII), and comparison to available standards suggested an oxidation state of Tc(IV). Further studies of tank waste samples by x-ray absorption spectroscopy (XANES or EXAFS) have not been reported, so the structure cannot be more precisely described at this time.

However, the nature of the apparent K_d measurement suggests a potential avenue for additional studies. The apparent K_d represents a weighted average of the sorption of all ^{99}Tc species in the sample. If the soluble complex has a non-zero affinity for the resin, then the % non-TcO$_4^-$ species in the wastes must be greater than that reported in Table 2. More significantly, the reduced species cannot have a true K_d value greater than the apparent K_d of the supernate sample as a whole. Initial estimates for the non-pertechnetate species' K_d in 101-SY and 103-SY were obtained by immediately running a second batch contact of the supernates from the first contacts, giving values <1 mL/g. Any postulated identification of the non-pertechnetate species should be corroborated by a comparable measured K_d value from actual or simulated wastes.

Stability of Non-Pertechnetate Species in Alkaline Solution

The coordination chemistry of technetium has been extensively characterized in near-neutral aqueous solution (9), mainly with the goal of generating pharmaceutically interesting complexes of ^{99m}Tc. In contrast, its reactivity in alkaline solution was largely unexplored prior to the discovery of Hanford non-pertechnetate problem. Because of their ability to stabilize reduced Tc in neutral to acidic solution, their strong binding affinity for simple cations and their abundance in some Hanford tanks, aminocarboxylate ligands such as ethylenediaminetetraacetic acid (EDTA), and nitrilotriacetic acid (NTA) were initially selected for complexation studies in strong base. Complexes of technetium with aminocarboxylates were prepared using literature methods, typically via sulfite reduction of TcO_4^- in dilute acid. Known complexes of EDTA, NTA, ethylenediamine-N,N'-diacetic acid (EDDA), iminodiacetic acid (IDA), and related organics were characterized using IR and UV-vis spectroscopy and combustion analysis.

Using these complexes, we attempted to address a series of questions: What classes of ligands are capable of stabilizing reduced technetium in Hanford wastes? What is the behavior of these complexes toward anion exchange? And finally, can these complexes be selectively oxidized to prepare the waste for disposal? While the first question might be addressed by spectroscopic studies, radioanalytical methods are ideally suited to study oxidation and separation.

In general, none of the aminocarboxylate complexes showed significant stability under alkaline conditions, as exemplified by the Tc(IV)-EDTA complex which has the dimeric solid-state structure $(H_2L)Tc(\mu-O)_2Tc(LH_2) \cdot 5H_2O$ (L = EDTA; 10). If stable, this complex is predicted to sorb weakly, or not at all, to anion exchange resins at low pH, where the four peripheral carboxylates are protonated and the complex has no net charge. At higher pH values, sorption should increase monotonically as the carboxylates are deprotonated.

To check this assumption, a stock solution (0.01 mol/L in ^{99}Tc) of the EDTA complex was diluted to 5 x 10^{-5} mol/L in solutions of pH ranging from 0 to 14. After standing for 24 hours to ensure equilibration, the ^{99}Tc distribution coefficient on Reillex-HPQ was measured (mass/volume ratio = 0.1 g/mL). The results in Figure 2 show the expected rise in K_d values as the pH is increased from 0 to 8. However, a sharp decline occurs at pH values above 8, indicating that the intact complex does not sorb onto anion exchangers, or alternatively that it decomposes to form a new non-pertechnetate species in highly alkaline solution. While the observed bleaching of the pink Tc(IV) solution color during equilibration might suggest reoxidation to Tc(VII), the observed K_d is consistently lower than the value of 3400 mL/g predicted for pertechnetate.

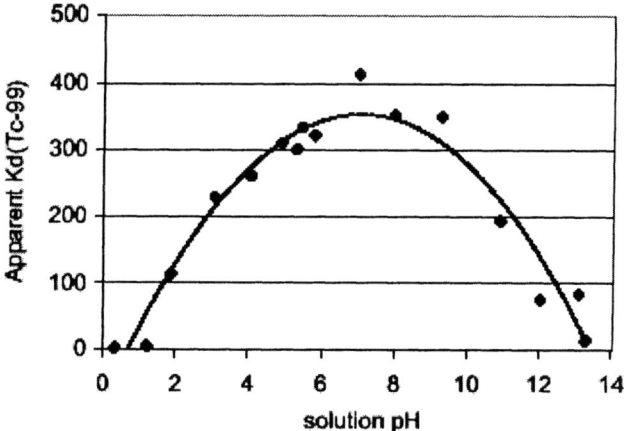

Figure 2. Apparent ^{99}Tc K_d of Tc(IV)-EDTA complex diluted to 10^{-5} mol/L.

In this case, radiochemical methods indicated an unexpected reaction, but UV-vis spectroscopy proved valuable in interpreting these data. Upon dilution of the Tc(IV)-EDTA complex in pH 12.8 solution, the characteristic dimer absorption band at 500 nm (ε = 2000 L mol^{-1} cm^{-1}; ref. *10*) bleached to 10% of its original intensity within 2 days, and eventually faded completely (Fig. 3). The rate of bleaching was not greatly affected by carefully purging air from the sample, suggesting an initial non-oxidative reaction with OH$^-$. Concurrent monitoring of the apparent ^{99}Tc K_d showed a gradual increase from the initial value of 20 mL/g, to 280 after 14 days. After 50 days, the K_d exceeded 1000 mL/g, consistent with decomposition to TcO_4^-.

In contrast, the solutions at pH 3.4, 5.6 and 10.9 reached an equilibrium absorption of about 60% their initial intensity, when air was excluded from the samples. The differential rates suggest acid- or base-catalyzed conversions, but

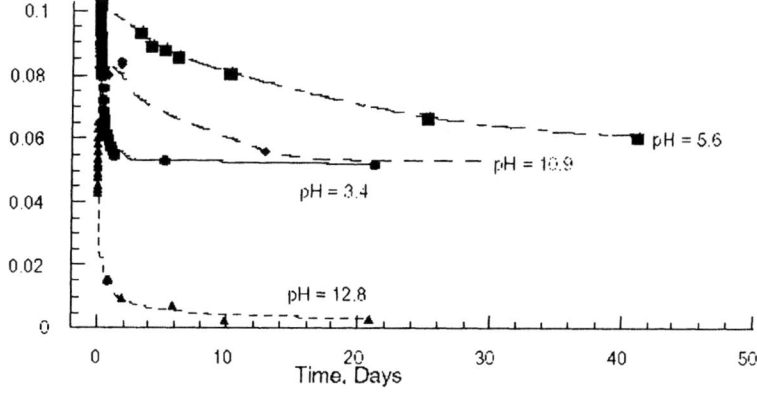

Figure 3. Bleaching of the 500-nm absorption of the Tc(IV)-EDTA dimer after dilution to 10^{-5} mol/L and pH adjustment.

the common equilibrium endpoint indicates a relatively simple reaction, probably a dimer-monomer equilibrium. The apparent K_d in the pH 10.9 sample remained constant at ~200 mL/g for 50 days, indicating that the product is stable with respect to oxidation, as long as it is maintained in a closed container. Collectively, these results suggest that stable technetium-EDTA complexes can exist in "alkali deficient" tanks such as 107-AN, but are unlikely to account for the non-pertechnetate fraction in high-pH tanks such as 101-SY and 103-SY. Furthermore, even at the lower pH of the 107-AN waste sample, it is mathematically impossible to account for the observed K_d of 5.2 mL/g by invoking a Tc-EDTA complex whose K_d is 200 mL/g.

Similar investigations of complex stability at both LANL and Lynntech, Inc, showed that Tc complexes of EDDA, IDA, and citrate have insufficient stability in strong base ([NaOH] > 0.5 mol/L) to account for the non-pertechnetate species in tank waste. It also proved difficult to synthesize most carboxylate and aminocarboxylate complexes directly in pH 12 solutions by reducing TcO_4^- with Sn(II). In general, the attempted syntheses using various complexants singly or in mixtures gave only hydrous TcO_2. However, two notable exceptions were found. First, a Tc-glycolate complex was moderately stable when dissolved in 2 mol/L NaOH, decomposing to TcO_4^- in about 2 days. A colorless glycolate complex could also be directly synthesized by Sn(II) reduction at pH 13, but within 18 hours showed 50% decomposition to TcO_2.

A more promising result was noted for gluconate, which may still be present in a number of Hanford tanks as the byproduct of sugar-based denitration, or from direct process addition (*1*). A stable pink Tc-gluconate complex is readily formed in 2.5 mol/L NaOH by reduction with sulfite or Sn(II), and shows no evidence of decomposition when stored in a closed container. When intentionally oxidized using hydrogen peroxide, the pink (512 nm) complex is converted to a yellow (400 nm) complex, which remained stable relative to TcO_4^- for more than 2 months.

Subsequent experiments were directed at synthesis of reduced Tc complexes in more realistic tank waste simulants. Simulant formulations for tanks 107-AN (developed by BNFL, Inc.) and 101-SY (developed by PNNL) are detailed in the Experimental section. The 107-AN simulant contains the complexants EDTA, HEDTA, glycolate, gluconate, citrate, NTA, IDA, formate, acetate, and oxalate, giving a total organic content of approximately 33 g/L. The 101-SY simulant contains a far lower total organic concentration and a more limited suite of organics, with only EDTA, HEDTA, citrate, NTA, IDA and gluconate present, but also has a higher pH.

The yellow to brown coloration of these simulants hinders characterization of dilute technetium complexes by UV-vis spectroscopy, so radiochemical evaluation of the speciation was essential. The simulants were spiked with $^{99}TcO_4^-$ at 2×10^6 Bq/L (approximately 0.01 mmol/L) and varying amounts of

Sn(II) chloride were added to reduce the pertechnetate. Figure 4 shows the effect of Sn(II) on the ^{99}Tc K_d, for the 107-AN simulant, compared to a control experiment using Sn(IV) to ensure that tin itself does not alter the pertechnetate sorption. The rapid decline in ^{99}Tc sorption onto Reillex-HPQ with increasing amounts of Sn(II) indicates that a non-extractable reduced form of ^{99}Tc can be prepared in simulated 107-AN tank waste. Consistent with observations for actual tank waste samples, the apparent K_d of the 0.01 mol/L Sn(II) solution gradually increased from 2 to 25 after one month in a capped (but not rigorously anoxic) bottle.

A similar result was obtained using the 101-SY simulant. Thus, addition of 0.01 mol/L Sn(II) to the 101-SY simulant gave an apparent ^{99}Tc K_d of only 0.84 mL/g. The stability of the reduced species was lower in this simulant (K_d increased to 131 mL/g after two weeks), probably due to the higher pH and/or lower organic content. Periodic measurements of dissolved ^{99}Tc showed no precipitation of TcO_2 in the simulants, indicating that the complexes decompose via soluble species only, and not hydrous TcO_2.

Although no ^{99}Tc complexes could be directly detected, it is instructive to compare the simulant results with the stabilities of Tc complexes in simple NaOH solutions. While both glycolate and gluconate gave indications of stable complexes in base, no glycolate was present in the 107-AN simulant. Both simulants contained gluconate, which also supported persistent reduced complexes in NaOH. Although possibly coincidental, the ^{99}Tc K_d values in the freshly reduced simulants were slightly less than 1 mL/g, which is comparable to the results of the second batch contact of the actual 101-SY and 103-SY samples.

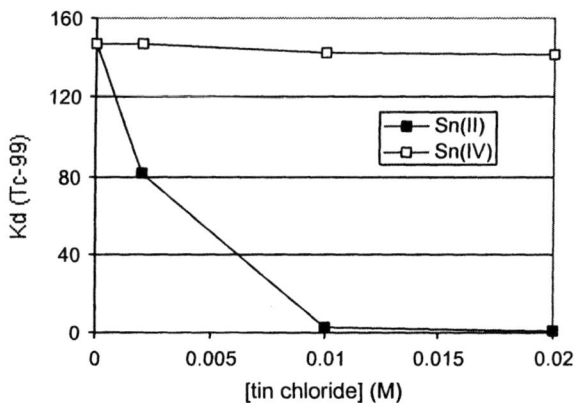

Figure 4. Effect of reducing agent $SnCl_2$, or control salt $SnCl_4$, on the apparent ^{99}Tc K_d in simulated 107-AN supernate.

Genesis of Non-Pertechnetate Species in Alkaline Solution

A final issue of interest is the mechanism of pertechnetate reduction in the tank waste environment. Understanding the mechanism would not only support identification of the non-pertechnetate species, but could also indicate which Hanford tank compositions are most likely to have supported reduction and thus may require special processing. It can be broadly assumed that the two key steps are reduction of the metal center and complexation by an organic ligand; in the absence of supporting ligands, reduction usually generates insoluble TcO_2. With the newly enhanced knowledge of alkaline technetium chemistry, the key question becomes: What reducing agents are present in the waste and what energy source causes their generation?

As in the studies described above, UV-vis spectroscopy is an excellent tool to monitor changes in technetium speciation, provided that the solution does not contain high concentrations of interfering species such as iron and nitrate. Other spectroscopic methods such as EPR and XAS are highly specific to the metal environment, and have been effectively applied in studies of TcO_4^- radiolysis in alkaline solution (7), but each individual experiment can be time-consuming and expensive. Taking advantage of the intrinsic radioactivity of technetium, radioanalytical methods provide a valuable complement to the other techniques by supporting rapid and inexpensive screening of a large parameter space. Experimental conditions of interest can be quickly identified and explored further using conventional spectroscopy (*e.g.*, for structure determination).

Given the abundance of beta and gamma emitting nuclides in the tank wastes, it is plausible that all reactivity is driven by radiolysis. However, it has also been shown that chemical catalysis by fission product metals such as ruthenium, rhodium, and palladium can promote formate decomposition and foaming during vitrification (*11*). Concurrent with our investigations, it was also discovered that palladium-catalyzed reactions were responsible for the unexpectedly rapid generation of benzene during an In-Tank Precipitation demonstration (*12*). These examples demonstrate the danger in ascribing all nuclear waste reactivity to radiolysis.

Studies at LANL focused on the effect of noble metal catalysts on technetium speciation in alkaline solution. Simple simulants containing TcO_4^- in 2.5 mol/L NaOH, along with chloride salts of Ru, Rh, and Pd, were mixed with organic complexants believed to exist in Hanford tanks. Since pertechnetate is quantitatively sorbed onto Reillex-HPQ from moderately strong base (*8*), a simple two-step analysis was developed for screening the catalytic reactivity. First, centrifugation removed any precipitates, typically TcO_2, and then the supernate was contacted with Reillex-HPQ resin. The proportion of ^{99}Tc activity contained in the precipitate and resin phases was calculated based on the attrition of solution beta activity after each step. Control experiments showed that any ^{99}Tc activity remaining in solution could be attributed to a reduced, soluble complex. However, technetium sorbed onto the resin could either be TcO_4^- or a soluble complex with an appreciable K_d, so the experiment actually determines the *minimum* extent of reduction in the sample.

Using this screening approach, it rapidly became clear that simply heating alkaline pertechnetate solutions to 65 °C, in the presence of noble metals and certain complexants, leads to quantitative reduction in a matter of days (*6*). Figure 5 shows that glyoxylate, glycolate and formate are excellent reducing agents, converting 1 mmol/L TcO_4^- solutions entirely to TcO_2 within 1 day. Aminocarboxylates showed varied behavior: hydroxyethyl(iminodiactetic acid) (HEIDA), hydroxyethyl(ethylenediaminetriacetic acid) (HEDTA), and EDTA supported slow but eventually quantitative reduction, while NTA had little effect on pertechnetate. Control experiments confirmed that neither organic complexants (50 mmol/L) nor noble metal salts (1 mmol/L) support reduction if the other component is not present. However, catalytic reduction using H_2 gas also converts TcO_4^- to TcO_2, even when no organics are present.

Here again, the behavior of Tc-gluconate mixtures was exceptional, forming a soluble reduced product instead of TcO_2 in the catalytic reaction. For example, in an alkaline solution of TcO_4^- (0.1 mmol/L) and gluconate (50 mmol/L) heated with the three catalyst metals, only 26% of the ^{99}Tc-containing product could be sorbed onto Reillex-HPQ resin, while the balance of the activity remained in solution, imparting a pink color. This result indicates a minimum of 74% yield of a reduced Tc-gluconate complex. Besides mirroring the results obtained for Sn(II) and sulfite reduction in strong base, the gluconate complex may be

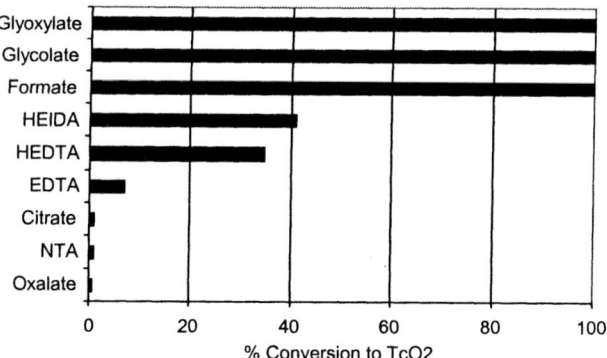

Figure 5. Catalytic reduction of TcO_4^- (1 mmol/L) to TcO_2 in the presence of organic complexants (50 mmol/L) after 1 day at 65 °C.

analogous to the Tc(IV)-diolate complexes reported as products of pertechnetate radiolysis in the presence of ethylene glycol, formaldehyde and glyoxylate (7).

To accurately determine the gluconate product yield and extend the investigation to related complexes, a method was needed to isolate TcO_4^- from other soluble species before counting. Paper chromatography proved to be a simple, general approach to quantitatively separate the TcO_4^- and soluble non-pertechnetate fractions. Using acetone as eluant, TcO_4^- ion migrates with the solvent front while the non-pertechnetate complex is stationary. TcO_2 can be separated and quantified by centrifugation, or by processing a second chromatogram with water to co-elute the two soluble compounds. Once separated, a Bioscan AR-2000 Imaging Scanner was used to locate and quantify the beta activity in each product fraction. The key limitation of this method is that multiple reduced complexes cannot be separated from one another.

Using this approach, the values in Table III were obtained for gluconate, sucrose, and tartrate, all of which were used in Hanford processing (1). All three carbohydrates are effective TcO_4^- reducing agents in the presence of Ru, Rh, and Pd catalysts, using the conditions reported above; no reduction occurred when the catalysts were absent. Of the three, gluconate supports the highest total reduction yield as well as the highest proportion of soluble, non-TcO_4^- complex.

By comparing the yield of reduced complex, determined using paper chromatography, with the total amount of ^{99}Tc sorbed onto Reillex-HPQ in the anion exchange experiments, it is also possible to calculate an approximate distribution coefficient for each of the reduced complexes. If more than one soluble complex is present, the apparent K_d represents a weighted average. Thus, Table 3 indicates that the gluconate complex is sorbed poorly (K_d < 1 mL/g), while the sucrose and tartrate complexes have significant affinity for Reillex-HPQ. Structurally, the poor uptake of the gluconate product from strong base implies that the complex contains no acidic protons, and by extension that all carboxylate groups are directly bonded to the Tc center. Recalling that the distribution coefficient of the non-pertechnetate species in the wastes must be no greater than the apparent K_d for the waste sample as a whole (*i.e.*, $\lesssim$ mL/g), and probably less than 1 mL/g (4), gluconate again appears to most closely mimic the actual speciation in the wastes. However, a definitive assignment clearly cannot be made until a more comprehensive survey of organics is completed and the product K_d values are measured in accurate waste simulants (13).

Equipped with the chromatography technique, it is convenient to begin scoping the full range of organic compounds and reaction conditions that support TcO_4^- reduction. When a series of concentrated solutions containing 10 mmol/L $^{99}TcO_4^-$ and 1 mol/L (or saturated below 1 mol/L) organic compounds in 2 mol/L NaOH were allowed to stand for 3 days at ambient temperature with no external energy source, reduction and complexation of Tc was revealed to be facile (Fig. 6). Even the relatively benign ligand EDTA gave evidence of reduction, although the product decomposed when exposed to air. Several 5- and

Table III. Anion Exchange and Chromatography Data for Soluble Complexes

Complexant	% TcO_2	% TcO_4^-	% complex	% sorbed	K_d (est)
Gluconate	3.3	11.8	84.9	5.4	0.6
Sucrose	35.5	29.4	35.2	51.2	11
Tartrate	18.4	43.2	38.5	92.9	135

NOTE: $[TcO_4^-] = 10^{-4}$ mol/L; [organic] = 0.05 mol/L. Mass/volume ratio = 0.1 g/mL.

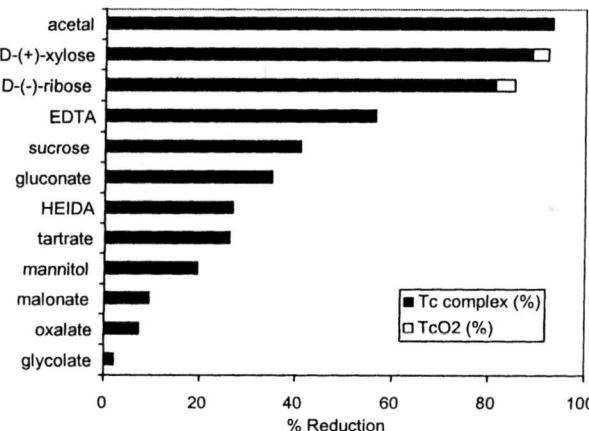

Figure 6. Autoreduction of TcO_4^- (0.01 mol/L) in concentrated ligand solutions.

6-carbon sugars and sugar derivatives supported extensive reduction, often producing intensely red or purple colored solutions. It is not yet clear whether this reduction is caused by radiolysis, or by a direct redox reaction between metal and ligand. In any event, the propensity toward autoreduction in these concentrated solutions is remarkable.

Conclusions

The routine use of radiochemical methods to evaluate technetium partitioning from simulated and actual Hanford tank wastes led directly to the discovery of non-pertechnetate species. Although the reduced complexes were further confirmed using spectroscopic methods, it is not clear that the problem would have been discovered in a timely fashion using those methods alone. Competition from other beta and gamma emitting nuclides remains the greatest challenge in analyzing ^{99}Tc in the tank waste matrix, but careful radiochemical methodology now supports routine analyses that are both accurate and reproducible.

Despite more than five years of study, a definitive assignment of the reduced species in the waste remains elusive, due to the continuing shortage of data on technetium reactivity in alkaline solution, as well as the complex matrix of the tank waste. However, the rapid screening capability provided by radioanalytical methods, occasionally augmented by spectroscopy, has indicated that aminocarboxylate ligands do not stabilize reduced technetium in strong base. Instead, the accumulated evidence from the synthesis, reactivity, and ion exchange studies all points toward gluconate, or a closely related carbohydrate, as the source of the very stable non-pertechnetate complexes found in Hanford tank wastes.

Experimental

Details of experimental methods are provided in references *2, 3, 4, 6* and *8*. Most experiments involved the use of radioactive materials and strong caustic solutions; appropriate safety precautions should be taken.

107-AN simulant composition (concentrations in mol/L): $Al(NO_3)_3$, 1.43×10^{-2}; H_3BO_3, 3.24×10^{-3}; $Ca(NO_3)_2$, 1.47×10^{-2}; $Ce(NO_3)_3$, 3.77×10^{-4}; $CsNO_3$, 1.40×10^{-4}; $Cu(NO_3)_2$, 4.74×10^{-4}; $Fe(NO_3)_3$, 3.03×10^{-2}; $La(NO_3)_3$, 3.28×10^{-4}; $Pb(NO_3)_2$, 1.87×10^{-3}; $Mg(NO_3)_2$, 1.03×10^{-3}; $MnCl_2$, 1.02×10^{-2}; $Nd(NO_3)_3$, 6.65×10^{-4}; $Ni(NO_3)_2$, 9.03×10^{-3}; KNO_3, 4.55×10^{-2}; $Sr(NO_3)_2$, 7.54×10^{-5}; $Zn(NO_3)_2$, 6.93×10^{-4}; $ZrO(NO_3)_2$, 7.67×10^{-4}; K_2MoO_4, 3.73×10^{-4}; $NaCl$, 3.11×10^{-2}; NaF, 7.00×10^{-3}; Na_2CrO_4, 3.38×10^{-3}; $NaCO_3$, 1.4; $NaOH$, 8.32×10^{-1};

NaNO$_2$, 1.33; Na$_2$HPO$_4$, 1.17 x 10^{-2}; Na$_2$SO$_4$, 8.59 x 10^{-2}; NaNO$_3$, 3.5; glycolic acid, 2.48 x 10^{-1}; sodium gluconate, 6.00 x 10^{-2}; citric acid, 4.49 x 10^{-2}; nitrilotriacetic acid, 2.98 x 10^{-3}; iminodiacetic acid, 4.54 x 10^{-2}; disodium EDTA, 1.95 x 10^{-2}; HEDTA, 7.78 x 10^{-3}; sodium formate, 2.31 x 10^{-1}; sodium acetate, 1.74 x 10^{-2}; sodium oxalate, 9.38 x 10^{-3}. Simulant formulation was supplied by Mike Johnson, BNFL, Inc.

101-SY simulant composition (concentrations in mol/L): Al(NO$_3$)$_3$, 4.15 x 10^{-1}; Ca(NO$_3$)$_2$, 4.20 x 10^{-3}; CsNO$_3$, 4.19 x 10^{-5}; Fe(NO$_3$)$_3$, 1.96 x 10^{-4}; Ni(NO$_3$)$_2$, 2.49 x 10^{-4}; RbNO$_3$, 4.20 x 10^{-6}; Zn(NO$_3$)$_2$, 5.00 x 10^{-4}; MoO$_3$, 4.20 x 10^{-4}; Na$_2$SO$_4$, 4.75 x 10^{-3}; Na$_2$HPO$_4$, 2.04 x 10^{-2}; Na$_2$SO$_4$, 4.75 x 10^{-3}; Na$_2$CO$_3$, 3.75 x 10^{-2}, NaOH, 3.78; KF, 3.38 x 10^{-2}; NaNO$_2$, 1.09; iminodiacetic acid, 3.05 x 10^{-2}; sodium gluconate, 1.25 x 10^{-2}; Na$_3$NTA, 2.50 x 10^{-4}; Na$_4$EDTA, 5.00 x 10^{-3}, Na$_3$HEDTA, 3.75 x 10^{-3}, Citric Acid, 5.00 x 10^{-3}. Simulant formulation was supplied by PNNL.

Acknowledgements

This work was supported by the U.S. Department of Energy under the SBIR (#DF-FG03-97ER82421) and EMSP (TTP #AL17SP23 and #AL17SP23) programs. Additional support was provided by the Los Alamos National Laboratory Director's Postdoctoral Fellowship program.

References

1. Kupfer, M. J.; Boldt, A. L.; Higley, B. A.; Hodgson, K. M.; Shelton. L. W.; Simpson, B. C.; Watrous, R. A.; LeClair, M. D.; Borsheim, G. L.; Winward, R. T.; Orme, R. M.; Colton, N. G.; Lambert, S. L.; Place, D. E.; Schulz, W. W. "Standard Inventories of Chemicals and Radionuclides in Hanford Site Tank Wastes," Report HNF-SD-WM-TI-740, U.S. Deparment of Energy, August 1997.
2. Schroeder, N. C.; Radzinski, S.; Ball, J. R.; Ashley, K. R.; Cobb, S. L.; Cutrell, B.; Whitener, G. "Technetium Partitioning for the Hanford Tank Waste Remediation System: Anion Exchange Studies for Partitioning Technetium from Synthetic DSSF and DSS Simulants and Actual Hanford Waste (101-SY and 103-SY) Using Reillex™-HPQ Resin," Report LA-UR-95-4440, Los Alamos National Laboratory, 1995.
3. Schroeder, N. C.; Radzinski, S. D.; Ashley, K. R.; Truong, A. P.; Szczepaniak, P. A. "Technetium Oxidation State Adjustment for Hanford Waste Processing," in *Science and Technology for Disposal of Radioactive*

Tank Waste, W. W. Schulz and N. J. Lombardo, eds., Plenum, New York, 1998.
4. Schroeder, N. C.; Radzinski, S. D.; Ashley, K. R.; Truong, A. P.; Whitener, G. D. *J. Radioanal. Nucl. Chem.*, **2001**, *250*, 271-284.
5. Blanchard, D. L, Jr.; Brown, G. N.; Conradson, S. D.; Fadeff, S. K.; Golcar, G. R.; Hess, N. J.; Klinger, G. S.; Kurath, D. E. "Technetium in Alkaline, High-Salt, Radioactive Tank Waste Supernate: Preliminary Characterization and Removal." Report PNNL-11386, Pacific Northwest National Laboratory, January 1997.
6. Bernard, J. G.; Bauer, E.; Richards, M. P.; Arterburn, J. B.; Chamberlin, R. M. *Radiochim. Acta*, **2001**, *89*, 59-61.
7. Lukens, W. W., Jr.; Bucher, J. J.; Edelstein, N. M.; Shuh, D. K. *Environ. Sci. Technol.*, **2002**, *36*, 1124-1129. (b) Lukens, W. W.; Jr.; Bucher, J. J.; Edelstein, N. M.; Shuh, D. K. *J. Phys. Chem. A*, **2001**, *105*, 9611-9615.
8. Ashley, K. R.; Whitener, G. D.; Schroeder, N. C.; Ball, J. R.; Radzinski, S. D. *Solvent Extr. Ion Exch.*, **1998**, *16*, 843-859, and references therein.
9. Baldas, J. "The Coordination Chemistry of Technetium," in *Advances in Inorganic Chemistry*, A. G. Sykes, ed., **1994**, *41*, 1-123.
10. (a) Bürgi, H.B.; Anderegg, G.; Bläuenstein, P. *Inorg. Chem.* **1981**, *20*, 3829-3834. (b) Linder, K. "Aminocarboxylate Complexes of Technetium," hesis, Massachusetts Institute of Technology, June 1986.
11. King, R. B.; Bhattacharyya, N. K.; Wiemers, K. D. *Environ. Sci. Technol.* **1996**, *30*, 1292-1299.
12. (a) Oji, L. N.; Barnes, M. J. "Batch Studies of Sodium Tetraphenylborate Decomposition on Reduced Palladium Catalyst." Report WSRC-TR-2000-00459, Savannah River Site, October 2000.
13. For a preliminary report of XANES results comparing a synthesized gluconate sample with actual tank waste, see: Shuh, D. K.; Burns, C. J.; Lukens, W. W. "Research Program to Investigate the Fundamental Chemistry of Technetium: 2001-2002 Progress Report," Environmental Management Science Program, Project ID 73778, July 2002.

Chapter 13

Thorium-229 for Medical Applications

Miting Du, Fred Peretz, Rose A. Boll, and Saed Mirzadeh

Nuclear Science and Technology Division, Oak Ridge National Laboratory, Oak Ridge, TN 37831–6385

Medical researchers are assessing the ability of several alpha-emitting radioisotopes to treat cancer and reduce tumor burden. In particular, ^{213}Bi ($t_{1/2}$ = 45.6 min) attached to tumor specific antibodies has been under clinical investigations for treatment of certain cancers. ^{213}Bi is a decay product of ^{225}Ac ($t_{1/2}$ = 10.0 d), itself an alpha emitter. Currently the domestically-produced supply of ^{225}Ac, however, is limited by the availability of the parent radionuclide, ^{229}Th ($t_{1/2}$ = 7340 y). ^{229}Th, in turn, is a decay product of highly fissile ^{233}U, which is currently stored at ORNL and is now scheduled for long-term storage. In this paper we describe the primary separation of ^{229}Th from ^{233}U stockpile and further purification of this radioisotope from trace levels of uranium and plutonium. The chemical separation process is primarily based on the well known chemical behavior of these actinides on anion-exchange resins in both nitric and hydrochloric acid solutions.

The use of the alpha emitters ^{213}Bi and ^{225}Ac is being explored by the medical research community for the treatment of a variety of cancers. An example is the use of ^{213}Bi for the treatment of acute myeloid leukemia proposed by Memorial Sloan-Kettering Cancer Center (*1*). In this application, the humanized antibody HuM195 is used to target a surface protein on a leukemia white blood cell, with ^{213}Bi linked to the antibody using a chelating agent (modified DTPA). Because of the high linear energy transfer and short range of the alpha particle emitted following the decay of ^{213}Bi, there is a high probability of killing a targeted cancer cell with minimal exposure of other parts of the body to a large dose of radiation. A supply of ^{229}Th is maintained at Oak Ridge National Laboratory (ORNL), and its ^{225}Ac daughter isotope ($t_{1/2}$ = 10.0 d) is extracted on a regular basis and supplied to medical research institutes. Because of its short 45.6 m half-life, ^{213}Bi must be separated from its parent isotope ^{225}Ac at the site of its use. With its 7,340 y half-life, ^{229}Th can serve as a source of ^{225}Ac for a long period of time (Figure 1). An adequate supply of ^{229}Th can support basic research and clinical trials for a number of proposed alpha particle-mediated radioimmunotherapy.

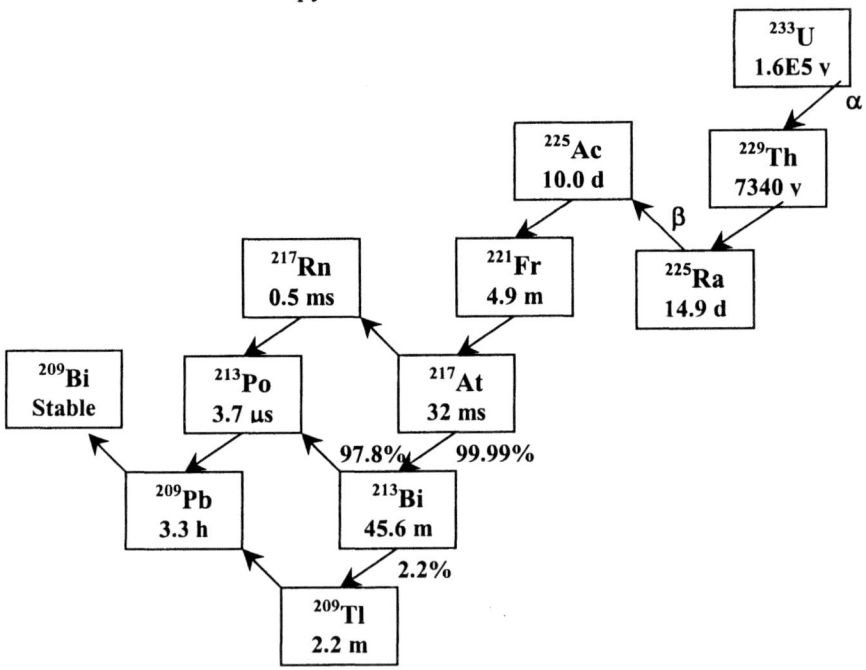

Figure 1: Th-229 Decay Chart (Neptunium Series)

There are two general pathways of obtaining ^{229}Th on a significant scale (millicurie levels). One is by neutron transmutation of ^{226}Ra target (successive neutron captures followed by β decays) (*2*). This approach typically requires a reactor with a high neutron flux and associated facilities for fabrication and post-irradiation processing of highly radioactive targets. The second pathway is to separate ^{229}Th produced by the alpha decay of ^{233}U. Alternatively, ^{225}Ra and ^{225}Ac can be produced directly by gamma-ray, proton and deuteron induced reactions on Ra targets (2).

Separation of ^{229}Th from ^{233}U is less complicated than the production by irradiating ^{226}Ra targets, and less expensive especially if available funding dictates a relatively small program, although, as one may expect, the safeguard of ^{233}U is rather problematic. About 450 kg of ^{233}U is stored at ORNL, of which about 250 kg contains relatively low levels of the contaminant ^{232}U. The ^{232}U contaminant in ^{233}U stock was probably produced by ^{233}U[n,2n] reaction and partly from the decay of ^{236}Pu, which in turn was produced by neutron transmutation of ^{232}Th target. Decay daughters of ^{232}U include ^{208}Tl (Figure 2), which emits a 2.6 MeV gamma-ray with high intensity. Thus, ^{233}U containing significant concentrations of ^{232}U must be handled in a shielded environment (Th separated from ^{233}U also contains the ^{228}Th decay daughter of ^{232}U, thus its

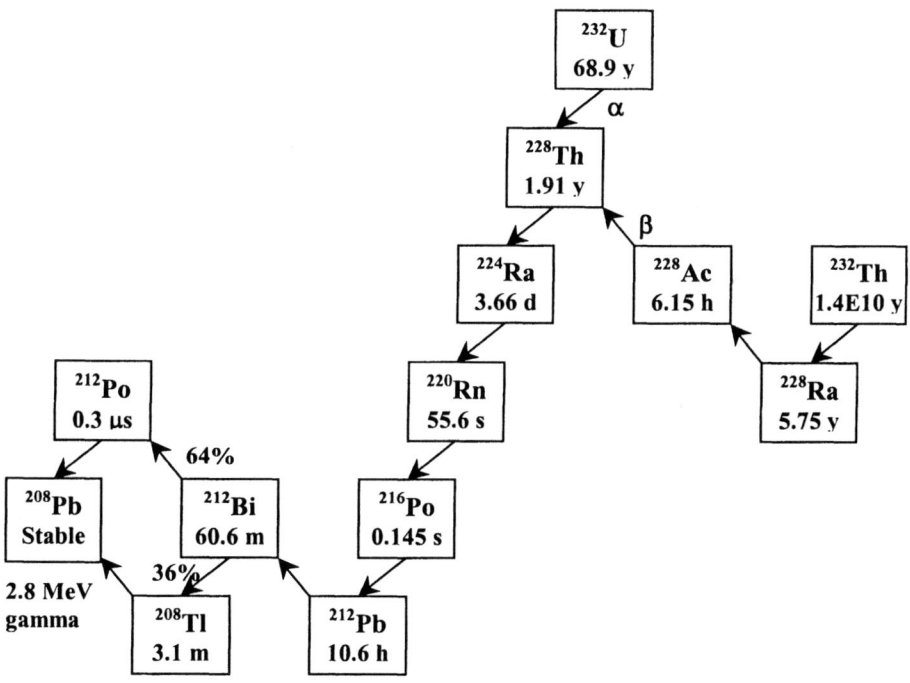

Figure 2: Th-228 Decay Chart (Thorium Series)

radioactivity is also related to the ^{232}U content of the original uranium). The ORNL ^{233}U (stored in ~1,000 separate packages), has ^{232}U concentrations ranging from ~2 to ~200 ppm, and it has been in storage for up to 35 years. During this priod, significant quantities of ^{229}Th has grown in (the ^{229}Th annual in growth rate is ~1 mCi per Kg of ^{233}U).

In April of 2000 the U.S. Department of Energy (DOE) made $1 million available over three years for the separation of Th from ^{233}U, to serve as feedstock for the production of additional ^{225}Ac. A set of 19 packages that were recently sent to ORNL from Mound Laboratory, containing 3.3 kg of ^{233}U with an estimated 72 mCi of ^{229}Th, were originally identified for separation. This material contains between 2 and 16 ppm of ^{232}U. As the project progressed, other packages of ^{233}U containing about 6 ppm ^{232}U were identified, and processing of the 16 ppm Mound material was deferred. The amount of ^{229}Th expected to be present in the revised package sequence remained essentially unchanged, but a variable amount of plutonium, mainly ^{239}Pu, was unexpectly identified, which was not shown on file for the packages (the ^{239}Pu may be a by-product in original ^{233}U production due to the impurity ^{238}U inside thorium). A typical process batch contains hundreds of grams of uranium, tens of milligrams of thorium and variable amount of plutonium ranging from milligrams to hundreds of milligrams. Thus, separation and purification of ^{229}Th from this material became a process of separating a few mgs of thorium and plutonium from a few hundred grams of uranium.

Process for ^{229}Th Separation and Purification

The flow sheet for separating ^{229}Th from ^{233}U consists of two major processes, Thorium Separation Process and Thorium Purification Process. The Thorium Separation Process identifies the steps for the separation of the thorium and plutonium from bulk of uranium, and for the conversion of uranium back into a stable oxide form for continued storage (Figure 3). The Thorium Purification Process of the flowsheet, lists the steps to further purify the ^{229}Th product from plutonium and residual uranium contaminants (Figure 4).

The Thorium Seperation Process begins by opening packages of ^{233}U oxide and confirming the uranium content by titrametric method using FeSO$_4$ as reductant (3), and mass analysis (ICP-MS). The oxide (~250 g) is then dissolved in 11 mol/L HNO$_3$ (2.8 mL per gram of U$_3$O$_8$), and filtered to remove undissolved impurities. The acid concentration of this solution is then adjusted to ~8 mol/L HNO$_3$ and is loaded onto a 200 mL column containing anion-exchange resin (tertiary amine-based Bio-Rad AG1-X8, which was later replaced with MP-1) (100-200 mesh, ~200 mL bed volume) (4). Nitric acid (7.5 mol/L) is used to elute the bulk of uranium (k_d ~ 14), while Th^{+4} (k_d ~ 300) and Pu^{+4} (k_d > 1000)

(5) are retained on the column. If significant quantities of Pu are present, a dark green band can be seen near the top of the resin bed. Both thorium ($k_d \sim 1$) and plutonium ($k_d \ll 100$) (5) are stripped off the column with 250 mL of 0.1 mol/L HNO_3 and then sent for process of Thorium Purification. The bulk of uranium eluted from the anion-exchange column was precipitated as hydroxide by the addition of conc. NH_4OH, filtered, washed with DI water, and then converted to U_3O_8 by heating in air to 800 °C. Initial batches, with less than 2 ppm ^{232}U, were processed in a set of glove boxes. As material containing 6 ppm ^{232}U was handled, most of the process was moved into a shielded manipulator hot cell in the same building.

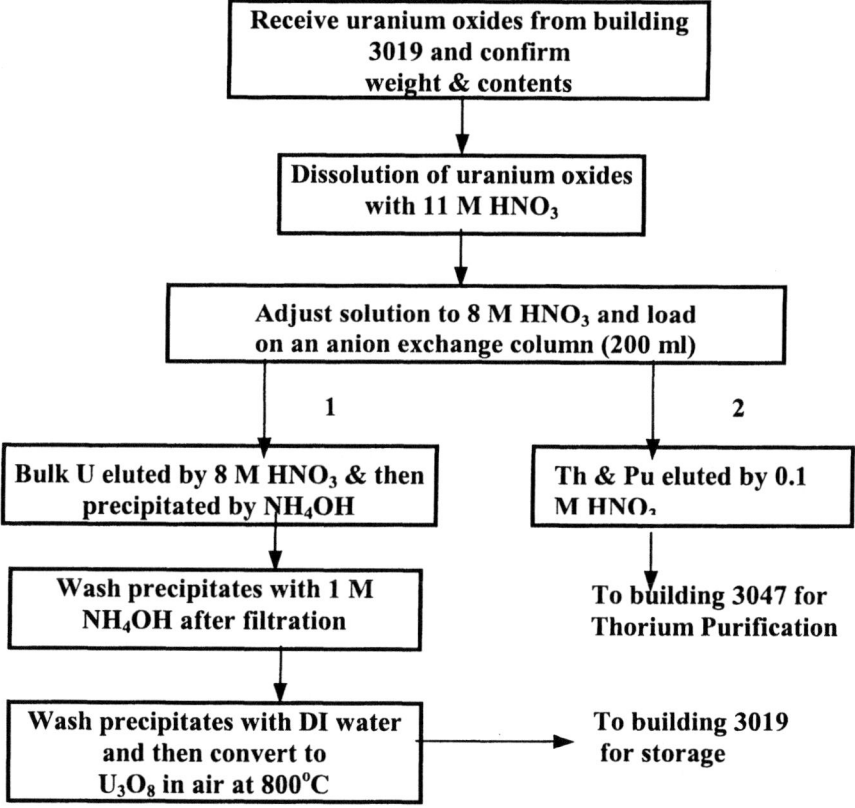

Figure 3: Thorium Separation Process Flow Chart

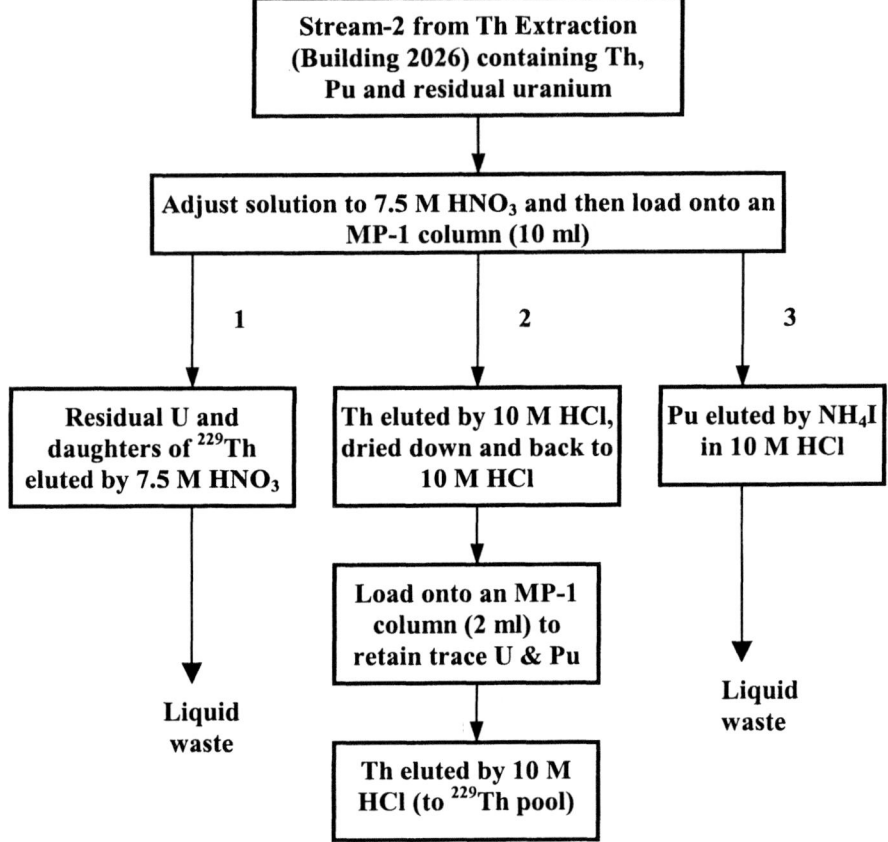

Figure 4: Thorium Purification Process Flow Chart

In the Thorium Purification Process (Figure 4), ^{229}Th is separated from plutonium and residual uranium impurities in two consecutive MP-1 resin columns. These columns are 10 and 2 mL bed volumes, and are much smaller than the 200 mL column used in the Thorium Separation Process. The 0.1 mol/L HNO$_3$ solution containing Th, U and Pu is evaporated to near dryness in a semi-closed distillation apparatus and is then redissolved in a minimum volume (~5 mL) of 7.5 mol/L HNO$_3$. This solution is loaded on a MP-1 anion exchange column (10 mL bed volume, 100-200 mesh, pre-conditioned with 7.5 mol/L HNO$_3$) and column is washed with an additional 30 mL of 7.5 mol/L HNO$_3$. Similar to the first process, the residual U is eluted with the 7.5 mol/L HNO$_3$ while Th and Pu(IV) are retained on the resin. Thorium is then selectively eluted with ~60 mL of 10 mol/L HCl, while Pu is retained on the resin (k_d ~1 and >

1000, respectively, (5, 6). Again, a visible Pu dark green band can be seen on the top of column; the Pu band turns white as HCl is added to the column. The plutonium retained on the column can be discarded as a solid waste, or it can be stripped off the column with 40 mL of 10 mol/L HCl + 0.05 mol/L NH$_4$I (I$^-$ will reduce Pu^{+4} to Pu^{+3}) (7). The thorium eluted in this step still contains trace quantities of U, which is removed in the second MP1 column. The Th fraction in 10 mol/L HCl is again evaporated to near dryness in a semi-closed system, 2-3 mL of 10 mol/L HCl is added and solution is evaporated again. Th, containing trace levels of U, is then dissolved in 3x1 mL of 10 mol/L HCl, and loaded on a very small MP-1 anion exchange column (~2 mL bed volume, 200-400 mesh, pre-equilibrated with 10 mol/L HCl). The column is washed with an additional 12 mL of 10 mol/L HCl (trace U is retains on the column). The column load and wash fractions containing the purified Th is combined and aliquots are taken for γ-ray and mass analysis.

Thorium Purification Process Alternative.

In cases where the amount of Th in the feed solution was greater than the amounts of U and Pu, the 10 mL MP-1 column can be used to retain U and Pu, instead of retaining Th. In this case, the evaporated thorium feed (from the Thorium Seaparation Process) is brought up in a minimum volume of 10 M HCl (rather than 7.5 M HNO$_3$) and is run through the 10 mL MP-1 column (pre-equilibrated with 10 M HCl) to retain U and Pu (VI) on the column, while Th is not retained and is eluted in the load and the wash solutions. The Th fraction then can be further purified by eluting through another small MP1/HCl column.

Results and Discussion

Table 1 shows the composition of the ten batches of ^{233}U processed during 2000-2002. The ^{233}U content of the load solution ranged from 150 to 340 g, and mass analysis indicated that the levels of ^{234}U, ^{235}U, ^{236}U and ^{238}U where <0.1% of the total U mass. The Th mass distribution primarily included ^{229}Th and ^{232}Th of varying fractions (columns 3 and 4 in Table 1) with a small contribution of ~0.5% from ^{230}Th and ppm levels of ^{228}Th (data not shown). The age of the batch and the level of ^{232}Th, left from the original ^{233}U production, primarily dictated the variations in the levels of ^{229}Th and total Th in the U batches. Note that ^{232}Th is the target for producing ^{233}U. Batch 02-05 contained almost 2 gram of ^{232}Th, and after purification it was stored separately. As pointed out earlier, although the ^{228}Th mass is almost negligible, its radiation contributes significantly to the total radiation field due to rather short half-life of ^{228}Th (2.8

y) and emission of a high energy and high intensity γ-rays. The level of total Pu in the U feed solution also varied significantly, ranging from 0.15 mg to 220 mg, with ^{239}Pu being the major component and with a ~50 fold less contribution from ^{240}Pu and ^{242}Pu.

Table 1. Composition of the ^{233}U feed stock[a]

Batch No.	^{233}U (g)	Th(total) (mg)	^{229}Th (mg)	Pu(total) (mg)
00-01	163.3	15.6	12.5	18.7
00-02	149.2	38.9	11.2	92.5
01-01	182.6	17.1	14.7	47.1
01-02	287.4	165	27.8	220
02-01	309.4	62.2	29.6	1.77
02-02	142.1	25.1	15.7	0.10
02-03	272.4	58.9	30.4	0.69
02-04	295.2	62.3	26.3	0.74
02-05	340.4	1719	27.0	1.45
02-06	284.4	56.1	35.3	1.93

[a] All masses are determined by mass spectrometry.

The results of thorium separation process, post AG1 or MP1 columns, are summarized in Table 2. Both total Th and the mass of ^{229}Th compares fairly with those in the feed solution. The ^{229}Th radioactivity after the Thorium-Separation process (measured by γ-ray spectroscopy) is given in column 3, and the ratio (R) of ^{229}Th activity from column 3 to the theoretical ^{229}Th activity based on the mass from column 2 is given in column 4. The correlation of data from mass and γ-ray spectroscopy ranged from 0.58 to 1.1. The levels of U and Pu isotopes in the Th fraction after the Thorium Separation process are given in columns 6 and 7 of the Table 2. As indicated, significantly higher decontaminations (~100 fold) from U and Pu were achieved by replacing AG1 resin with MP1 resin. Batch 02-05, which has gone through the Thorium Seperation process (employing a 200 mL bed volume of MP-1 resin), contained 31.2 mg of ^{229}Th, 1.7 g of ^{232}Th, 2.7 mg of residual U, and 1.72 mg of Pu. This mixture was fed into a 10 mL MP-1 column in 10 mol/L HCl, and mass analysis of the Th fraction (after the second 10 mL column) indicated that Pu was effectively removed, and only ~0.02 mg of U remained in the Th product. This trace quantity of U was then effectively removed in a 2 mL MP-1/ 10 M HCl column.

Table 2. Composition of purified ^{229}Th post Thorium Extraction process

Batch No.	^{229}Th (mg)[a]	^{229}Th (mCi)[b]	R [c]	Th (total) (mg)[a]	U (total) (mg)[a]	Pu (total) (mg)[a]
AG1X8 - 7.5 M HNO$_3$, 200 mL						
00-01	12.5	2.9	1.1	12.8	64.3	22.9
00-02	11.2	2.1	0.88	20.0	38.9	92.5
01-01	14.7	3.0	0.96	17.1	419	106
01-02	32.8	6.8	0.97	165	306	219.6
MP1 - 7.5 M HNO$_3$, 200 mL						
02-01	55.5	6.4	0.54	62.2	1.87	1.77
02-02	21.5	4.6	1.0	25.1	1.4	0.1
02-03	35.3	6.1	0.81	58.9	5.01	0.69
02-04	31.2	4.7	0.71	62.3	2.68	0.74
02-05	27.4	3.4	0.58	1,722	2.70	1.72
02-06	35.2	4.7	0.63	55.9	2.06	2.28

[a] All masses are determined by mass spectrometry.
[b] Thorium-229 activities are determined by γ-ray spectroscopy.
[c] R is the ratio of ^{229}Th activity from column 3 to the theoretical ^{229}Th activity based on the mass given in column 2 (assumed a value of 0.213 mCi/mg for the theoretical specific activity of ^{229}Th).

Plutonium Valence Adjustment

Adjusting the plutonium valence is an important step for an effective separation of plutonium from other actinides, since Pu(IV) is retained with Th(IV) on the anion-exchange resin in strong nitric acid elutions while Pu(III) and Pu(VI) are not retained under the above conditions. However, if Pu recovery is not required then no Pu valence adjustment is required because the Pu(III) and Pu(VI) ions that may be present in the original feed material will elute with the bulk uranium in the Thorium Seperation Process (see above). As discussed earlier, the Pu(IV), which is carried over with the Th in the Thorium Seperation section, subsequently was separated from Th in the Thorium Purification Process described above.

AG-1 vs. MP-1 Resin.

Under our experimental conditions wherein the 7.5 mol/L HNO_3 load solution was nearly saturated with U and contained only small amount of Th, the MP-1 resin provided much better separation than the AG-1 resin. Uranium tailing was significantly less in MP-1 resins, resulting a cleaner separation between U and Th. The major difference in performance of these reins are due to the pore size of these resins. MP-1 resins are macroporous (macroreticular) with irregular structures while AG-1 resins are microporous with rather smoth spherical surfaces (the Bio-Rad AG-1 and MP-1 resins have similar chemical structures -- backbones consisting of cross-linked polystyrene divinylbenzene copolymer and identical quaternary ammonium functional groups). As we demonstrated, the use of MP-1 resins not only resulted in a significant reduction in U contamination of the Th product, but it also resulted in some timesaving because the separation was accomplished with a smaller volume of eluent, which needs to be evaporated in the subsequent step.

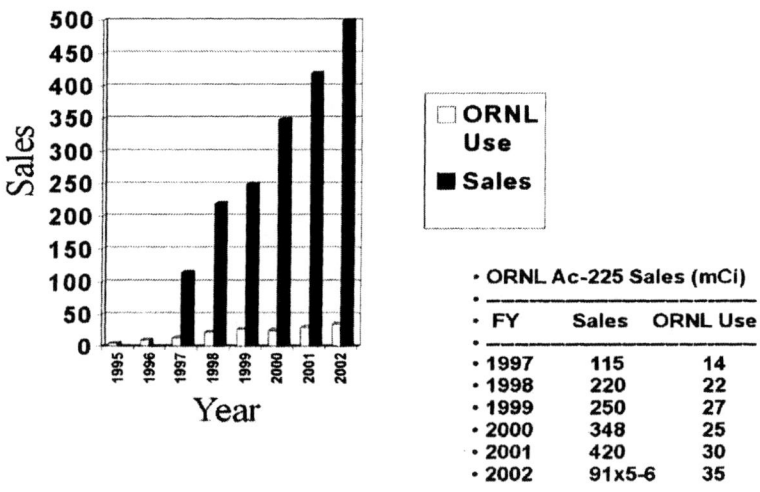

Figure 5: ORNL Actinium-225 Sales

The separation of ^{229}Th was initiated at ORNL prior to 1995 (8). A total of ~90 mCi of ^{229}Th was recovered from the legacy waste materials left over from the processing of hundreds of kilograms of ^{233}U during the 1970s. To date, the direct separation of ^{229}Th from ^{233}U stockpile described here added another ~60 mCi of ^{229}Th to the inventory. With our current stock of ^{229}Th, we have been able to supply ~500 mCi of ^{225}Ac in 2002 (Figure 5). It appears that the current demand for ^{225}Ac is about twice the amount currently supplied. This demand could increase sharply if clinical trials in the U.S., Europe, or Australia proceed into Phase III. Efforts are underway to identify funding sources to continue the separation of ^{229}Th from the ORNL inventory of ^{233}U.

Acknowledgements

The authors would like to recognize the efforts of the chemical and health physics technicians and the supporting technical staff at ORNL Buildings 2026, 3047, and 3019 for their efforts in support of this project. In particular, John Keller and Scott White should be recognized for their development and implementation of the thorium separation process in Building 2026.

References

1. Jurcic J. G., Larson S. M., Sgouros G., McDevitt M. R., Finn R. D., Divgi C. R., Ballangrud A. M., Hamacher K. A., Ma D., Humm J. L., Brechbiel M. W., Molinet R., and Scheinberg D. A. *Blood*, **2000**, *100/4*, accepted
2. Mirzadeh, S. *Applied Radiation and Isotopes*, **1998**, *49*, 345-349.
3. Davies, W., Gray, W. *Talanta*, **1964**, *11*, 1203-11.
4. Boll R. A., Mirzadeh S., Kennel S. J., DePaoli D. W., Webb O. F. *Journal of Labelled Compounds*, **1997**, *XL*, 341.
5. Korkisch, J. *Handbook of Ion Echnage Resins: Their Applications to Inorganic Analytical Chemistry*: CRC Press: Ann Arbor, MI, 1989, Vol. II
6. Kluge, E., Lieser, K.H. *Radiochimica Acta*, **1980**, *27*, 161.
7. Maiti, T. C., Kaye, J. H., Kozelisky, A. E. *J. Radioanal. Nucl. Chem.*, **1991**, *161*, 533.
8. Boll R. A., Mirzadeh S. and Kennel S. J. *Radiochimica Acta*, **1997**, *79*, 145.

Interdisciplinary Applications

Chapter 14

Radioanalytical Chemistry in the Courtroom

J. David Robertson

Department of Chemistry and Research Reactor, University of Missouri, Columbia, MO 65211

Two examples of the use of particle-induced X-ray emission (PIXE) analysis to forensic science are presented. In the first, PIXE was used as a non-destructive method to search for signatures of gunshot residue on clothing. In the second, PIXE was used to investigate the authenticity of gold coins from the New World.

Introduction

The first application of radiochemical methods to forensic analysis is, most probably, found in the popular anecdote told of Georg Hevesey who was awarded the Nobel Prize in Chemistry in 1943 "for his work on the use of isotopes as tracers in the study of chemical processes." The story is told that Hevesey used a dose of ^{212}Pb for a bit of detective work at his austere boarding house where his frugal landlady continued to serve the same old pudding, disguised by the addition of fruits and nuts, for more than one week. The first application of forensic activation analysis, published in 1963, was the determination of arsenic in human hair in relation to possible case of poisoning (*1*) and the first U.S. court case involving activation analysis occurred in March of 1964 (*2*). From this point forward, there was much optimism of the tremendous sensitivities of high-flux neutron activation analysis for the detection for large numbers of elements in very small samples for solving problems in forensic science. From the literature and other sources of information, it is evident that the use of neutron activation, and other radioanalytical methods, in forensic analysis has gradually diminished over the last two decades. For example, in the early 1970s, one forensic activation analysis lab alone was processing more than 1,000 gunshot residue hand swabs per month but a 1996 survey of 80 forensic science labs found that only 2% of those labs surveyed still used activation analysis for gunshot residue anlaysis (3). This decline is due, in part, to the emergence of powerful, non-nuclear techniques for trace element analysis over the last three decades. Not only do these competing techniques perform well in respect to analytical characteristics, but they also have the advantages that they may often be carried out in house, enable a substantial throughput with short turn-around times, and do not require access to radioactive materials, a nuclear reactor, or particle accelerator. There are, however, still instances in which the unique characteristics of radiochemical analytical methods lead to their application in forensic analysis. Two such cases are described in this work.

Particle-Induced X-ray Emission (PIXE)

The technique of using an accelerated particle beam for X-ray emission analysis was first introduced at the Lund Institute of Technology in 1970 (4). Like other X-ray spectroscopic techniques used for elemental analysis, PIXE utilizes the X-rays that are emitted from the atoms in a sample when that sample is exposed to an excitation source. The energies of the X-rays are characteristic of the elements from which they are emitted and the number of X-rays of a given energy is proportional to the mass of that corresponding element in the sample. The use of a high-energy proton beam as an excitation source offers several advantages over other X-ray techniques including a higher rate of data accumulation across the entire periodic table and better overall sensitivities, especially for the lower atomic number elements. In the case of electron excitation, the better sensitivity is due to a lower bremsstrahlung background and, in the case of X-ray fluorescence analysis, the better sensitivity is due to the lack of a background continuum across the entire spectrum. Of course, the chief disadvantage of PIXE is the reason that it is considered a radioanalytical technique; it requires the use of a particle accelerator. For the interested reader, detailed descriptions of the theory and application of PIXE can be found in references (5) and (6). In this work, I describe the application of PIXE to gunshot residue and coin analysis. Other forensic applications of PIXE include the analysis of bone (7), ink (8), paint chips (9), and hashish (9).

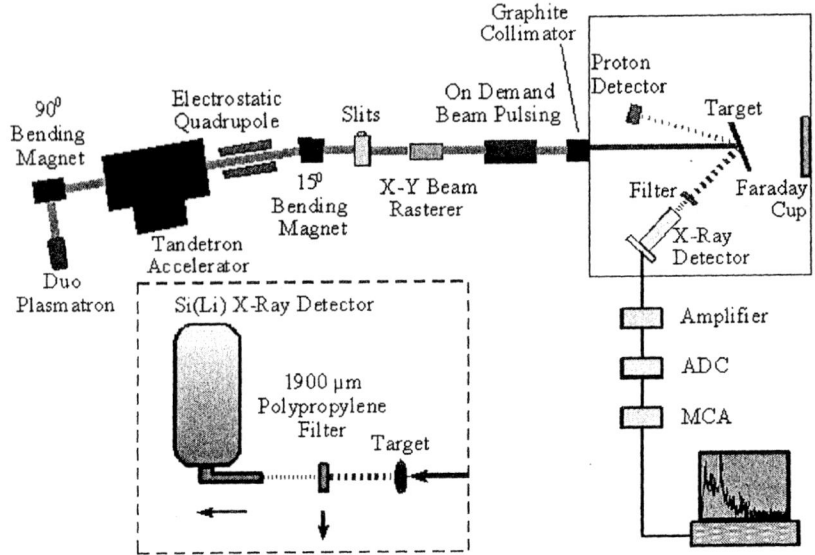

Figure 1. Schematic diagram of PIXE analysis system

A schematic diagram of the PIXE analysis system used in this work is shown in Figure 1. Protons enter the target chamber by passing through a 7-μm thick Kapton window and X-rays exit through a 2.5-μm thick Mylar window at 45° relative to the beam. A 30 mm^2 Si(Li) detector with a full-width-at-half-maximum resolution of 150 eV at Mn K$_\alpha$ records the X-rays. The proton beam, which is at an angle of 23° relative to the sample surface, is continuously swept over the target. The area of irradiation is selected by a series of graphite collimators and can range from 0.3 to 16 mm in diameter. The sample chamber is flushed with helium at atmospheric pressure to reduce sample heating and charging and the sample is typically irradiated for 15 minutes. For 10 minutes of the 15-minute irradiation, the detector is placed 3.5 cm from the target, a 1900 μm polypropylene filter is placed between the sample and the detector, and the sample is irradiated with a 2.1 MeV proton beam. The filter reduces the count rate by absorbing both bremsstrahlung and lower energy X-rays. For the remaining 5 minutes of analysis, the polypropylene filter is removed, the detector is moved further away from the target (7.5 cm), and the sample is irradiated with a 1.6 MeV proton beam. At each bombarding energy, a relative measure of the number of protons striking the target is made by measuring the number of protons that backscatter from the Kapton window into the surface barrier detector. The two X-ray spectra are then combined and the thick-target PIXE analysis is performed with a modified version of the GUPIX (*10,11*) PC-based software package.

A key advantage of PIXE is that the matrix effects associated with the analysis are well understood. As a result, PIXE can provide accurate analytical results in widely varying matrices without the use of "well-matched" standards. For example, the PIXE results presented in Tables I and II for two very different matrices, a pressed pellet of NIST-SRM1577 Bovine Liver and a disk of NIST-SRM1107 Naval Brass, were acquired with the same calibration curve. The calibration curve for these thick-target analyses was obtained by the analysis of thin-film gravimetric standards. It should be stressed, however, that accurate PIXE analysis of complex, inorganic matrices requires knowledge or measurement of both the low- and high-Z elements. Using NIST-SRM1413 High Alumina Sand as an example, the PIXE results for Fe (0.17 wt. %) change by nearly ten percent if Ca (0.53 wt. %) is not included in the matrix in the PIXE analysis. Fortunately, this analytical bias can be readily overcome with a complete PIXE analysis in which the same spot is analyzed multiple times with different energies and filters or in which two or more X-ray detectors are used to record the full PIXE spectrum simultaneously. To achieve more uniform detection limits across the entire periodic table, we elect to use the dual-energy approach even though it requires more analysis time than a single irradiation with two detectors. Example dual-irradiation PIXE spectra from activated carbon and a rare-earth catalyst are shown in Figure 2.

Gunshot Residue (GSR) Analysis of Jeans

My first experience with forensic analysis began with a request from the Pinellas County Florida Sheriff's Department to look for gunshot residue on a pair of jeans. Their two key criteria in selecting PIXE were the excellent sensitivity of the technique for a surface analysis of high-Z materials (barium, antimony and lead) on an "organic" matrix and the fact that the measurement could be made without dissolving or destroying the jeans. One of the largest applications of PIXE has been, and continues to be, the multi-element analysis of air particulate collected on the surface of filters (*6*). Previous applications of PIXE to the analysis of GSR can be found in references (*9, 13, and 14*).

The case began when the suspect telephoned the authorities and reported that he had come home to find his wife in bed with a fatal gunshot wound from her own gun. After questioning the husband at the scene, the officer in charge asked if he would be willing to accompany them to the station to answer a few more questions and take a "hand swipe" test. The suspect agreed and was placed in the back of a patrol car and left unattended. Approximately twenty minutes later, one of the officers at the scene noticed that the husband was "vigorously"

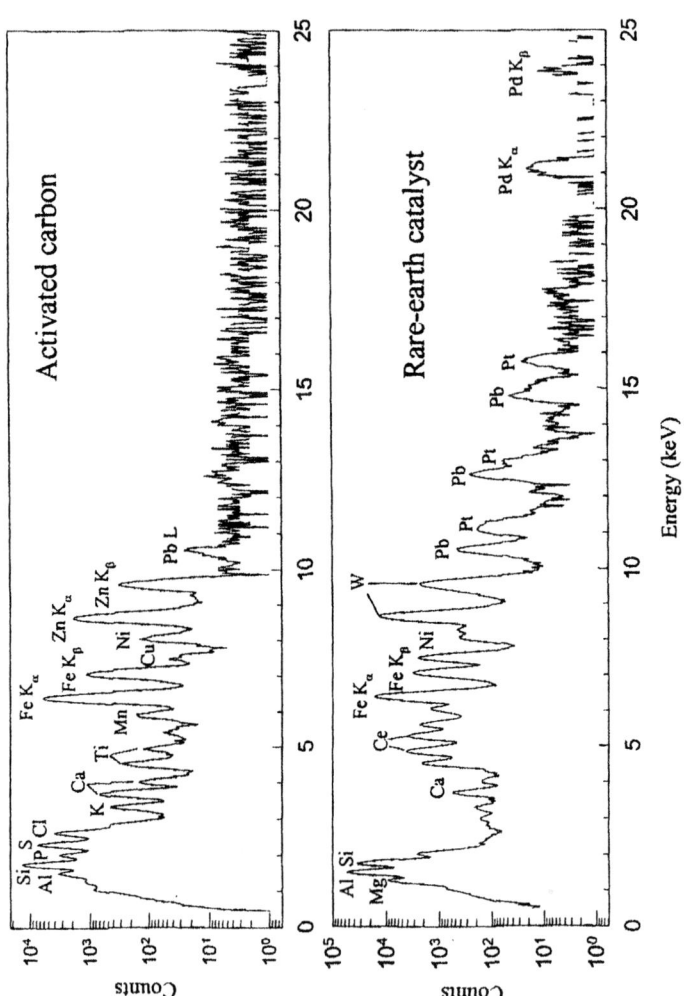

Figure 2. Example PIXE spectra using dual-energy irradiation.

Table I. PIXE Results for NIST SRM 1577 Bovine Liver

Element	PIXE Result	Certified Value
Na	0.27 ± 0.03 %	0.243 %
P	1.17 ± 0.08 %	(1.13 ± 0.12 %)
S	0.72 ± 0.05 %	(0.79 ± 0.10 %)
Cl	0.27 ± 0.03 %	(0.27 ± 0.01 %)
K	1.00 ± 0.07 %	0.97 %
Ca	164 ± 15	124
Mn	10.6 ± 0.5	10.3
Fe	264 ± 15	268
Cu	191 ± 12	193
Zn	141 ± 9	130
Se	1.2 ± 0.3	1.1
Br	8.2 ± 1.1	(9.1 ± 0.9)
Rb	16.6 ± 1.4	18.3

NOTE: The PIXE results are the average and standard deviation of 6 analyses. Units are µg/g or weight percent (%). Values in parentheses are recommended values (*12*).

Table II. PIXE Results for NIST SRM 1107 Naval Brass

Element	PIXE Result	Certified Value ()
Fe	396 ± 10	370
Ni	1260 ± 30	980
Cu	60.3 ± 0.6 %	61.21 %
Zn	38.3 ± 0.4 %	37.34 %
Sn	0.94 ± 0.04 %	1.04 %
Pb	0.13 ± 0.01 %	0.18 %

NOTE: The PIXE results are the average and standard deviation of 3 analyses. Units are µg/g or weight percent (%).

wiping his hands back and forth on the upper portion of his jeans. At this point, the husband's hands were restrained, he was taken to the station, and his hands were swabbed with a standard gunshot residue test kit. The request for PIXE analysis of the jeans was made when the crime lab found some evidence of GSR, but at levels below their threshold limits for positive GSR by atomic absorption spectrometry. We were never informed which GSR elements were observed or what positive threshold values the Florida lab used. The median positive threshold values in atomic absorption from a recent survey of 80 forensic laboratories in 44 U.S. states and two Canadian Provinces are 0.5 µg/mL for lead, 0.4 µg/mL for barium, and 0.05 µg/mL for antimony (*15*).

The Pinellas County Florida Sheriff's Department requested that we perform the work by randomly analyzing spots on the upper portion of the front of the jeans. The fact that six spots were analyzed was driven primarily by their budget. While not included in the original request, we emphasized the need for a set of control samples and insisted on also analyzing several spots on the upper portion of the back of the pants. The maximum spot size of 16 mm in diameter was selected in order to examine as much of the surface of the pants as possible. Lead was found in all six spots on the front of the pants with an observed range of 71 to 141 ng/cm^2 and antimony was found in four of the six spots with an observed range of 96 to 438 ng/cm^2. In contrast, lead and antimony were not observed in any of the spots on the back of the pants. The limits of detection for lead and antimony, calculated on the basis of 3 standard deviations of the background area over one full-width-at-half-maximum of the principal X-ray peak centroid, were 5 ng/cm^2 and 50 ng/cm^2, respectively. Barium, the third element frequently used to test for GSR (*3*), was not observed in any of the spots. The 95% confidence level detection limit for barium was, however, several hundred ng/cm^2.

Upon receiving our report, the first question from the prosecutor in the case was "does this prove that the suspect fired the gun?" Yet, precisely because the probability that the PIXE results indicate that the suspect fired the gun in question can only be answered through a set of carefully controlled experiments that were not performed, the report made no mention of the likelihood that the observed lead and antimony came from GSR. In follow-up conversations, I reiterated that we could only attempt to answer this question through a series of additional, controlled experiments with the gun in question. The prosecutor elected at this point, primarily because of the economics associated with trying the case, to proceed with a report that merely stated the observation of lead and antimony on the front of the pants and the absence of these elements on the back of the pants.

My next involvement in the case was a pre-trial deposition in which the defense attorney for the client had an opportunity to question our results and report. While the beginning of the deposition focused on a description of the technical details of the measurement, the defense attorney, like the prosecutor, spent a good deal of time asking me to speak to the probability that the suspect fired the gun in question. Each time I politely reminded the attorney that I was an analytical chemist, not a forensic scientist, and that the probability question could not be answered with our limited results. To his credit, the defense attorney then asked if the observed lead and antimony could have come from other sources. I readily admitted that this was a distinct possibility and suggested several environmental sources for lead and antimony. The last part of the deposition was spent on of the absence of barium in our results and a detailed technical discussion of why the detection limits for barium were poorer than those for lead and antimony.

My final involvement in the case was testifying as an expert witness for the prosecution in the murder trial. Aside from a series of questions by the prosecutor about my experience with PIXE in order to qualify me as an expert in the eyes of the court, the line of questioning at the trial was similar to that of the pre-trial deposition. Of special note, however, was a series of questions from the defense attorney at the end of my testimony in which he repeatedly asked for "my opinion" that the results indicate that the suspect fired the gun. Each time I replied that I was not qualified to offer an opinion and that I could not speak to the probability question from our results. I later learned from the prosecutor in the case that the jury found the defendant guilty based, in part, on our analytical results.

Case of Suspect Coins

In the second example, a special investigator for the Florida State Attorney's Office contacted the lab to find out if PIXE could be used to perform a non-destructive analysis of gold coins. The investigator was referred to us by the Pinellas County Florida Sheriff's Department. In this case, the key criteria for selecting PIXE appeared to be the non-destructive nature of the measurements and our ability to perform the measurements in a timely fashion. Because the special investigator was required by the Florida Department of State to remain with the "control" gold coins provided by the Division of Historical Research at all times, we were required to complete the work in one day. In this particular application, PIXE has been applied to the analysis of ancient silver coins

(*16,17*), Roman coins of the republican and imperial age (*18*), ancient Chinese coins (*19*), and gold coins of the eighteenth and nineteenth centuries (*20*).

In the mid 90s, the Florida State Attorney's Office began to suspect the authenticity of gold coins being marketed and sold as historical artifacts recovered from the Spanish galleon the Nuestra Señora de Atocha. Spain's trade with her colonies followed a well-established system. Beginning in 1561 and continuing until 1748, two fleets a year were sent from Spain to the New World. The ships brought supplies to the colonists and were then filled with silver, gold, and agricultural products for their return to Spain. For protection against pirates, each fleet was equipped with two heavily armed guard galleons. The Nuestra Señora de Atocha, or Atocha, was designated as the rear guard galleon of the Tierra Firme Fleet that arrived in Havana on its way back to Spain in late August, 1622. On September 4th, the fleet left Havana and sailed due north towards Florida. That evening, the fleet encountered a major storm and, while much of the fleet was driven past the Dry Tortugas and into the relatively safe waters of the Gulf of Mexico, the Atocha and four other ships were lost. During the storm, the Atocha was smashed violently on a coral reef and sunk. Although initially located in 55 feet of water, salvage attempts over the next ten years proved futile as a second hurricane struck the region before equipment could be brought from Havana to complete the salvage operation. The crew of Mel Fisher discovered the Atocha on July 20, 1985 in 55 feet of water west of Key West, Florida.

According to the special investigator, the authenticity of coins being marketed and sold as recovered from the Atocha was first called into question by several coin experts asked to provide appraisals by individuals who purchased them. The coins, which contained approximately $300 worth of gold, were being marketed and sold for several thousand dollars each. According to these coin experts, the imprint on the coins was too good. In their opinion, the edges of the relief were much too regular and sharp for the coins to have been produced by a hammer and die. Prior to 1733, coins were manufactured by cutting coin blanks from crudely cast bars of refined metal and hand hammered between engraved dies. Often, the irregular surface thickness and circumference of the blanks prevented well-defined strikes. An example of the relief on the suspect coins can be seen in Figure 3.

Not willing to risk making their case only on the opinion provided by a visual inspection from coin experts, the Florida State Attorney's Office contacted our lab and requested a chemical analysis of Spanish gold coins known to have been produced in the New World colonies in 1714 and the suspect coins. The results of the PIXE analysis of a 1, 2, 4 and 8 Escudo provided by the Florida Department of State and a suspect coin are given in Table III. As in the previous case, the first question asked by the special investigator when the first suspect coin was found to have a higher ratio of gold to silver than the control samples was "does this mean that the coin is a fake?" Even in the informal setting of the laboratory, I was careful to state that "I was not an expert in the composition of ancient coins and that his office would have to seek out experts who could answer this question." The other similarity to the previous case was that the special investigator decided, for budgetary reasons, to only request the analysis of a single suspect coin at until he determined if this was the type of difference one would expect in composition if the coins were manufactured from "modern" gold.

Figure 3: Suspectcoin market as recovered from the Nuestra Señora de Atocha

Approximately one year after performing this work, we were contacted by the special investigator and informed that the individuals marketing the coins had reached an agreement with the Florida State Attorney's Office to return the money to any individual that purchased one of the coins in question. The principle individual in question claimed that he had been misled by his partner in the operation as to the authenticity of the coins and was, in exchange for the plea agreement, not prosecuted. According to the special investigator for the case, the main motivator for the individuals in question to accept the plea agreement was the results of the chemical analysis. Of note is the fact that, while the attorneys for the individuals were provided our analytical report, they never contacted us to question either the results or the conclusions that the State Attorney's Office drew from the results.

Table III. PIXE Results for Spanish Gold Coins

Coin	Au (wt. %)	Ag (wt. %)	Fe ($\mu g/g$)	Cu ($\mu g/g$)
1 Escudo	92.9	6.0	2582	363
2 Escudo	91.2	8.8	4270	290
4 Escudo	91.2	8.2	4376	651
8 Escudo	90.4	8.4	4858	784
Suspect	96.3	3.9	3224	723

NOTE: The control Escudo were provided by the Florida Department of State Division of Historical Research

Summary

In 1972, V.P. Guinn presented a paper at the 2nd International Conference on Forensic Activation Analysis outlining the "growing pains" that forensic activation analysis was experiencing on its way to becoming a well accepted method of analysis in the eyes of the law (21). He outlined a series of cases that revealed problems in forensic activation analysis, not from the standpoint of its use in the investigative stages of criminal cases, but from the standpoint of the proper presentation and interpretation of such results in court. Any analyst who becomes involved in the analysis of forensic samples would do well to accept and adhere to the guidelines that he prescribed. In particular, it is clear from my experience, that the analyst must approach each case in an impartial manner – regardless of whether the analyses are to be performed for the prosecution or for the defense. In addition, if the analyst interprets the results, the interpretation must be based upon a large body of background data or carefully controlled experiments. While the client (prosecutor or defense) is likely to press for a quick or definite interpretation, providing one without the appropriate background or controls could, as in the formative years of forensic activation analysis, lead to travesties of justice and loss of credibility for the technique.

References

1. Savel, P. *Ann Pharm. France*, **1963**, *21*, 303.
2. U.S. vs. Anderson *et al.* (Brooklyn, N.Y., March 1964).

3. Singer, R.L.; Davis, D.; Houck, M.M. *J. Forensic Sci.*, **1996**, *41(2)*, 195-198
4. Johansson, T.B.; Akselsson, K.R.; Johansson, S.A.E. *Nucl. Instr. Meth. Phys. Res.* **1970**, 84, 141.
5. Johansson, S.A.E.; Campbell, J.L. *PIXE A Novel Technique for Elemental Analysis*, Wiley: New York, NY, 1988.
6. *Particle-Induced X-ray Emission Spectrometry (PIXE)*; Johansson S.A.E.; Campbell J.L.; Malmqvist K.G., Eds.; Chemical Analysis Vol. 133; Wiley: New York, NY, 1995.
7. Warren, M.W.; Falsetti, A.B.; Kravchekno, I.I.; Dunnam, F.E.; Van Rinsvelt, H.A.; Maples, W.R. *Forensic Sci. Int.*, **2002**, *125(1)*, 37-41.
8. Vogt, C.; Becker, A.; Vogt, J. *J. Forensic Sci.*, **1999**, *44(4)*, 819-831.
9. Sen, P.; Mehta, G.K. *IEEE Trans. Nucl. Sci.*, **1983**, *NS-30(2)*, 1302-1304.
10. Maxwell, J.A.; Campbell, J.L.; Teesdale, W. *J. Nucl. Instr. Meth. Phys. Res*, **1989**, *B43*, 218.
11. Campbell, J.L.; Higuchi, D.; Maxwell, J.A.; Teesdale, W.J. *Nucl. Instr. Meth. Phys. Res.* **1993**, *B77*, 95-109.
12. *Compilation of Elemental Concentration Data for NBS Clinical, Biological, Geological, and Environmental Standard Reference Materials*; Gladney, E.S.; O'Malley, B.T.; Roelandts, I.; Gills, T.E.; Natl. Bur. Stand. Spec. Publ. 260-111; Washington, DC, 1987.
13. Sen, S.; Varier, K.M.; Mehta, G.D.; Rao, M.S.; Sen, P.; Panigrahi, N. *Nucl. Instrum. Methods*, **1981**, *181*, 517-521.
14. Sen, P.; Panigrahi, N.; Rao, M.S.; Varier, K.M.; Sen, S.; Mehta, G.K. *J. of Forensic Sci.*, **1982**, *27(2)*, 330-339.
15. Singer, R.L.; Davis, D.; Houck, M.M. *J. Forensic Sci.* **1996**, *41(2)*, 195-198.
16. Uzonyi, I.; Bugoi, R.; Sasianu, A; Kiss, A.; Constantinescu, B; Torbagyi M. *Nucl. Instr. Meth. Phys. Res*, **2000**, *B161-3*, 748-752.
17. Smit, Z.; Kos, P. *Nucl. Instr. Meth. Phys. Res*, **1984**, *231*, 416-418.
18. Dacca, A.; Prati, P; Zucchiatti A.; Lucarelli, F; Mando, P.A.; Gemme, G.; Parodi, R.; Pera, R. *Nucl. Instr. Meth. Phys. Res*, **2000**, *B161-3*, 743-747.
19. Lin, E.K.; Wang, C.W.; Yu, Y.C.; Cheng, W.C.; Chang, C.H.; Yang, Y.C.; Chang, C.Y. *Nucl. Instr. Meth. Phys. Res*, **1995**, *B99*, 394-398.
20. Ferreira, G.P.; Gil, F.B. *Archaeometry*, **1981**, *23(2)*, 189-197.
21. Guinn, V.P. *J. Radioanal. Chem.* **1973**, *15*, 389-398.

Chapter 15

Study of Traces of Tritium at the World Trade Center

Thomas M. Semkow[1,2], Ronald S. Hafner[3], Pravin P. Parekh[1], Gordon J. Wozniak[4], Douglas K. Haines[1], Liaquat Husain[1,2], Robert L. Rabun[5], and Philip G. Williams[6]

[1]Wadsworth Center, New York State Department of Health, Albany, NY 12201
[2]School of Public Health, University of Albany, State University of New York, Albany, NY 12201
[3]Fission Energy and Systems Safety Program, Lawrence Livermore National Laboratory, Livermore, CA 94551
[4]Nuclear Science Division, E. O. Lawrence Berkeley National Laboratory, Berkeley, CA 94720
[5]Tritium Engineering Department, Westinghouse Savannah River Company, Aiken SC 29808
[6]Physical Biosciences Division, E. O. Lawrence Berkeley National Laboratory, Berkeley, CA 94720

Traces of tritiated water (HTO) were detected at the World Trade Center (WTC) ground zero after the 9/11/01 terrorist attack. A water sample from the WTC sewer, collected on 9/13/01, contained 0.164±0.074 (2σ) nCi/L of HTO. A split water sample, collected on 9/21/01 from the basement of WTC Building 6, contained 3.53±0.17 and 2.83±0.15 nCi/L, respectively. These results are well below the levels of concern to human exposure. Several water and vegetation samples were analyzed from sites outside ground zero, located in Manhattan, Brooklyn, Queens, and the Kensico and Croton Reservoirs. No HTO above the background was found in those

© 2004 American Chemical Society

samples. Tritium radioluminescent (RL) devices were investigated as possible sources of the traces of tritium at ground zero. It was determined that the two Boeing 767 aircraft that hit the Twin Towers contained a combined 34 Ci of tritium at the time of impact in their emergency exit signs. There is also evidence that many weapons from law enforcement were present and destroyed at WTC. Such weaponry contains by design tritium sights. The fate and removal of tritium from ground zero were investigated, taking into consideration tritium chemistry and water flow originating from the fire fighting, rain, as well as leaks from the Hudson River and broken mains. A box model was developed to describe the above scenario. The model is consistent with instantaneous oxidation of the airplane tritium in the jet-fuel explosion, deposition of a small fraction of HTO at ground zero, and water-flow controlled removal of HTO from the debris. The model also suggests that tritium from the weapons would be released and oxidized to HTO at a much slower rate in the lingering fires at ground zero.

1. World Trade Center

The World Trade Center was built in New York City during the 1960s through the 1980s. It contained seven buildings designated as WTC 1 through WTC 7. The most prominent were the 110-floor Twin Towers, WTC 1 - The North Tower built in 1970, and WTC 2 - The South Tower built in 1972. The WTC was owned and operated by the Port Authority of New York and New Jersey (PANYNJ). It is important to this investigation that several federal law enforcement agencies were located at the WTC (*1,2*). US Customs and the Bureau of Alcohol Tobacco and Firearms (ATF) were housed in WTC 6, also called the US Customs House. US Secret Service, US Department of Defense, Central Intelligence Agency, and the New York City Office of Emergency Management (OEM) had offices in WTC 7.

The original, 1776 Manhattan shoreline crossed the WTC complex in the north-south direction. The present-day land to the west of the complex is actually a fill (*3*). Since WTC 1 and 2 had to have foundations down to the bedrock, the required engineering solution was achieved by constructing the so-called Bathtub. It is surrounded by the Slurry Wall, 510 ft × 980 ft, 70-ft deep, and 3-ft thick (*4*). The Slurry Wall prevented leaks from the Hudson River. Besides the foundations of the buildings, the Bathtub contained a Concourse and a six-level basement underground. On the lowest B6 level there was a tunnel and a station for the Port Authority Trans-Hudson (PATH) train, providing commuting from and to New Jersey under the Hudson River.

2. The Terrorist Attack

On September 11, 2001 at 8:46 AM, a Boeing 767-223ER aircraft operated by American Airlines as Flight 11 hit WTC Tower 1 causing jet fuel explosion and fire. At 9:02 AM, a second aircraft, a Boeing 767-222 operated by United Airlines as Flight 175, hit WTC Tower 2. Both flights originated in Boston, so the aircraft were full of fuel, estimated at 10,000 gallons each (*5*). WTC 2 collapsed after 56 min, followed by WTC 1 which lasted 102 min.

The collapse of the Towers has been studied in detail (*5,6,7,8*). The floor trusses of the Towers were supported by the steel perimeter columns, while the central columns supported the elevator shafts. If there had been no fire, the Towers would have not collapsed. However, due to the fires, when temperature reached 1500F (816C), the steel support systems lost their strength, causing the structures to collapse. By some estimates, the temperatures could have locally reached 1800C from burning of the aluminum bodies of the airplanes. At this temperature, hydrogen gas is evolved from burning of the concrete, which fuels further burning. The reasons that WTC 2 collapsed first included the higher speed of the aircraft at collision (586 mph) compared to the speed of the aircraft colliding with WTC 1 (494 mph), as well as noncentral and lower point of impact in the case of WTC 2. The collapsing Towers destroyed other WTC buildings, and the debris compacted and destroyed much of the Bathtub. The debris from WTC 1 plunged through the center of WTC 6, creating a pit stretching down to the basement of the Bathtub (*9*). At 5:20 PM, WTC 7 collapsed due to a weakening of its steel support structure caused by a diesel fuel fire. The fuel was stored in the tank for emergency power generation for the OEM (*10*). The WTC area is referred to as ground zero.

Authorities determined that 2795 people died in the attack on the WTC (*5*), including 157 people onboard the aircraft, 343 New York City Fire Department firefighters, 23 officers from the New York City Police Department, 37 officers from the PANYNJ Police Department, and 3 officers from the New York Office of Court Administration (*11,12*).

3. Tritium Measurements

Tritium is produced naturally in the atmosphere from the reactions of cosmic ray protons and neutrons with N and O nuclei, as well as by ternary fission in geological formations (*13,14,15*). However, the bulk contribution to environmental accumulation comes from the nuclear testing in the atmosphere, nuclear fuel cycle, and some from consumer products. The global present-day inventory of tritium in the environment is 19 EBq, only 1.3 EBq of which is attributed to natural production (*16*). The levels of tritium in the environment have been decreasing steadily, due to its decay with a half-life of 12.3 years, since the ban on atmospheric nuclear testing. Tritium occurs in the environment primarily as tritiated water, and much less as organically bound tritium. Typical present-day concentrations of HTO in water in the US are 0.1-0.2 nCi/L (*17*).

We became interested in the subject of tritium at WTC because of the possibility that tritium RL devices could have been present and destroyed at WTC. Three groups of environmental samples were analyzed for tritium as HTO, to confirm or disprove this hypothesis. The 1st group consisted of the samples collected by the EPA not specifically for tritium analysis. They were analyzed for tritium after this investigation had started. The 2nd group was analyzed for tritium before this investigation started, and was collected by the New York City Department of Environmental Protection (samples 23-35 at the request of EPA). The 3rd group consisted of the samples collected especially for this investigation.

Water was distilled once from the environmental stationary water samples, and twice from the vegetation samples. 10 ml of such distillate was mixed with 13 ml of Instagel XF cocktail (Packard) in a borosilicate glass vial and measured on an ultralow-background liquid scintillation counter TRI-CARB, model 3170TR/SL by Packard. The samples from groups 1 and 3 were measured for 200 min, while from group 2 for 100 min. The tritium end-point beta energy is 18.6 keV. We used the energy window 1-13 keV to maximize signal to background ratio. The background rate was about 2 cpm. The efficiency of the instrument was calibrated using HTO standards as a function of the tSIE quench index. The environmental samples had a tSIE value around 230, corresponding to efficiency in the range 0.20-0.25.

The results are given in Table I. Samples 1, 6, and 7 are from ground zero and they are all positive. The rest of the results in Table I are upper limits. Sample 1, measuring 0.164±0.074 nCi/L, is from the WTC sewer, collected three days after the attack, and is just above the detection limit. Samples 6 and 7 of about 3 nCi/L are split samples from WTC 6, basement B5, collected 10 days after the attack. Thus, tritium was detected in these samples from ground zero, but the concentrations are very low. In fact, 3 nCi/L is about 7 times less than the EPA limit in drinking water of 20 nCi/L (*18*). No health implications are known or expected at such low concentrations (*14*). As a consequence, no additional ground-zero samples were judged to be necessary.

Table I. The results of tritium analysis in New York State.

Gr. no	Samp. no	Coll. date[a]	Samp. type	Sampling location	Activity[b] (nCi/L)
1	1	9/14	water	WTC storm sewer	0.164±0.074
1	2	9/17	water	Manh., 55 Broadway, 32 fl., roof tank	<0.13[c]
1	3	9/18	water	Manh., 111 Broadway, 22 fl., roof tank	<0.13[c]
1	4	9/18	water	Manh., 45 Wall St., 30 fl., roof tank	<0.13[c]
1	5	9/18	water	Manh., 7 Hanover Sq., 29 fl., roof tank	<0.13[c]
1	6	9/21	water	WTC 6, basement B5	3.53±0.17
1	7	9/21	water	same	2.83±0.15
2	8-22	9/11,12	water	Kensico and Croton Reservoirs[d]	<0.11-<0.19
2	23-35	9/15	water	South Manhattan water distribution	<0.12-<0.15[c]
3	36	10/25	grass	Albany	<0.12[e,f]
3	37	10/27	grass	Brooklyn, Brooklyn Heights	<0.21[e,f]
3	38	10/27	water	Brooklyn, Govanus Canal	<0.11
3	39	10/27	grass	Brooklyn, Govanus Park	<0.091[e]
3	40	10/27	water	Brooklyn, English Kills	<0.11
3	41	10/27	water	Brooklyn, Prospect Park	<0.090
3	42	10/27	grass	same	<0.093[e]
3	43	10/27	water	Brooklyn, Marine Park	<0.11
3	44	10/27	grass	same	<0.090[e]
3	45	10/27	water	Brooklyn, Paerdegat Basin	<0.090
3	46	10/27	water	Brooklyn, Coney Island	<0.11
3	47	10/27	grass	same	<0.092[e]
3	48	10/27	water	Queens, Forest Park	<0.090
3	49	10/28	water	Poughkeepsie	<0.11
3	50	10/28	grass	same, with weeds	<0.17[e,f]
3	51	11/4	leaves	Manhattan, Battery Park	<0.12[e,f]

FOOTNOTES: a) In 2001; b) 2σ errors or limits; c) System closed or restricted to atmospheric deposition; d) New York City raw water reservoirs in Westchester County; e) Activity given per volume of water extracted from the vegetation; f) Problems with chemiluminescence and color quench, measured with instrumental luminescence correction. The upper limits for samples 9 and 22 are higher because the efficiency was lower due to higher quench (lower tSIE index), the detection limit being inversely proportional to the efficiency.

Samples 2-5 are from roof tanks in South Manhattan near ground zero. Since these tanks are vented, there was a possibility of some atmospheric contamination, although restricted. Samples 23-35 are from the New York City water distribution system in South Manhattan, which is closed to the atmospheric deposition. None of these samples show any tritium present, as expected. There was also a possibility that some HTO would have been transported with the fire plume during the first several days after the attack and deposited downwind. The wind direction was approximately northwest during 9/11 and 9/12 (*19*). Therefore, we did environmental sampling in Brooklyn, Queens, and South Manhattan, which are downwind from ground zero. The sample numbers are 37-48 and 51 in Table I. They were taken about seven weeks after the attack. All the results were zero within the detection limits, which is consistent with the low levels of HTO detected at ground zero.

4. Tritium Radioluminescent Devices

The difference between tritium RL devices and CRT tubes is that, in the former, β particles from tritium decay, rather than accelerated electrons, generate light in the phosphor (*15*). ZnS is the most widely used phosphor and is activated by an impurity. ZnS-Ag gives a green glow, with a decay constant of 0.2 μs. ZnS-Cu gives blue-green light, and ZnS-(Cu,Mg) gives yellow-orange light (*20*). There are two basic types of RL devices: i) gaseous tritium light sources (GTLS) sealed in borosilicate glass tubes, internally coated with the phosphor, and ii) tritium chemically incorporated into a polymer such as polystyrene and mixed with the phosphor. There is no tritium leakage from GTLS, unless broken. There is a small diffusion of tritium from polymers. GTLS are used as airport runway lights at remote airports (Alaska); emergency EXIT and other signs in buildings; emergency EXIT signs, handles, and aisle markers in airplanes; as well as sights in weaponry and markings in time devices. The polymer-based RLs are used in emergency signs and as paints in watches. When GTLS tubes age, they acquire a small percentage (<2%) of HTO due to radiolytic reactions with the phosphor binder (*15,21,22*).

Typical emergency EXIT signs in buildings contain from several to several tens of Ci of molecular tritium. The maximum recommended tritium activity by ANSI standard is 50 Ci (*23*). The activity of tritium is regulated by the Nuclear Regulatory Commission (NRC), per request of a manufacturer. For instance, Mb-Microtec AG registered with NRC sealed RL devices containing up to 50 Ci of tritium (*20*). The typical content of tritium per device is 10 Ci. The tritium emergency signs in airplanes have a regulatory limit of 10 Ci (*24*).

GTLS are used extensively in weaponry and are standard equipment in military as well as law enforcement. Of interest to this work are gun sights containing GTLS capsules, either cylindrical or spherical, which facilitate aiming at night. There are two categories of interest: scopes and night sights. The content of tritium depends on the configuration as well as on the manufacturer. Trijicon Inc. uses 100 mCi in scopes and three capsules of 18 mCi each (54 mCi total) in night sights (25). Innovative Weaponry Inc. uses 54 mCi in their PT night sights (26). Meprolight Ltd. uses between 30 to 54 mCi per set of night sights (27).

Tritium in timing devices is used as GTLS or polymer paint. NRC regulations limit tritium content per timepiece to 25 mCi for paint (28) and 200 mCi for GTLS (29). The ISO standard recommends for paints a maximum average activity of 5 mCi per lot, and 7.5 mCi per isolated instrument (30). The US military standard recommends the maximum activity for a GTLS device as 25 mCi (31). A major manufacturer of GTLS-containing watches is Mb-Microtec AG, who offers the watches to the US market under the brand name Luminox. The watches are licensed with NRC under NR-0446-D-103-E for up to 100 mCi of tritium; however, the watches on the market contain up to 41 mCi of tritium (32). Luminox makes dive watches for the US Navy and aviator watches for the US Air Force. Consumer models are available. These types of watches are expensive, available through specialty stores only and are, therefore, not widely worn.

Less expensive and more popular watches use paint containing tritiated polymer, in a plastic casing. A major manufacturer of tritiated paint is Rc Tritec AG. The typical range of tritium activity per timepiece is 0.8-2.7 mCi (33). However, a non-radioactive photoluminescent material, Super-LumiNova, has been developed by Nemoto & Co., based on mixed aluminum oxides and activated with a rare earth element (20). It is characterized by high intensity and long afterglow, and is used in more than 95% of luminescent watches currently manufactured, instead of tritium paint (33).

5. Sources and Fate of Tritium at WTC

As described in Section 3, HTO was detected at ground zero at low concentrations. Several sources of tritium were considered and analyzed, as consistent with the experimental data: i) EXIT signs in the buildings, ii) emergency signs on the airplanes, iii) fire and emergency equipment, iv) weaponry, and v) timepieces.

Presence of RL EXIT signs in the buildings would have implied large available source of tritium. We were informed by PANYNJ authorities that there were no tritium signs at the WTC, only photoluminescent ones (34). This is entirely consistent with our observations.

It was determined by the Federal Aviation Administration that Boeing 767, Serial Number 21873, operated by United Airlines, Tail Number 767-222 N612UA, was delivered in February, 1983, with 43.2 Ci of tritium in emergency signs (35). The 43.2 Ci of tritium was contained in four EXIT signs (10 Ci each) and four slide/raft handles (0.8 Ci each). The same activity of tritium was present upon the April, 1987 delivery of the second Boeing 767, Serial Number 22322, Tail Number 767-223ER N334AA, operated by American Airlines. Since neither of these aircraft were modified after delivery (35,36), the total activity from the aircraft was 34 Ci at the time of attack, when the radioactive decay of tritium has been accounted for.

The tritium from the airplanes was released at the two points of impact with the Towers. Conversion of molecular tritium (T_2) to HTO in the atmosphere is normally negligible: the formation of HTO through chemical kinetics is extremely slow (37). Rather, the conversion to HTO in atmospheric transport goes through a stage of deposition of molecular tritium to soil, followed by a microbial and exchange oxidation in soil. HTO is then directly reemitted, or taken up by plants first and then reemitted into the atmosphere. The combined process results in measured conversion rates of between 10^{-5} and 10^{-3} for downwind distances of up to 15 km.

However, at each of the two points of impact there was an explosive release. Considering the jet fuel explosion and high-temperature fires at the WTC, T_2 was efficiently oxidized to HTO, based on weapons-testing data (38), as well as laser heating experiments (39). This oxide immediately vaporized due to the intense heat. Most of the HTO would be transported in the vapor phase with the wind, since the weather was dry on 9/11/01 (19). One cannot accurately determine how much HTO condensed on building surfaces and deposited on the ground with the collapse of the buildings, but this would have been a small fraction of the 34 Ci available. One indication is the low 0.164±0.074 nCi/L from the WTC sewer, collected two days after the attack. Since the initial source was small, it is consistent that the environmental samples collected downwind over seven weeks after the attack contained no tritium (Section 3).

It is important to compare this small release of tritium in the fire with two other incidents caused by fire and involving the release of molecular tritium. One incident involved a fire in a community building at Council, Alaska, on 9/6/87, where 12 RL light panels for airport runway marking were stored, totaling 3000 Ci of tritium (40). It was a free-burning fire, which consumed the building in 1 hr. Tritium assessment was done 11 days after the accident. The remaining GTLS tubes were mostly undamaged but disfigured, indicating that all tritium had escaped. No air-borne tritium was detected. All tubes were carefully wiped on surfaces, and the HTO activity from the wipes amounted to 6.5×10^{-8} of that originally present. No HTO was found in bioassay or environmental samples. The release scenario at the WTC from the airplanes is consistent with this accident. However, the Twin Towers collapsed before their complete burning, so the fraction of tritium deposited at the WTC might be larger. Another incident, involving containers with tens of thousands Ci of tritium, was a fire on a C-124 airplane on the ground at the Wright-Patterson Air Force Base, Dayton, OH, on 10/12/65 (*41*). That fire was actively extinguished. Elevated levels of HTO were found in bioassay samples, on emergency and fire equipment, clothing, in the debris, as well as in the soil and water from nearby samples. In comparison with the Alaska incident, the active fire fighting contributed to capture of some of the HTO on site.

After the WTC buildings collapsed, fire fighting and rescue operations continued. The fires at ground zero were smoldering for months after the attack (*42*). It was determined that 3 million gallons of water were hosed on site in the fire-fighting efforts between 9/11 and 9/21 (the day of the tritium measurement; samples 6 and 7 in Table I) (*43*). In addition, there were two episodes of rain during the same 10-day period: on 9/14 and 9/20,21 (*19*), totaling 0.9 million gallons of water in the Bathtub area. Considering the neighboring areas, we take 1 million gallons from the rain. Therefore, a total of 4 million gallons of water percolated through the debris in the first 10 days and collected at the bottom of the Bathtub. The percolating water efficiently dissolved that part of the airplane HTO, which was deposited in the building collapse, and carried it to the bottom of the Bathtub.

An engineering assessment determined that there was a water leak into the Bathtub, adding to the rain and hose water. The main leak was from the Hudson River via two WTC cooling water outfall lines, while the incoming lines were shut down (*44*). There were reported leaks from broken water mains (*3,45*). There were also problems with the water table due to a hole in the damaged Slurry Wall along Liberty Street (*46*). The combined water from rain and hoses, as well as the leaks, collecting at the bottom of the Bathtub, transferred into the PATH train tunnel. Water then flowed under the Hudson River to the Exchange Place Station, Jersey City, NJ, since it is lower in elevation than WTC B6 level (*3,44*), where it was pumped out. Other pumps were installed (after 9/21) along Liberty Street to stabilize the Slurry Wall, which had moved (*46*). Based on the pumping records, a total of 30 million gallons of water passed through the Bathtub between 9/11 and 9/21 (*4,47*). Therefore, 26 million gallons were from the leaks. Even on 10/8/01 there was still some water flowing to New Jersey (*45*). HTO that collected at the bottom of the Bathtub was removed with the water flow. The 9/21 HTO sample, reportedly collected from basement B5, sampled that dynamic system close to the bottom of the Bathtub.

It was concluded that fire and emergency equipment could not have been a source of tritium, since such equipment does not typically use tritium RL devices, at least for the type of emergency response conducted at the WTC. Weaponry was another likely source of tritium. As described in Section 1, several federal and state law enforcement agencies were housed at WTC, in buildings 6 and 7. ATF had two vaults filled with tactical weapons and guns (*1,48,49*). The ATF vaults were in WTC 6, where our samples 6 and 7 were measured. A total of 63 police officers died in the attack (*12*). They may have been carrying pistols equipped with tritium night sights. In fact, many guns have been recovered from the debris (*48,49,50*), some of them in good condition. It would take 20 equipped weapons destroyed, 50 mCi each, to give approximately 1 Ci of tritium (Section 4).

Considering the 2795 victims in the attack, tritium watches could have been another source of tritium. Tritium paint watches were less likely, since they contain much less tritium and are generally no longer manufactured in modern watches (Section 4). However, GTLS-type watches, although expensive, could have been worn by more affluent public of Lower Manhattan. In addition, the military-style watches may have been worn by the emergency law enforcement personnel who perished. It would take 40 GTLS watches, 25 mCi each, to give 1 Ci of tritium activity. The GTLS watches can be obtained in specialty stores only. No specialty watch stores were located at WTC (*51*). Some watches, but not necessarily tritium, were recovered from the debris with only minor damage (*50*). Probability-wise, weapons were definitely present at WTC, and the law-enforcement types contain tritium night sights by design; tritium watches were probably present, but in numbers difficult to determine.

The mechanism of tritium release from the weapons or watches would have been much different than from the airplanes. Some devices could have been catastrophically destroyed in the buildings collapse; however, surprisingly, many were recovered with only minor damage. In addition, GTLS weapon sights are well-encapsulated in metal protective shields. Many devices would have been subjected to smoldering fires of much lower temperature than the explosive and high-temperature fires up in the Towers (with the exception possibly of the WTC 7 fire). At such temperatures, GTLS tubes would soften and disfigure, slowly releasing tritium. Some of that tritium would diffuse from the debris and be dispersed in the air, while some would remain trapped in the debris. While oxidation of molecular tritium is slow in the air, tritium is known to adsorb on surfaces and exchange with the adsorbed monolayer of water to form HTO due to a catalytic action of the surface (*15,52,53,54*). At elevated tritium concentrations, radiolytic and hot-atom chemistry effects also assist in the oxidation (*22*). Consequently, some molecular tritium released in the debris would convert to HTO and be swept with the hose and rain water down to the basement of the Bathtub, sharing the fate of HTO from the airplanes, but on a much slower time scale. This mechanism resembles leaching of HTO from landfills containing tritium RL devices (*55*).

6. Modeling of Water Flow and Tritium Removal from Ground Zero

A 3-Box model was developed to quantify water flow and tritium removal, depicted in Fig. 1. Box 0 describes the debris, from which HTO is assumed to be transferred to the flowing water at a rate λ. The Bathtub is divided into 2 boxes.

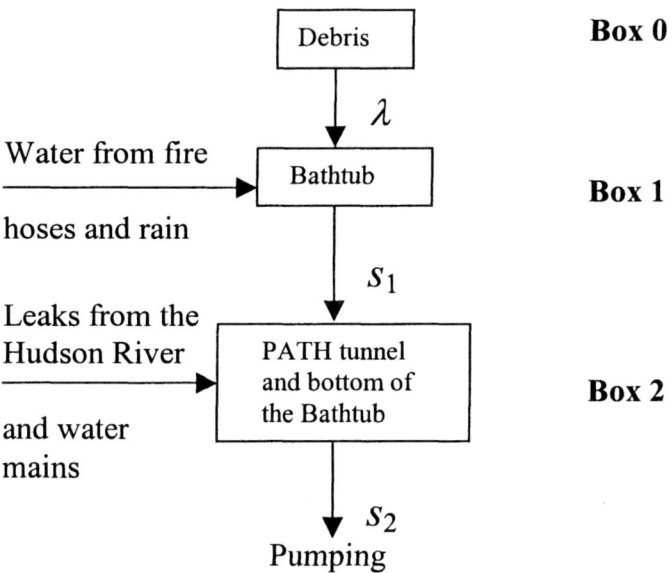

Figure 1. Model of water flow and tritium removal from the WTC.

Box 1 consists of 6/7 of the void volume of the Bathtub (Concourse plus levels B1-B5), through which 1.51×10^7 L of the hose and rain water flowed in 10 days. Therefore, an experimental flow rate $f_1=1.51\times10^6$ L/day through Box 1. Considering that the Bathtub was at least 50% destroyed and filled with the debris from the buildings (*56*), its air porosity could be assumed as 0.26 (a value for close packed spheres). For such porosity, the void volume can be calculated as $V_1=2.21\times10^8$ L. It would take $V_1/f_1=146$ days for such a volume to fill Box 1. The experimental volume of water would not have even reached the bottom of the Bathtub in 10 days. We conclude that the water could not have filled the air porosity completely. Water would flow through filled small pores, however, it would flow only on the surfaces of larger voids. A reasonable estimate of the water-filled volume can be made by equating it to the experimental volume of water that is known to have flowed in 10 days, $V_1=1.51\times10^7$ L. This yields a water percolation time constant $s_1=f_1/V_1=0.1$/day for Box 1.

The B6 level with the PATH tunnel is taken as Box 2. One could estimate an upper limit for the water-filled porosity as 0.1. Its exact value is less important for the model, since the experimental water flow rate through Box 2, $f_2=1.14\times10^7$ L/day $\gg f_1$. This yields $V_2 = 1.42\times10^7$ L, and the water percolation time constant $s_2=f_2/V_2= 0.803$/day for Box 2.

The following differential equations describe tritium propagation through the three boxes at ground zero, for a general tritium source:

$$\frac{dA}{dt} = -\lambda A, \tag{1a}$$

$$\frac{dc_1}{dt} = \frac{\lambda}{V_1} A - s_1 c_1, \tag{1b}$$

$$\frac{dc_2}{dt} = \frac{f_1}{V_2} c_1 - s_2 c_2, \tag{1c}$$

where t is the time, A is the total tritium activity in the debris, and c_1, c_2 are HTO concentrations in Boxes 1 and 2, respectively. Equations (1a-1c) are linear, 1st order, of which (1b) and (1c) are inhomogeneous. They can be solved by standard methods (57), resulting in the following equation describing the measured HTO concentration c_2 at the bottom of the Bathtub:

$$c_2 = \frac{s_1 \lambda A_0}{V_2}\left(\frac{e^{-\lambda t}}{(s_1 - \lambda)(s_2 - \lambda)} + \frac{e^{-s_1 t}}{(\lambda - s_1)(s_2 - s_1)} + \frac{e^{-s_2 t}}{(\lambda - s_2)(s_1 - s_2)}\right), \tag{2}$$

where A_0 is the total HTO activity from a general tritium source at time zero.

Equation (2) needs to be simplified for the specific sources, using the approximation $e^{-s_1 t} \gg e^{-s_2 t}$, which is based on the experimental condition $s_2 > s_1$. For the source term from the airplanes, tritium was in HTO form in the debris (Section 5) and its removal would be controlled solely by the flow rate (the slowest process) in Box 1, rather than by the transfer rate λ. We thus set $\lambda \gg s_2$, and obtain from Eq. (2)

$$A_0 = c_2 f_2 (1/s_1 - 1/s_2) e^{s_1 t}. \tag{3}$$

Using c_2=3.18 nCi/L, t=10 days, and the values s_1 and f_2 given above, we obtain A_0=0.86 Ci from Eq. (3). Taking the total tritium activity of 34 Ci from the two airplanes implies an upper limit for the HTO deposition fraction of 2.5%. This fraction, although the right order of magnitude, is high by a comparison with the two tritium fire incidents described in Section 5, indicating that the airplane source alone was insufficient.

For the tritium source term from the weapons subjected to the smoldering fires, tritium removal would be controlled by the transfer rate from the debris (the slowest process). We thus set $s_1 \gg \lambda$, and obtain from Eq. (2)

$$A_0 = \frac{c_2 f_2}{\lambda[1 - e^{-(s_1-\lambda)t}]}. \qquad (4)$$

Equation (4) cannot be solved uniquely with one value of c_2. There is, however, a constraint $s_1 \gg \lambda$. Taking $\lambda = 0.1 s_1$ would imply $A_0 = 6.1$ Ci. Such activity of tritium could be generated by 120 equipped weapons, 50 mCi each. It is thus an entirely reasonable scenario, however it is to too high since it would require a complete destruction of 120 weapons and a quantitative tritium capture as well as a conversion to HTO. Taking λ larger would invalidate experimental conditions for this source. Taking λ even smaller would further overestimate this tritium source. Therefore, such a mechanism alone was not sufficient and another tritium source must have been present, which originated from the airplanes.

7. Conclusions

34 Ci of tritium were released from the emergency tritium RL signs onboard the two Boeing 767s, on impact with the Twin Towers at the WTC. The measurements and modeling are consistent with a prompt creation of HTO in the jet-fuel explosion and fire, deposition of a small fraction of HTO at ground zero, and water-flow controlled removal from the site. The modeling implies that the contribution from the aircraft alone would yield the HTO deposition fraction of 2.5%. This value is too high by a comparison with other incidents involving fire and tritium. Therefore, the source term from the airplanes alone is too small to explain the measured concentrations, and another missing source is needed.

There is evidence that weapons belonging to federal and law-enforcement agencies were present and destroyed at the WTC. Such weapons contain tritium sights by design. The exact activity of tritium from the weapons was not determined. The data and modeling are consistent with the tritium source from the weapon sights (plus possibly tritium watches) in the debris, from which tritium was slowly released in the lingering fires, followed by an oxidation and removal with the water flow. Our modeling suggests that such a scenario would require a minimum of 120 equipped weapons destroyed and a quantitative capturing of tritium, which is too high, since many weapons were found with only minor damage and tritium sights are shielded in a metal. Therefore, such a mechanism alone is not sufficient to account for the measured HTO concentrations. This indicates that the weapons/watches are consistent with the missing source, which would have complemented the airplane source.

Acknowledgments

The following individuals are highly acknowledged for their substantial contributions to the overall success of this work: D. McChesney, EPA Region 2, for collecting group-1 samples; V. Murray and R. Pirritano, New York City Department of Environmental Protection, for collecting group-2 samples; E. Regilski and C. Schwenker, New York State Department of Health, for their help in the sample analysis and measurements; D. Dischino, Bristol-Myers Squibb Corporation, for scientific discussions on tritium; F. Lombardi, Port Authority of New York and New Jersey, for information about emergency EXIT signs at the WTC, as well as estimation of the total water flow rate at the WTC; T. Headley, Fire Department of New York, for estimation of water flow rate due to fire fighting at the WTC; J. Cashdollar and N. Sabatini, Federal Aviation Administration, for the estimation of tritium activity onboard the aircraft; J. Gaver, US Department of Energy, and C. Skinner, Princeton Physics Plasma Laboratory, for their expertise on tritium chemistry at high temperatures; R. Pause and A. Verschoor, New York State Department of Health, for critical reading of the manuscript.

References

[≈]**Disclaimer:** The statements and conclusions in this paper are those of the authors and not necessarily those of the New York State Department of Health State, State University of New York, Lawrence Berkeley National Laboratory, Lawrence Livermore National Laboratory, or Westinghouse Savannah River Company. The mention of commercial vendors, products, and news media in connection with the work reported herein is not to be construed as either actual or implied endorsement of such vendors, products, or media.

[&]**Corresponding author:** Thomas Semkow, Wadsworth Center, New York State Department of Health, Empire State Plaza, Albany, NY 12201-0509. Email: semkow@wadsworth.org.

1. Miller, J. Digging for gold, guns and drugs. *American Broadcasting Corporation*, abcnews.go.com, October 3, 2001.
2. Cummings, S. WTC complex: 6WTC, 7WTC. *Patriot Resource*, www.patriotresource.com, 2002.
3. Overbye, D. Engineers tackle havoc beneath Trade Center. *New York Times*, www.nytimes.com, September 18, 2001.
4. Lombardi, F.J. Response at ground zero - role of engineers. *National Engineers Week 2002*, Fishkill, NY, February 19, 2002.

5. Associated Press. Engineers study WTC collapse. *New York Times*, www.nytimes.com, April 30, 2002; WTC attacks cost NY over $33 billion, *ibid.*, November 13, 2002.
6. Ashley, S. When the Twin Towers fell. *Scientific American*, www.sciam.com, October 9, 2001.
7. Lipton, E.; Glanz, J. First Tower to fall was hit at higher speed, study finds. *New York Times*, www.nytimes.com, February 23, 2002.
8. *World Trade Center building performance study: data collection, preliminary observations, and recommendations.* McAllister, T., Ed.; Report FEMA 403; Federal Emergency Management Agency: Washington, DC, 2002.
9. Slobin, S. W.T.C.: Clearing the debris. *New York Times*, www.nytimes.com, October 13, 2001.
10. Glanz, J.; Lipton, E. Burning diesel is cited in fall of 3rd tower. *New York Times*, www.nytimes.com, March 2, 2002.
11. Lipton, E.; Glanz, J. A nation challenged: forensics; DNA science pushed to the limit in identifying the dead of Sept. 11. *New York Times*, www.nytimes.com, April 22, 2002.
12. *List of officers missing or killed in the line of duty during the attacks in New York;* International Union of Police Associations, www.iupa.org, 2001.
13. Okada, S.; Momoshima, N. *Health Phys.* **1993**, *65*, 595-609.
14. Hill, R.L.; Johnson, J.R. *Health Phys.* **1993**, *65*, 628-647.
15. Traub, R.J.; Jensen, G.A. *Tritium radioluminescent devices. Health and safety manual;* Report PNL-10620 (UC-610); Pacific Northwest National Laboratory: Richland, WA, 1995.
16. Bennett, B.G. *Health Phys.* **2002**, *82*, 644-655.
17. *Environmental radiation data;* Report 81; EPA-402-R-97-004; US Environmental Protection Agency: Washington, DC, 1997.
18. *Maximum contaminant levels for radionuclides;* US Code of Federal Regulations; 40 CFR Part 141.66.
19. *Local climatological data;* National Oceanic and Atmospheric Administration; US Department of Commerce: Washington, DC, September 2001.
20. Lulu, B. Luminous watch hands. *Power Reserve Inc.*, www.timezone.com, November 19, 1998.
21. Niemeyer, R.G. *Tritium loss from tritium self-luminous aircraft exit signs;* Report ORNL-TM-2539; Oak Ridge National Laboratory: Oak Ridge, TN, 1969.
22. Wermer, J.R. *Assessment of commercial self-luminous exit signs containing tritium: Final Report - Task 1;* Report WSRC-TR-0311; Westinghouse Savannah River Company: Aiken, SC, 1995.

23. *Classification of Radioactive self-luminous light sources;* American National Standard ANS/HPS N43.4; American National Standards Institute/Health Physics Society: McLean, VA, 2000.
24. *Luminous safety devices for use in aircraft: Requirements for license to manufacture, assemble, repair or initially transfer;* US Code of Federal Regulations; 10 CFR Part 32.53.
25. *Trijicon Inc.*, www.trijicon.com, 2002.
26. *Innovative Weaponry Inc.*, personal communication, March 18, 2002.
27. Kinder, K. *Kimber of America Mfg. Inc.*, personal communication, March 19, 2002.
28. *Certain items containing byproduct material;* US Code of Federal Regulations; 10 CFR Part 30.15.
29. *License applications for certain items containing byproduct material;* US Code of Federal Regulations; 10 CFR Part 32, RIN 3150-AF76.
30. *Radioluminescence for time measurement instruments – specifications;* International Standard 3157; International Organization for Standardization: Geneva, Switzerland, 1991.
31. *Watch, wrist: general purpose;* US Military Specification MIL-W-46374F; Naval Publications and Forms Center: Philadelphia, PA, 1991.
32. Benziger, J. *Mb-Microtec AG*, personal communication, March 19, 2002.
33. Zeller, A. *Rc Tritec AG*, personal communication, March 18, 2002.
34. Lombardi, F.J. *Port Authority of New York and New Jersey*, personal communication, December 10, 2001.
35. Sabatini, N.A. *Federal Aviation Administration*, US Department of Transportation, personal communication, March 26, 2002.
36. Cashdollar, J. *Federal Aviation Administration*, US Department of Transportation, personal communication, April 23, 2002.
37. Pan, P.Y.; Rigdon, L.D. *Tritium oxidation in atmospheric environment;* Report LA-UR-96-2953; Los Alamos National Laboratory: Los Alamos, NM, 1996.
38. Gaver, J.M. *Savannah River Site*, US Department of Energy, personal communication, January 30, 2002.
39. Skinner, C.H.; Gentile, C.A.; Carpe, A.; Guttadora, G.; Langish, S.; Young, K.M.; Shu, W.M.; Nakamura, H. *J. Nucl. Mater.* **2002**, *301*, 98-107.
40. Jensen, G.A.; Martin, J.B. *Investigation of fire at Council, Alaska: A release of approximately 3000 curies of tritium;* Report PNL-6523; Pacific Northwest National Laboratory: Richland, WA, 1988.
41. *Health physics sampling results, C-124 incident;* Report AFR 11-30; Bio-environmental Engineering Branch, US Air Force Hospital Wright-Patterson: Wright-Patterson Air Force Base, OH, 1965.
42. Duenes, S. Hot spots linger at ground zero. *New York Times*, www.nytimes.com, October 2, 2001.

43. Headley, T. *Fire Department of New York*, personal communication, February 12, 2002.
44. Post, N.M. Engineers plan assessment and new support for WTC foundation perimeter walls. *Engineering News Record*, www.enr.com, September 17, 2001.
45. Cho, A.; Post, N.M. Crews shore damaged subways. *Engineering News Record*, www.enr.com, October 8, 2001.
46. Post, N.M. WTC 'Bathtub' stabilization begins. *Engineering News Record*, www.enr.com, October 29, 2001.
47. Lombardi, F.J. *Port Authority of New York and New Jersey*, personal communication, January 28, 2002.
48. World Trade Center rubble holds buried secrets. *WPVI*, abclocal.go.com, Philadelphia, PA, October 5, 2001.
49. Gardiner, S.; Hurtado, P. Billions in stocks, bonds found in WTC rubble. *Seattle Times*, seattletimes.nwsource.com, October 6, 2001.
50. Koppel, T. Ground zero. *American Broadcasting Corporation*, Nightline, abcnews.go.com, March 11, 2002.
51. NYNEX Directory for New York Metro Area. *NYNEX*, April, 1995 – March, 1996.
52. Ono, F.; Tanaka, S.; Yamawaki, M. *Fusion Tech.* **1992**, *21*, 827-832.
53. Dickson, R.S.; Miller, J.M. *Fusion Tech.* **1992**, *21*, 850-855.
54. Antoniazzi, A.B.; Shmayda, W.T.; Surette, R.A. *Fusion Tech.* **1992**, *21*, 867-871.
55. Hicks, T.W.; Wilmot, R.D.; Bennett, D.G. *Tritium in Scottish landfill sites;* Report 014-2; Galson Science: Rutland, UK, 2000.
56. Post, N.M. Half of WTC 'Bathtub' basement damaged by Twin Towers' fall. *Engineering News Record*, www.enr.com, October 8, 2001.
57. Margeneau, H.; Murphy, G.M. *The mathematics of physics and chemistry;* Krieger Publishing Co.: Huntington, NY, 1976.

Chapter 16

Tritium-Helium-3 Age-Dating of Groundwater in the Livermore Valley of California

Gail F. Eaton, G. B. Hudson, and J. E. Moran

Lawrence Livermore National Laboratory, 7000 East Avenue, L–231, Livermore, CA 94550

Tritium (^{3}H) is an ideal tracer for water since it is part of the water molecule. Its 12-year half-life is well suited for tracking the movement of young groundwater. Traditional methods for measuring tritium employ sensitive decay counting often preceded by isotopic enrichment of tritium using electrolysis. The introduction of ^{3}He mass spectrometry provides both a ^{3}H measurement technique and a radiometric dating technique for analyzing young groundwater. By measuring both the parent ^{3}H and daughter product ^{3}He, we can estimate the age of groundwater; that is, the time since it was isolated from the atmosphere. These ^{3}H-He groundwater ages are useful in constraining groundwater flow paths, flow rates, and in identifying recharge areas. We present examples of these methods for groundwater samples from the Livermore Valley in California.

© 2004 American Chemical Society

Introduction: Helium In-Growth Groundwater Age-Dating Technique

Tritium (^{3}H) is a naturally occurring, though very low abundance, radioactive isotope of hydrogen. The average concentration of ^{3}H in surface waters today ranges from between about 1 to 10 tritium atoms per 10^{18} total hydrogen atoms. Tritium has a half-life of 12.32 years (*1*). Several units are commonly used to describe tritium abundances. The Tritium Unit (TU) corresponds to 1 atom of ^{3}H per 10^{18} hydrogen atoms, and represents an isotope ratio as opposed to a concentration. Concentrations are commonly given in activity units such as pCi/L (2.09 × 10^4 atoms ^{3}H/g) or Bq/L (1 Bq = 27.0 pCi). Tritium is produced naturally in the stratosphere by spallation reactions involving cosmic rays and atoms in the atmosphere (e.g., N, O, Ar), and enters the hydrologic cycle through meteoric precipitation. As a result of atmospheric nuclear weapons testing between 1945 and 1963, a relatively large amount of tritium (^{3}H) was introduced into the hydrologic cycle, and is referred to as the "bomb-pulse." This bomb-pulse of tritium, which peaked in 1963, is several orders of magnitude above the background concentration of naturally produced tritium. Since then, the ^{3}H concentration in precipitation has steadily decreased due to radioactive decay and mixing with tritium depleted ocean water. In addition to nuclear testing, nuclear reactors also produce and release tritium to the environment, but only in very small amounts. Tritium is also widely disseminated in luminous displays (10 Ci of ^{3}H in exit signs) and wristwatches (0.3 Ci of ^{3}H).

The concentration of tritium in groundwater is affected both by radioactive decay and mixing. The presence of tritium in groundwater is an excellent indicator that some portion of the water recharged into subsurface aquifers less than ~50 years ago. However, more precise statements of age are difficult to make due to large-scale mixing and dispersion of tritium in the subsurface. However, if we measure the daughter product of tritium decay, helium-3 (^{3}He), we can greatly improve our ability to use ^{3}H for age-dating groundwater (*2, 3*). Helium-3 is a stable noble gas isotope that is very rare in nature: approximately 7 parts per trillion (ppt) in air, and about 5 × 10^{-17} mole ^{3}He/mole H$_2$O. Since ^{3}He is not chemically active, or sorbed onto aquifer materials, it remains in solution after ^{3}H decay and moves nearly identically to the water parcel in which it was born. Thus, for groundwater, the sum of the ^{3}H and ^{3}He concentrations is a constant in time and equals the initial ^{3}H concentration. Knowing both the current and initial ^{3}H concentration, we can write the equation:

$$\exp(-ln2 \times T/12.32) = {}^3H/({}^3H + {}^3He_{tritium}) \qquad (1)$$

where T (in years) is the time since the water parcel was isolated from the atmosphere and $^3\text{He}_{tritium}$ began to accumulate. The subscript "tritium" on $^3\text{He}_{tritium}$ refers to ^3He from ^3H decay (tritiogenic ^3He), which is generally only a fraction of the ^3He present in the sample. This equation can be rewritten as:

$$T(\text{yrs}) = 17.8 \times ln(1 + {}^3\text{He}_{tritium}/{}^3\text{H}) \qquad (2)$$

It is important to note that since ground water samples are often a mixture of water molecules with an age distribution that can span a wide range, the reported groundwater age represents a "mean age" for the sample.

Since $^3\text{He}_{tritium}$ is typically a very small fraction of the total ^3He present in the sample, we use the other air-derived noble gases present in the sample (^4He, Ne, Ar, Kr, and Xe) to calculate and then subtract the air-derived ^3He from that produced solely from tritium decay (*4, 5*). Under most circumstances, the air-derived ^3He can be determined to about 1%, so if $^3\text{He}_{tritium}$ is 20% of the total ^3He, then we can determine $^3\text{He}_{tritium}$ with an uncertainty of 5%. This uncertainty associated with the subtraction of air-derived ^3He corresponds to a 1-2 year uncertainty in the measured age.

Laboratory Procedures

Each ^3H-^3He age determination requires two measurements: one for the concentration of tritium, and the other for the concentration of the dissolved noble gases in a groundwater sample. Previously, tritium measurements were made using a small water sample (30 mL), which required a minimum 30-day accumulation period for ^3He in-growth. However, with increased demand for groundwater age dates, it became necessary to increase sample through-put while maintaining accurate, high-quality measurements. Several experiments were conducted where varying sample sizes and accumulation times were tested. The procedure outlined below was adopted as a result of these experiments, which concluded that 500 mL of water would accumulate enough ^3He in 14-20 days to make an accurate determination of ^3H concentration in groundwater samples.

Tritium Procedure

To make tritium concentration measurements, a 1-liter water sample is collected in a Pyrex media bottle. Then, 500 grams from this water sample is loaded into a 1 liter, stainless-steel bottle, and attached to a gas-handling manifold (Figure 1). Since the tritium concentration is determined by measuring the rate at which ^{3}He grows into the water sample, all ^{3}He must be removed from the sample prior to accumulation of tritiogenic ^{3}He (6). This degassing process involves several cycles of heating, cooling, and pumping. The samples are cooled with ice to reduce the water vapor pressure and headspace gases are evacuated (Figure 2). Samples are then heated with the valves closed to re-equilibrate the water and the headspace void. Samples are once again chilled with ice, and headspace gases are again pumped away. This cycle is repeated four times, and with each cycle approximately 98% of the He is removed so that after the final step virtually no ^{3}He remains (<100 atoms) in the sample. Approximately 1 gram of water is lost in this process.

Figure 1. Photo showing a 1 liter stainless steel vessel containing 500 grams of water for helium accumulation.

A sample with 1 Bq/L of ^{3}He (27 pCi/L) produces about 4×10^5 atoms (2×10^{-18} gram) of ^{3}He in ten days. Therefore, after an accumulation period of 14-20 days, there is enough ^{3}He present in the bottle from ^{3}H decay to be measured by mass spectrometry. Prior to mass spectrometer analysis, the samples are prepared by first heating the bottles (to ensure that the sample and headspace are in equilibrium), then freezing the water inside the stainless steel bottles to 200K (dry-ice temperatures). Once the water is frozen, the headspace gases are expanded into a gas-handling manifold. Residual reactive gases (N_2,

CO, H_2) are removed using a hot (600K) metal alloy (primarily Ti) getter. Then the remaining noble gases are collected, with the Ar, Kr, and Xe being sorbed onto an activated charcoal trap at a temperature of 77K, followed by the sorption of He and Ne onto another activated charcoal trap at 15K. The temperature of the trap containing the He and Ne is raised to 35K, and only the He is released into a VG5400 mass spectrometer, where the isotopes ^{3}He and ^{4}He are measured. The ^{4}He measurement is used to determine that no atmospheric He is present (^{3}He/^{4}He = 1.38×10^{-6} in air) in the sample. During these analyses, the mass spectrometer is operated in static mode (vacuum pumps closed), and has sufficient mass resolution to completely separate the hydrogen peaks ($^1H^2H^+$ and $^1H^{3+}$) at mass 3 from ^{3}He. Ions of ^{3}He are detected with a 17-stage electron multiplier (background of 1 count per 1000 seconds). The overall sensitivity of the mass spectrometer corresponds to 1 ion detected per second per 10^5 helium-3 atoms present in the sample. The total background of the system is approximately 5000 atoms of ^{3}He (0.05 counts/second). Data for ^{3}He is accumulated for 1000 seconds. The ^{3}He in-growth procedure is calibrated using samples with known amounts of tritium. Tritium detection limits are a function of both sample size and accumulation time, but are typically around 0.25 pCi/L for our standard accumulation periods (Figure 3).

Figure 2. Samples being chilled with ice just prior to degassing. The samples go through four heating and chilling cycles before ^{3}He accumulation begins.

Dissolved Gas Procedure

Sampling for dissolved gases requires a method that retains the gases in the sample and prevents atmospheric exposure. We use soft copper tubing and metal pinch clamps that provide a gas-tight seal for a 10 gram sample of water. The sample is collected by allowing water to flow through the copper tube (from the groundwater source – usually a down-hole pump), gently tapping the tube to remove any air bubbles, and then sealing clamps.

In the laboratory, each sample is attached to a 250 cm^3 receiving volume, which is attached to a gas-handling manifold (Figure 4). The pinch clamp closest to the receiving volume is removed and the water expands into the volume. The receiving volume is heated to degas the sample, then frozen to 200K (dry-ice temperatures), freezing the water completely to separate the gas from the water. The sample gas is then released from the receiving volume, and known amounts of ^{22}Ne, ^{86}Kr and ^{136}Xe are introduced to the sample gas as internal standards.

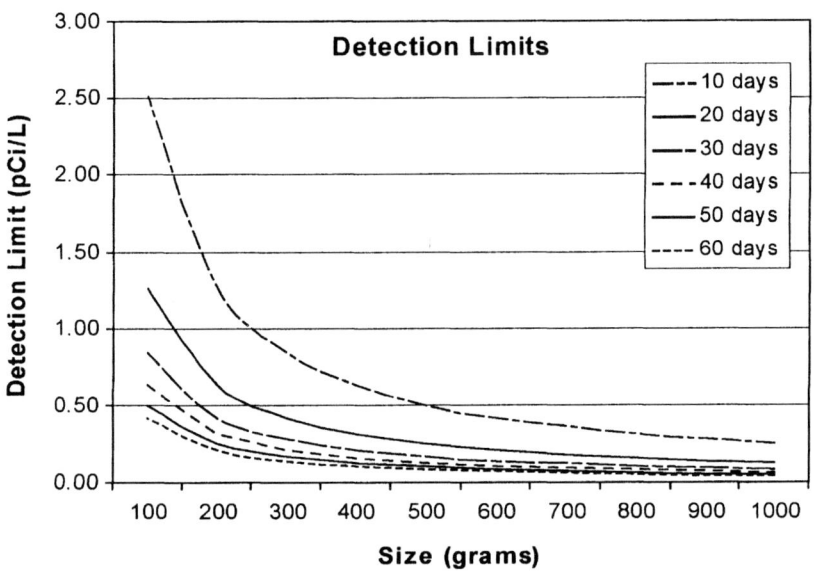

Figure 3. Plot showing detection limits as a function of sample size and accumulation time. For a sample size of 500 grams of water with a 20-day accumulation time, the tritium detection limit would be 0.25 pCi/L.

Gases are purified and separated using cryogenic sorption as described in the tritium procedure. In order to measure the Ne concentration and determine the He aliquot size needed to make the helium isotope measurements, a small portion of the He and Ne are analyzed using a quadrupole mass spectrometer.

Then, an appropriate amount of He is analyzed using the VG5400 mass spectrometer previously described in the tritium procedure section. The separated Ar fraction is measured by sensitive manometry, while the Kr and Xe are measured using the quadrupole mass spectrometer. The ^{3}He/^{4}He ratio is measured to an accuracy of approximately 0.75%, and the Ne, Ar, Kr and Xe concentrations are measured to an accuracy of about 2%. The system is calibrated using samples of air and samples of water in equilibrium with air at known temperature and pressure.

Figure 4. Copper tube samples for analyses of dissolved noble gases in groundwater samples. The receiving volume in contained in dewars of dry-ice.

Application of the ^{3}H-^{3}He dating method to Groundwater Basins

Age dating of groundwater is a useful tool for understanding groundwater recharge and rate of flow in groundwater basins, and has been applied in several studies of basin-wide flow and transport (*4, 7, 8, 9*). While individual ^{3}H-^{3}He ages are interesting, the real power of the method becomes apparent when we are able to make many age determinations within a single flow-field. In the simplest case, the spatial derivative of the age-field, as determined by a large number of measurements, results in a clear determination of the flow field. However, our ability to be successful in determining a flow-field depends on the availability of samples. The mostly widely available source of groundwater samples comes from municipal water supply wells. This type of groundwater well introduces one significant complication into the interpretation of the data: high-volume wells generally produce water from a significant range of depths beneath the surface. This means that any sample collected will represent a mixture of groundwaters with diverse ages, which to some extent integrate across the depth of the aquifer. Thus when we interpret the ^{3}H-^{3}He ages, we must explicitly consider this age distribution.

Livermore Valley Groundwater Basin

The Livermore Valley is a highly populated region located approximately 45 miles east of San Francisco. Tritium age dating of groundwater samples collected from wells in the Livermore Valley was completed in 2001 using the ^{3}H-^{3}He in-growth method. Mean groundwater ages vary from 9 years to greater than 50 years for Livermore Valley Basin wells. The lack of a clear spatial pattern in the mean groundwater ages determined for water samples from Livermore Valley basin wells indicates that recharge is distributed over a large area (Figure 5).

In this instance, the flow field determined from mean ages from wells in the Livermore Valley shows areas of more rapid recharge (younger aged water in the eastern side of the valley) and areas where the aquifers are extensively confined (older water in the western side of the basin). The apparent mean age reported for a well is actually the mean age of the water that still has detectable tritium. There may also be a component of water recharged before about 1955 that no longer contains detectable tritium. Additional information about the distribution that results from the mixture of ages present in water drawn from a well can be determined through closer examination of the relationship between initial tritium (tritium concentration at the time of recharge) and the mean age. The concentration of tritium at the time of recharge is fairly well constrained from measurements of tritium in precipitation at several sites in North America (Figure 6). A coarse estimate of the fraction of pre-modern water that is drawn from a well can be derived from the difference between the measured tritium (decay-corrected according to the mean age of the water sample) and the 'initial' tritium that should have been recorded at the time of recharge. Curves representing additions of equal fractions of pre-modern water are shown below the initial tritium curve in Figure 7. A groundwater sample for which the measured age gives a decay-corrected tritium value that falls on or near the curve, is not significantly diluted with a component of 'older', tritium-free water. Samples that fall below the 'initial tritium' curve contain a fraction of water that recharged before 1955 ('pre-modern').

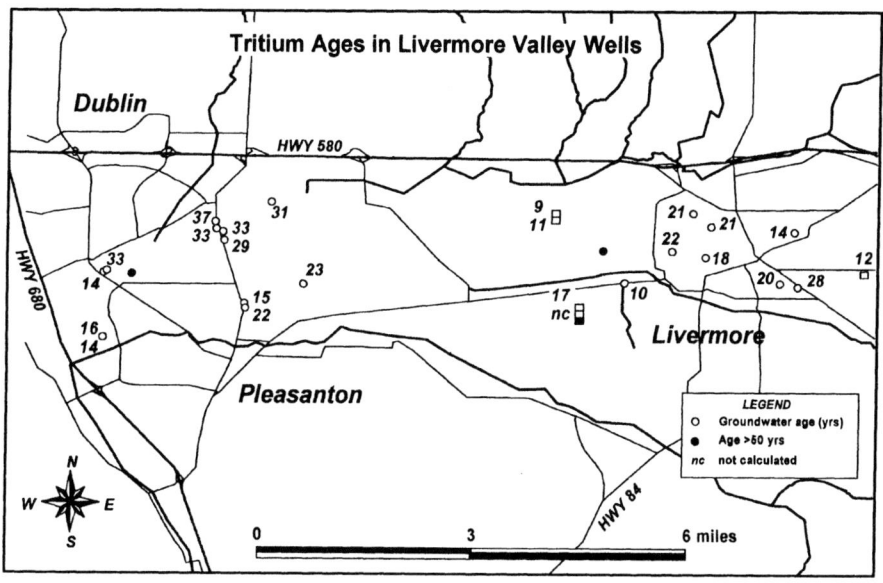

Figure 5. Map showing tritium ages for groundwater in the Livermore Valley, California.

Many of the Livermore Valley samples fall well below the initial tritium curve, indicating that a large component of older water reaches these wells. Several of the wells from the western side of the basin produce water that has an estimated fraction of greater than 90% pre-modern. Hence the mean ^{3}H-^{3}He age that is reported actually represents a broad age distribution with a large proportion of pre-modern groundwater. A few samples shown in Figure 7 plot above the initial tritium curve. These waters likely have a small component of tritium from a local, anthropogenic source. In the case of Livermore Valley groundwater, the likely source is the low level emissions to the atmosphere from Lawrence Livermore National Laboratory (*10*).

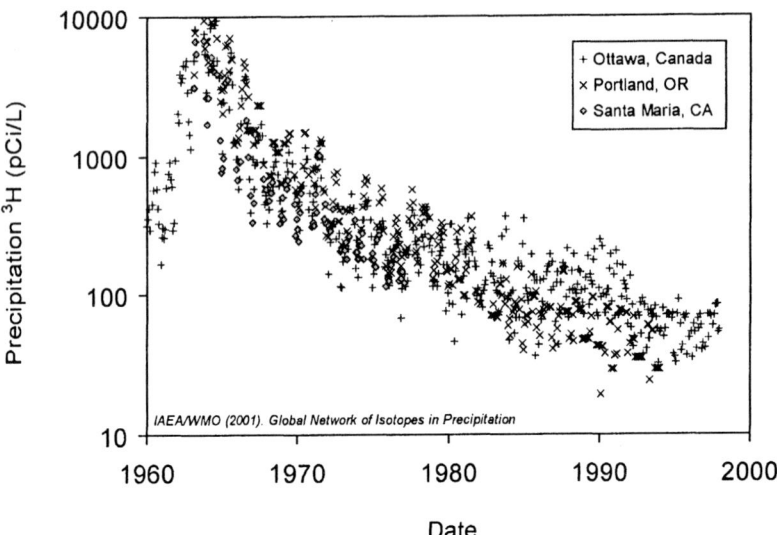

Figure 6. Plot of initial tritium versus mean age. Tritium that was present at the time of recharge is well known from measurements of tritium in precipitation at three sites in North America.

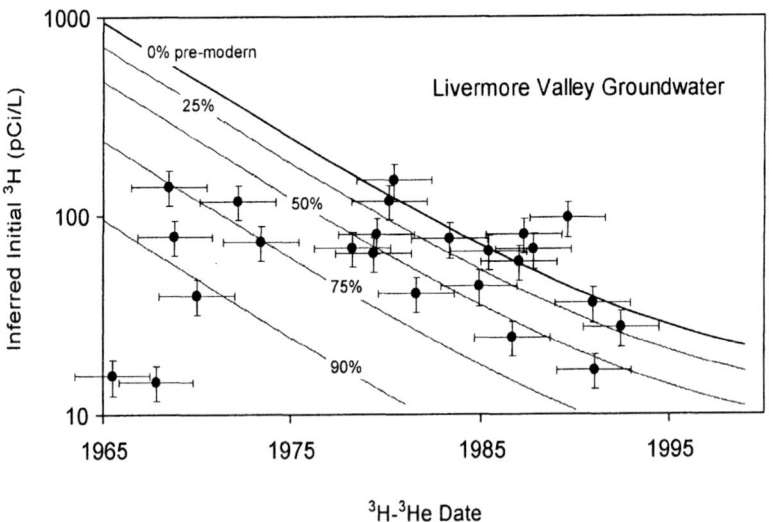

Figure 7. Plot of inferred initial tritium versus mean age for Livermore Valley wells. Samples that fall below the 'initial tritium' curve contain a fraction of water that recharged before 1955.

Conclusions

The use of larger sized accumulation vessels proved to be highly effective in producing quality tritium mean ages over shorter periods of time. This increased sample throughput has allowed more detailed studies of groundwater basins, their flow-fields, and their susceptibility to contamination.

The ^{3}H-^{3}He in-growth method is extremely effective in determining groundwater flow-path in the Livermore Valley of California, and showing areas where aquifers are fairly confined since they contain a larger fraction of pre-modern water.

References

1. Lucas, L. L.; Unterweger, M. P. *Journal of Research of the National Institute of Standards and Technology* **2000**, *105*, 541-549.
2. Schlosser, P.; Stute, M.; Dorr, H.; Sonntag, C.; Munnich, O. *Earth, Planetary Science Letters* **1988**, *89*, 353-362.
3. Poreda, R. J.; Cerling, T. E.; Solomon, D. K. *Journal of Hydrology* **1988**, *103*, 1-9.
4. Ekwurzel, B.; Schlosser, P.; Smethie, W. M.; Plummer, L. N.; Busenberg, E.; Michel, R. L.; Weppernig, R; Stute, M. *Water Resources Research* **1994**, *30*, 1693-1708.
5. Aeschbach-Hertig, W.; Peeters, F.; Beyerle, U.; Kipfer, R. *Water Resources Research* **1999**, *35*, 2779-2792.
6. Clarke, W. B.; Jenkins, W. B.; Top, Z. *International Application of Radiation Isotopes* **1976**, *27*, 515-527.
7. Hudson, G. B.; Moran, J. E.; Eaton, G. F. *Lawrence Livermore National Laboratory Internal Report* **2002**, UCRL-AR-151447, p 59.
8. Moran, J. E.; Hudson, G. B.; Eaton, G. F.; Leif, R. *Lawrence Livermore National Laboratory Internal Report* **2002**, UCRL-AR-148831, p 31.
9. Szabo, Z.; Rice, D. E.; Plummer, L. N.; Busenberg, E.; Drenkard, S.; Schlosser, P. *Water Resources Research* **1996**, *32*, 1023-1038.
10. Gallegos, G. M., et al. *Lawrence Livermore National Laboratory Internal Report* **2001**, UCRL-50027-01, p 375.

The work was performed under the auspices of the U.S. Department of Energy by University of California Lawrence Livermore National Laboratory under contract No. W-7405-Eng-48.

Chapter 17

Automation of Radiochemical Analysis: From Groundwater Monitoring to Nuclear Waste Analysis

Oleg B. Egorov, Matthew J. O'Hara, R. Shane Addleman, and Jay W. Grate

Pacific Northwest National Laboratory, 902 Battelle Boulevard, Richland, WA 99352

In the current analytical practice determination of nongamma-emitting radionuclides relies on manual radiochemical analysis methods performed in centralized laboratories. Manual methods are not well suited to support nuclear waste processing and long-term environmental monitoring applications. Our research is directed at the development of chemical and instrumentation approaches towards automated radiochemical analysis and radionuclide-specific sensing. Our overall approach is based on integrating selective separation chemistries and radiation detection within a single functional sensor or analyzer device. This paper will present examples of selective radiometric sensing of beta- and alpha-emitters in water. Analysis of total ^{99}Tc and actinides in aged nuclear waste samples will be discussed as examples of automated analytical procedures and analyzer instruments.

Introduction

The production of nuclear weapons materials and storage of nuclear wastes at U.S. Department of Energy (DOE) sites has lead to radioactive contamination of soil and groundwater by several routes of intentional and unintentional release. Monitoring of the radionuclide contaminants in various environmental matrixes (including groundwater) represents an important element in support of remediation and long term stewardship activities. Rapid radiochemical analysis methods are also required in support of waste processing and stabilization.

Direct, isotope-specific determination of pure beta- and alpha-emitting radionuclides in liquid samples represents a substantial challenge due to short radiation ranges, rapid energy dispersion, and limited resolution capabilities of radiation detection instrumentation. With the exception of a limited number of radionuclides that can be analyzed directly by high resolution gamma-spectroscopy (e.g., ^{134}Cs, ^{137}Cs, ^{154}Eu, ^{155}Eu, ^{60}Co, ^{54}Mn, ^{125}Sb, and ^{241}Am), non-destructive radiochemical analysis is either exceptionally problematic or not possible.(1)

Radiometric detection of beta emitters (e.g., ^{99}Tc) by liquid scintillation or gas proportional counting has limited isotope discrimination capability due to the continuous nature of beta energy spectrum and limited energy resolution of the detectors. Alpha particles are emitted with characteristic energies, and high-resolution alpha spectroscopy using passivated ion implanted silicon (PIPS) diode detectors is possible. However, a number of important radionuclides have unresolvable energies (e.g., ^{241}Am/^{238}Pu, and ^{237}Np/^{234}U).(2) Chemical separations are required prior to preparation of thin counting sources by electroplating or co-precipitation.

Due to continuing improvements in sensitivity and detection limits, inductively coupled plasma mass spectrometry (ICP MS) is becoming an increasingly important method for the analysis of longer lived radionuclides.(3-6) Compared to classical radiometric detection techniques, ICP MS may potentially offer lower detection limits, less tedious sample preparation, and shorter analysis times. Nevertheless, in many instances, direct ICP MS analysis of liquid samples may be subject to isobaric, molecular, and spectral interferences. Chemical separations are often necessary for unambiguous and reliable quantification of the individual radionuclides. In addition, analyte preconcentration and matrix removal will typically improve sensitivity and detection limits.

Chemical separation steps are required when detection by radiometric or mass spectrometric methods cannot distinguish radioisotopes that may be present simultaneously in the sample. Individual, group, or radionuclide/matrix separations represent an important part of the overall radiochemical analysis scheme. In the case of environmental measurements (e.g., groundwater monitoring), sample preconcentration may be required to achieve low regulatory detection limits.*(7-10)*

In conventional radiochemical analysis, chemical separations are carried out by a variety of classical and chromatographic methods, including precipitation, liquid-liquid extraction, and ion exchange. Sequential combinations of these methods are often employed. As the result, manual radiochemical analysis methods are costly and time consuming. Moreover, in many applications, it is desirable to be able to perform measurements at-site or *in situ* using analyzer or sensor instruments. Radionuclide sensors and analyzers must perform all the functions previously carried out as sequential manual steps in the laboratory, while achieving this rapidly and efficiently in an automated device.

Our overall approach towards automated radionuclide analysis is based on integrating modern selective separation chemistries and radiation detectors within a single functional sensor device or an analyzer instrument.

Radionuclide Selective Sensors for Water Monitoring

Functional Requirements and General Approach

Compared to gamma-emitting radionuclides, nongamma-emitting contaminants present significant fundamental challenges for detection and identification in a field deployable sensor device. Because of low-level detection requirements and the need for isotope-specific detection, radiometric detection represents a detection method of choice. However, beta and especially alpha particles are characterized by short ranges and rapid energy dispersion in condensed media. For example, the ranges in water for beta particles emitted by ^{90}Y (E_{max}=2282 keV), ^{90}Sr (E_{max}= 546keV), and ^{99}Tc (E_{max}=294 keV) are 1.1 cm, 1.8 mm, and 750 µm, respectively. The range of a 5.5 MeV alpha particle emitted by ^{241}Am is only 47 µm in water.*(11)* Consequently, radioactive species are detected only if they are in close proximity (i.e., within the particle range) to a transducing material or device. If spectral information is to be obtained from particles with characteristic energies, such as alpha particles, it is especially important to minimize the path length through condensed media to the detector. Moreover, because of the energy dispersion, the energy resolution of the radioactivity measurement may not be adequate to distinguish individual radioisotopes, even if the charged particle has been emitted with a discrete characteristic energy. In environmental monitoring, analytes must be determined to extremely low detection limits set by regulatory requirements. Preconcentration and separation methods and long signal accumulation or counting times are necessary in order to meet these detection requirements. It is for these reasons that nongamma-emitting radionuclide contaminants are normally analyzed in laboratories using manual procedures. This approach is clearly inapplicable to on-site and especially *in situ* measurements of beta and alpha-emitting radionuclide contaminants.*(12)*

For a radioactivity detector to detect and identify these contaminants in a field deployable device, the species of interest must be spatially localized at the detector surface or within a detector volume of well-defined geometry. For long term *in situ* measurements, the sensor device must preferably work with chemically untreated samples and without generating secondary waste.

Following the definition of a chemical sensor given by Janata et. al *(13)*, radionuclide sensors and radionuclide sensor instruments can be defined as "*devices that provide information on a specific radionuclide content of the sample in real time or nearly real time fashion. They consist of a radioactivity detector and may include a chemically selective layer.*"

We are currently investigating two general sensor concepts for sensitive and selective determination of beta and alpha emitters in water: a) preconcentrating minicolumn sensors and b) chemically selective diode detectors. These sensor configurations are schematically illustrated in Figure 1.

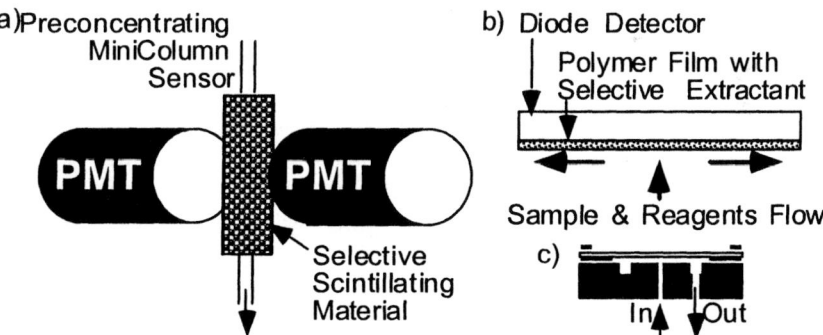

Figure 1. Schematic diagrams of the radionuclide sensor concept a) Preconcentrating minicolumn sensor using scintillation detection, b) Diode detector modified with selective thin film, and c) Schematic of a "fountain" flow cell for diode sensor.

Preconcentrating Minicolumn Sensors

The preconcentrating minicolumn concept for a radionuclide sensor is based on the use of a dual-function column bed designed to incorporate 1) selective separation chemistry to capture and retain the radionuclides of interest; and 2) scintillating properties so that the energy of radioactive decay of the captured species can be converted into a measurable light output. One approach is to use dual function materials that combine selective sorptive and scintillating properties within a single bead. Alternatively, scintillation and chemical selectivity can be attained in a composite sensor column by using a homogeneous mixture of sorbent and scintillator beads. The selectivity is determined mainly by the separation chemistry used to preferentially capture and preconcentrate the analyte of interest. Dual function sensor beads or the mixture of sorbent and scintillator beads are arranged in a minicolumn located between two photomultiplier tubes of a scintillation detection system. Automated fluidics is used to deliver sample and reagent solutions to the sensor column. This scintillating minicolumn sensor configuration for preconcentration and detection is shown schematically in Figure 1 a). This configuration meets all the functional requirements for sensors described in the previous section. The packed bed column format provides for efficient fluidic processing of the sample for preconcentration. When necessary, the sensing material can be regenerated or renewed.*(14, 15)*

Dual Functionality Sensor Materials

Dual functionality extraction chromatographic materials can be made scintillating by incorporating organic scintillator fluors into the bead structure. This can be accomplished by: 1) co-depositing scintillating fluors with the extractant into the bead pores; 2) diffusing the scintillator fluors into the bead structure prior to impregnating with extractant; or 3) introducing scintillator fluors into the bead structure during polymerization.(14,15) Pertechnetate-selective sensor materials were prepared by co-depositing a mixture of liquid anion exchanger tricaprylylmethylammonium chloride (Aliquat™-336), 2,5-diphenyloxazole (PPO) primary scintillator, and 1,4-bis(2-methylstyryl)-benzene (bis-MSB) secondary scintillator in the pores of macroreticular acrylic support beads.(14) Pertechnetate uptake properties of this sensor material are illustrated in Figure 2. Strong uptake at low acid and neutral pH range is favorable for analyte preconcentration from groundwater. The instrumental pulse height spectra of ^{99}Tc obtained using the selective sensor material indicated that luminosity of the sensor material is lower than the liquid scintillator, but the detection efficiency remains sufficiently high (absolute efficiency 56% using

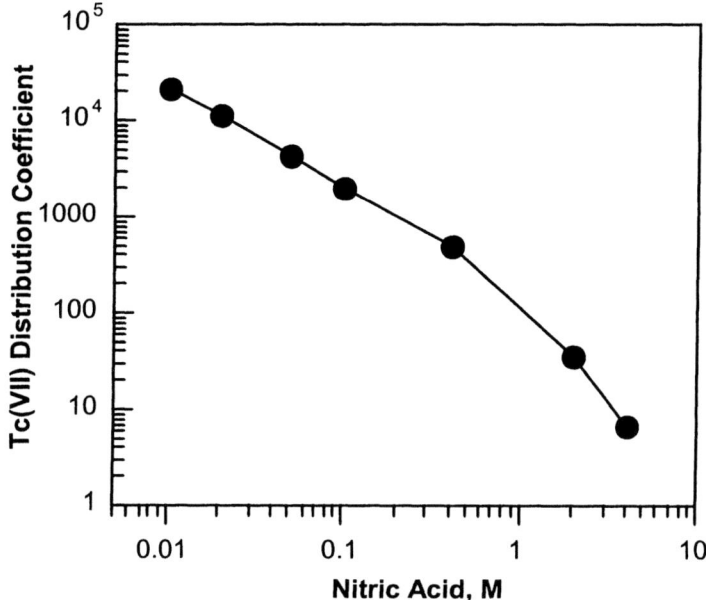

Figure 2. Pertechnetate uptake properties of scintillating sensor material as a function of nitric acid concentration.

static liquid scintillation spectrometer) for practical analytical applications.*(14)* Similar uptake and luminosity (absolute detection efficiency 30-50%) characteristics were observed for the Aliquat-336 based sensor materials prepared by diffusing scintillator fluors into the bead matrix or by incorporating scintillator fluors into the bead during the polymerization process.*(15)*

As shown in Figure 3, injection of an aliquot of ^{99}Tc standard in dilute acid results in the analyte capture and measurable scintillation light output. In this example, the analyte capture was quantitative and the signal persists as the sensor column is washed with additional dilute acid. The absolute detection efficiency for ^{99}Tc using flow through scintillation detection equipped with the sensor cell was 45%. Because the sensor material exhibits high binding affinity towards Tc(VII), large sample volumes can be preconcentrated using a miniature sensor column.

Injection of a radioactive species that does not exhibit affinity to the sensor

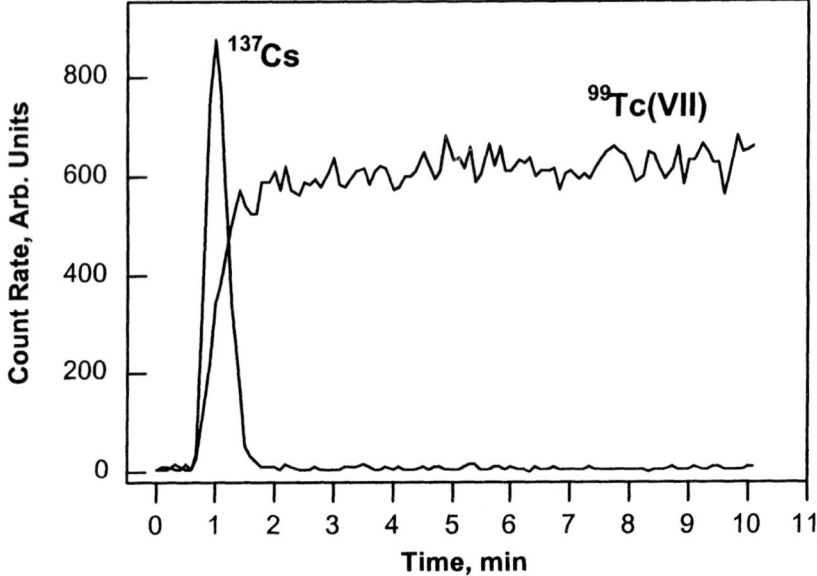

Figure 3. Detector signals illustrating response of the ^{99}Tc-sensor to the injection of sample solution containing target analyte (^{99}Tc(VII)) and unretained species (^{137}Cs).

material results in a transient peak signal as shown in Figure 3 using ^{137}Cs as a model analyte. Unretained interfering species can be removed from the system using a small volume of wash solution. In this manner, the sensing method is selective towards the target analyte.

Uptake data for ^{99}Tc in Figure 2 indicates that Tc elution and sensor regeneration is possible using solution of nitric acid (6mol/L). However, the scintillating properties of the sensor material prepared via co-precipitation of the fluors and extractant was not stable at this acid concentration. Sensor materials prepared via fluor diffusion or immobilization during polymerization exhibited improved stability in acid media. Typically, several sample load/sensor regeneration cycles are possible with these materials without significant loss of detection efficiency. Nevertheless, these materials exhibit tendency towards sporadic chemoluminescence and long term stability over extended periods of operation remains to be problematic.

Composite Bed Sensors

The limited long-term stability of sensor materials based on impregnated fluors and extractants prompted us to explore an alternative approach for developing preconcentrating minicolumn sensors. We determined that scintillation properties and chemical selectivity could be attained in a composite sensor column by creating a uniform mixture of sorbent and non-porous scintillator beads. Using the composite bed approach, dual functionality is attained in a sensor column, rather than in a single bead. Sorbent beads selectively capture the radionuclide of interest and localize it in an intimate proximity to the scintillator bead. Relatively efficient scintillating functionality in a composite bed is feasible because the radiation range is comparable to or greater than the diameter of the sorbent bead. Because of the relatively long range of the beta particles, this approach appears to be particularly suitable for detection of beta-emitters such as ^{99}Tc and ^{90}Sr. Absolute detection efficiencies as high as 34% for ^{99}Tc and 55% for ^{90}Sr were obtained for the composite sensor columns prepared using selective solid-phase extraction sorbents and plastic scintillator beads. These detection efficiencies are comparable to the detection efficiencies obtained using sensor materials based on impregnated scintillator fluors.

The composite bed approach allows the use of solid-phase extraction sorbents that typically can not be readily converted to scintillators by impregnation techniques. For ^{90}Sr sensing, we were able to take advantage of the high uptake characteristics of the silica gel based Sr-selective Analig™ material (IBC Advanced Technologies) in a composite bed configuration (55% absolute detection efficiency for the 1:1 sorbent/scintillator ratio). Pertechnetate-selective composite bed sensors were obtained using strongly basic anion exchange sorbent which exhibits high affinity and selectivity towards pertechnetate in neutral solutions. The detector signal in Figure 4 corresponds to a ^{99}Tc(VII) standard loading experiment using composite bed consisting of a strongly basic anion exchange material (Biorad AGMP 1, particle size 20-50 µm) mixed with plastic scintillator beads (Bicron BC-400, particle size 100-250 µm). The sorbent-to-scintillator weight ratio in the mixed bed was 1:30.[12] Sensor responses indicate that the injected pertechnetate is captured by the sensor column, resulting in a continuous signal which persists as the sensor is washed with additional dilute nitric acid. The count rate corresponded to 34% absolute detection efficiency for the ^{99}Tc(VII) species captured on the sensor bed. A subsequent wash with 2mol/L nitric acid resulted in prompt removal of the analyte without the loss of scintillation efficiency. In this manner, rapid regeneration of the composite bed is possible using small volumes of 2mol/L nitric acid.

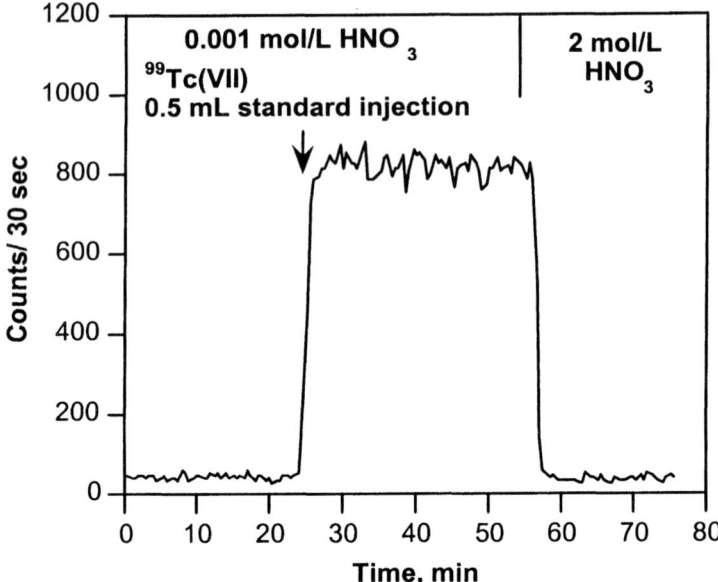

Figure 4. Detector response illustrating loading and regeneration of the Tc(VII) selective composite bed sensor. Solution flow rate is 0.5 mL/min.

Quantification Approaches

Using the column sensor approach, the radionuclide species are selectively captured and concentrated within the radiation detector. These sensor properties form the basis of the analytical measurement based on quantitative analyte capture. In this scenario, the sample volume delivered to the sensor column is selected to be below the chromatographic breakthrough volume of the sensor column. The maximum sample volume, V_{sm}, that can be delivered without the chromatographic breakthrough loss is determined by the column, size, analyte capacity factor, and chromatographic efficiency of the sensor column:

$$V_{sm} = k' V_{fc} \left(1 - 3\sqrt{\frac{H}{L}}\right) \tag{1}$$

where, k´ is the analyte capacity factor, V_{fc} is the free column volume of the sensor column, H is the column plate height and L is column length.

By stopping the flow after sample delivery, the analyte residence time within the sensor system can be increased indefinitely and counting time can be selected as necessary to achieve the required detection limit. Background subtracted or net count rate (in counts per second or cps) obtained during the stopped-flow counting interval, $C_{n,\,cps}$ is related to the sample activity $A_{Bq/mL}$ as follows:

$$C_{n,cps} = VE_e A_{Bq/mL} \qquad (2)$$

where, V is the sample volume and E_e is the effective analytical efficiency (analyte recovery times absolute detection efficiency). The E_e represents a sensor calibration parameter that can be determined from a calibration curve or by using the standard addition technique. In order to enable analysis of consecutive samples, the sensor must be either regenerated or renewed after each measurement. Quantitative analyte capture approach using the standard addition technique for sensor calibration was successfully demonstrated in the analysis of the acidified groundwater samples from the Hanford site. The ^{99}Tc analysis results obtained using the sensor system were in excellent agreement with the established values.*(14)*

In order to facilitate radionuclide selective sensing without the need for sensor regeneration or renewal, we have recently developed a sensing concept based on equilibrating the entire sensor minicolumn with the sample solution. In this scenario, the sample is delivered through the column until the entire sensor column reaches dynamic equilibrium with the sample solution and complete column breakthrough conditions are achieved. Under these dynamic breakthrough conditions no further analyte preconcentration occurs. The analyte concentrations in the sample before and after the sensor column are equal.

Chromatographic theory indicates that under dynamic equilibrium conditions, analyte concentration on the sensor column is proportional to the analyte concentration in the sample. The net count rate observed after sensor equilibration, $C_{e,cps}$, can be related to the sample activity, $A_{Bq/mL}$ as follows:

$$C_{e,cps} = DV_s E_e A_{Bq/mL} = E_o A_{Bq/mL} \qquad (3)$$

where D is the distribution coefficient, V_s is the volume of the stationary sorbent phase in the sensor column, E_e is the effective analytical efficiency, and E_o is the effective calibration parameter. Similar to the quantitative capture approach, the effective calibration parameter in the equilibration approach can be established from the calibration curve or by using the standard addition technique.

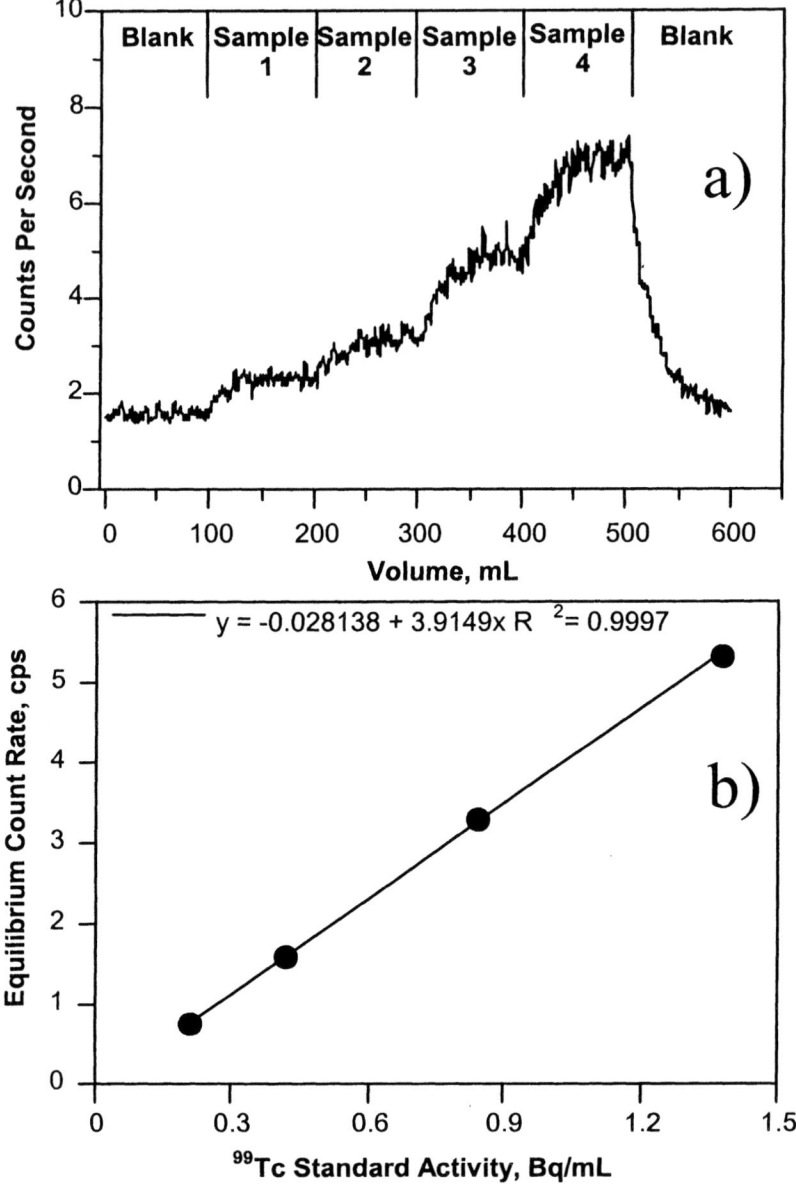

Figure 5. Detector signals (plot a) and calibration curve (plot b) illustrating the equilibration sensing approach for $^{99}Tc(VII)$.

The calibration parameter will remain constant unless sensor degradation or fouling occurs In the equilibration sensing scenario, the sensor column does not require regeneration or renewal, as the signal will change proportionally to the new sample concentration upon equilibration with the next sample. Consequently, the equilibration approach forms the basis for reagentless radionuclide selective sensing, which is particularly well suited for the development of sensor probes for *in situ* long term environmental monitoring.

The equilibration sensing approach is illustrated in Figure 5 a) using ^{99}Tc(VII) as the analyte. In this example, we used an anion exchanger-based composite sensor column and 100-mL aliquots of ^{99}Tc(VII) standards of increasing activities prepared in 0.02 mol/L nitric acid.*(12)* Delivery of approximately 80 mL of sample solution resulted in sensor equilibration. As can be seen in Figure 5 a), delivery of the additional sample solution did not result in a further increase of the count rate. Delivery of samples with higher ^{99}Tc(VII) activity resulted in increased equilibrium count rates. Figure 5 a) also indicates that sensor response returns to the baseline level after equilibration with a Tc-free solution. As shown in Figure 5 b), the net equilibrium count rate gave a linear calibration curve when plotted against sample activity. These results indicate that under equilibrium conditions, the sensor response is proportional to the analyte concentration.

Reagentless ^{99}Tc(VII) Sensing in Hanford Groundwater

The feasibility of ^{99}Tc detection in Hanford groundwater was investigated using two Hanford groundwater samples spiked with ^{99}Tc(VII) at 3.3×10^{-2} Bq/mL and 0.13 Bq/mL levels.*(12)* The composite sensor bed consisted of AGMP-1 anion exchange beads and BC-400 scintillator beads mixed at 1:30 weight ratio. The volume of the sensor bed was 50 µL. Following sensor column equilibration with the unspiked groundwater, two spiked samples were sequentially delivered to the sensor column. The volume of the sample used for sensor equilibration was 1.9 L. Each sample delivery cycle was followed by a 6-hour stopped-flow counting interval. Detector responses corresponding to the groundwater sensing experiments are shown in Figure 6. Delivery of approximately 900 mL of groundwater sample was necessary to achieve sensor equilibration. A large equilibration volume required in the analysis of the unacidified groundwater (pH~8) is due to a very strong uptake of the ^{99}Tc(VII) by the strongly basic anion exchanger used in the sensor column (distribution coefficient ~2.5×10^5 mL/g in Hanford groundwater). Sample activity of 3.3×10^{-2} Bq/mL corresponds to the regulatory drinking water standard level. The magnitude of the sensor signal observed in the analysis of this sample indicates that ^{99}Tc can be reliably detected at levels below regulatory drinking water requirements.

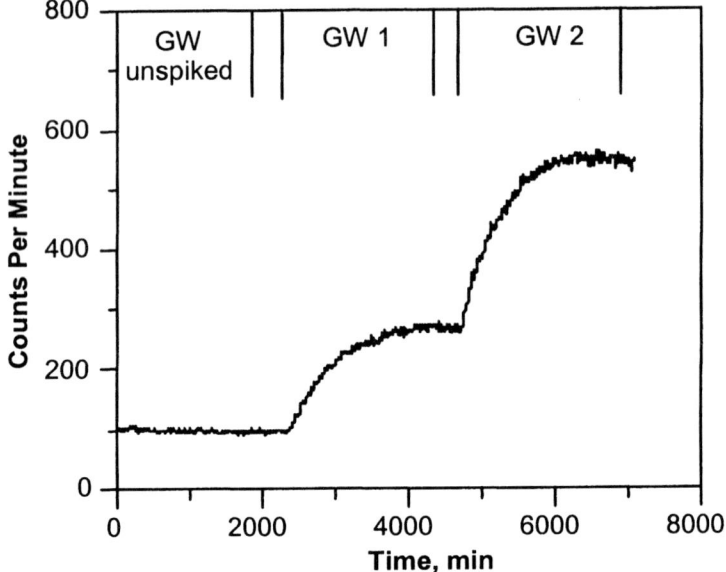

Figure 6. Detector signals corresponding to reagentless equilibration ^{99}Tc sensing in chemically untreated groundwater (GW). Sample activity is 3.3×10^{-2} Bq/mL (GW1) and 0.13 Bq/mL (GW 2). Sample volume is 1.9 L.

This sensor concept has been recently successfully implemented in a prototype of a fully automated ^{99}Tc sensor probe in a configuration compatible with a well casing for down-hole monitoring.

Feasibility of ^{90}Sr Sensing via ^{90}Y Cherenkov Radiation

High-energy beta emitters can be detected and quantified by measuring Cherenkov radiation produced in the adjacent media without the use of scintillators.[16] For high-energy beta emitters such as ^{90}Y (maximum E_β=2.28 MeV), the Cherenkov detection efficiency in water matrix can be as high as 65% using standard scintillation counting equipment. Detection efficiency for medium and low energy beta-emitters is insignificant (e.g. <1% for ^{90}Sr or ^{99}Tc). In this manner, Cherenkov detection is nearly selective for the high-energy beta-emitters, and does not require the incorporation of scintillating fluors into the selective material.

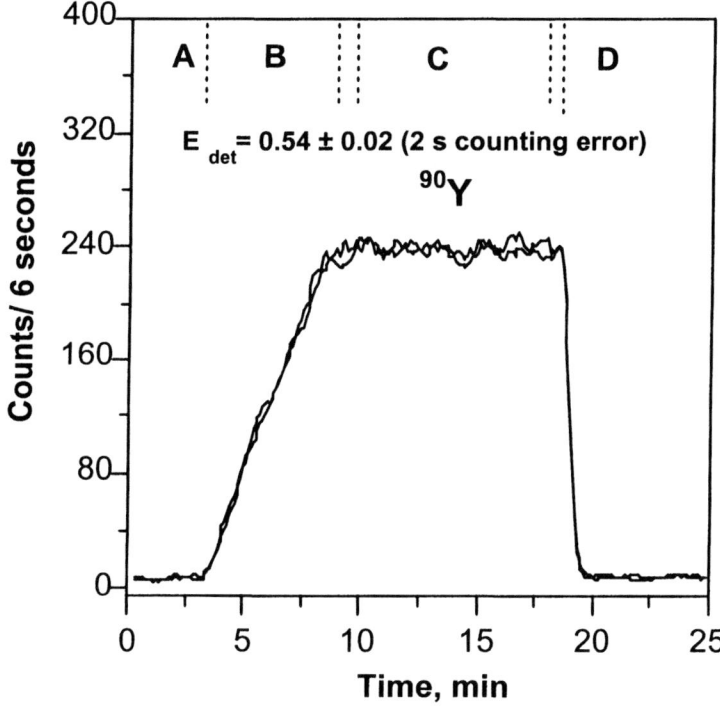

Figure 7. Detector signals illustrating the feasibility of ^{90}Y Cherenkov detection without the use of scintillating fluors; A) background count; B) 10 mL $^{90}Sr/^{90}Y$ standard in 0.01mol/L HCl is loaded on the column; C) sensor wash with 0.01mol/L HCl (40 mL); D) sensor regeneration using 4mol/L HCl.

The Cherenkov detection approach can be used to quantify ^{90}Sr by detecting its daughter ^{90}Y, which exists in radioactive equilibrium with ^{90}Sr. Bis(2-ethylhexyl) phosphoric acid (HDEHP), immobilized on inert polymeric support was used to capture ^{90}Y in a preconcentrating minicolumn sensor. The uptake properties of this material for Y ions was determined in Hanford groundwater, yielding distribution coefficients of 6.6 x 10^6 and 1.5 x 10^6 mL/g for Hanford groundwater acidified to pH 2 and pH 1.3, respectively. Lower uptake values were observed for the unacidified groundwater (distribution coefficient 2.5 x10^5 mL/g) and are attributed to the suppressing effect of the common divalent cations (Ca^{2+}, Mg^{2+}) which are abundant in Hanford groundwater. In general, HDEHP extractant exhibits cation exchange properties and has high affinity towards metal ions in neutral-low acid matrixes. Batch uptake results indicate that high levels of Y preconcentration are possible using small sensor columns.

Yttrium uptake drops rapidly with increasing acid concentration and sensor columns can be regenerated using acid (4-6 mol/L concentration) as the eluent.

The plots in Figure 7 demonstrate ^{90}Sr sensing via ^{90}Y Cherenkov detection using a 589 μL volume sensor column and 10 mL of ^{90}Sr/^{90}Y standard in 0.01M HCl. The absolute detection efficiency was 54±2% for the captured ^{90}Y using Cherenkov radiation on the sorbent column without the use of the scintillators. The sensor column was successfully regenerated using 4mol/L HCl as an eluent solution.

The feasibility of ^{90}Y capture and detection in groundwater was established using acidified (pH1.3) Hanford groundwater spiked with ^{90}Sr/^{90}Y standard. No decrease in detection and loading efficiency was evident after 1 L of groundwater sample had been delivered through the sensor bed. This sensor approach relies on secular equilibrium between ^{90}Sr and ^{90}Y that must exist in a sample prior to analysis.

Chemically Selective Diode Detectors

The second class of sensors for selective radionuclide sensing is based on the use of semiconductor diode detectors operated in contact with a liquid sample.[17, 18] Deposition of a selective sorbent thin film on the surface can collect and concentrate radionuclides in close proximity to the diode surface. This sensor configuration is shown schematically in Figure 1 b). Incorporation of this type of sensor in a "fountain cell" flow cell (Figure 1 c) provides for efficient processing of sample solutions over the planar selective film. Alternatively, the packaged diode can be used for batch contact measurements in cup geometry. Diode detectors offer three key advantages for radionuclide sensing. Diode detectors offer energy resolution that is intrinsically superior to that of scintillation detection. The background noise levels of diode detectors for alpha detection are inherently lower than those for scintillation detectors by orders of magnitude. Third, diodes designed for alpha particle detection provide excellent discrimination against beta/ gamma radiation.

We obtained working actinide-selective thin diode coatings by modifying a PVC polymer with actinide selective organic extractants such as bis(2-ethylhexyl)phosphoric acid (HDEHP) and bis(2-ethylhexyl)methane-diphosphonic acid (DIPEX) used as plasticizers. Initial batch contact uptake studies indicate that selective thin films exhibit high distribution coefficients for the uptake of actinides from groundwater matrixes acidified to pH 2. For example, Am and U distribution coefficients exceed 10^6 mL/g in Hanford groundwater at pH 2. High analyte uptake chemistries are critical to enable sufficient analyte preconcentration in thin-film geometry.

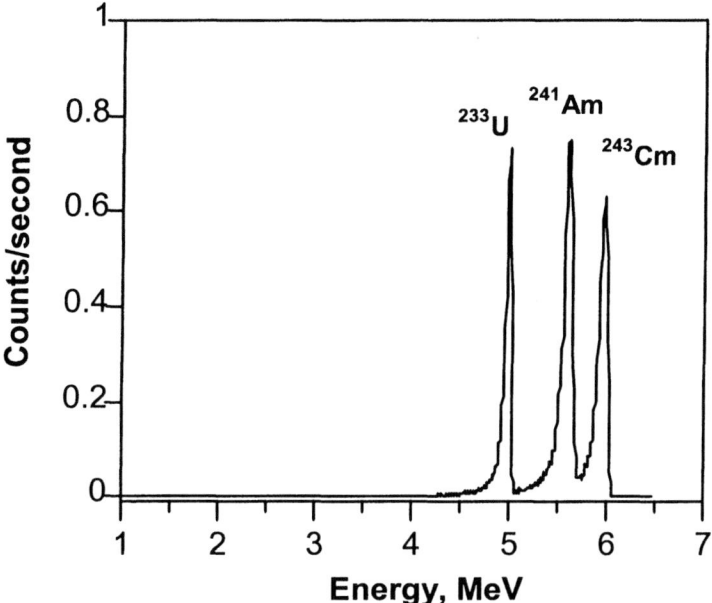

Figure 8. Alpha energy spectra illustrating feasibility of chemically modified diode detectors for in-situ detection of alpha-emitters in solution. Spectra obtained in a static contact regime. Sample volume is 3 mL. High-resolution alpha spectroscopy in solution was demonstrated using sample containing 16.6 Bq/mL of ^{241}Am, ^{243}Cm and ^{233}U. Selective film thickness is 0.5 μm.

Modern PIPS diode detectors were packaged in a flow cell for operation in solution. We were successful in developing diode packaging methods that resulted in minimal noise in the alpha region and reliable diode operation in solution. Diode leakage current and alpha noise were not significantly affected after surface modification and packaging for use in solution. We determined that selective films of reproducible and controlled thicknesses could be readily deposited on the surface of diode detectors using a spray-coating technique. Initial batch contact diode sensor results, shown in Figure 8, indicate that good energy resolution (≤100 keV FWHM) can be obtained for <0.5 μm thick films. However, energy resolution is degraded significantly for >2 μm thick films. Nevertheless, the alpha particle energy resolution for 3-5 μm thick films exceeds practical energy resolution capabilities of scintillation detection. The feasibility of flow-through sensing using a diode packaged in a "fountain cell" is shown in Figure 9 using an ^{241}Am standard as the sample. In this experiment, the diode was coated with a HDEHP-plasticized film of 3 μm thickness.

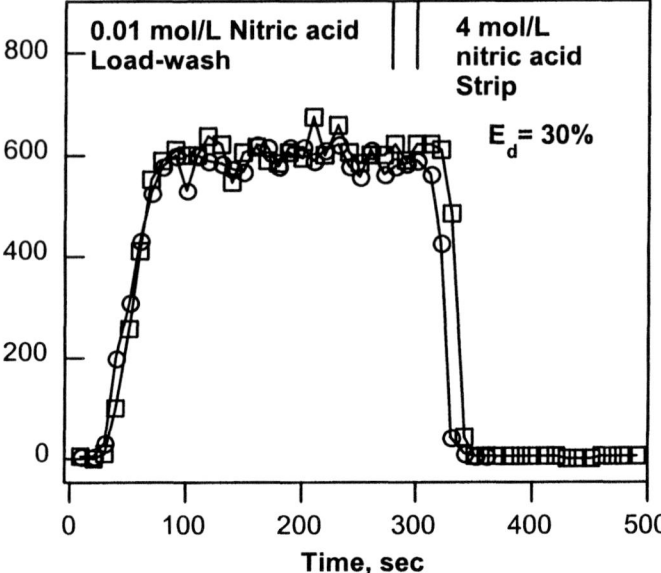

Figure 9. Duplicate detector signals showing flow-through operation of the chemically selective diode detector using 0.2 mL of ^{241}Am standard in 0.01mol/L nitric acid. Selective film thickness is 3 μm. Flow rate is 0.7 mL/min.

Diode detector signals in Figure 9 indicate that the analyte was captured and preconcentrated on the surface of the diode. The analyte remains in the diode coating during additional wash with dilute acid. The integrated count rate corresponds to ~30% absolute detection efficiency for the ^{241}Am captured on the diode surface. The diode sensor regeneration was feasible using 4 mol/L nitric acid for the analyte elution.

Given the remarkable robustness of modern PIPS diodes in contact with solution, we believe that use of more robust thin films based on solid-phase extraction materials for analyte capture may ultimately lead to the development of practical probes for monitoring alpha emitters in aqueous samples.

Automated Total ^{99}Tc Analyzer Instrument

In the United States, chemical processing of the low activity waste (LAW) at Hanford prior to vitrification requires removal of ^{90}Sr, TRU, ^{137}Cs and ^{99}Tc.

An on-line/at-line analytical method is desirable to enable rapid detection of total ^{99}Tc in the effluent streams from the technetium removal process.

Typical process streams are caustic brines with complex and varying chemical and radiological composition. In the case of ^{99}Tc, the radionuclide speciation is dependent on the source of the waste material. For instance, in waste matrixes with high organic content, a substantial amount of total ^{99}Tc is present in the reduced, non-pertechnetate form of unknown chemical composition. Currently, there is a lack of analytical methodologies suitable for the development of a process analyzer or sensor instruments capable of detecting total ^{99}Tc in chemically and radiologically complex matrixes.

Currently, preferred methods of total ^{99}Tc analysis are ICP MS or classical wet radiochemical measurements performed off-line in centralized analytical laboratories. The ICP MS technique is capable of unambiguous detection of the total ^{99}Tc in complex matrixes at the required detection levels. However, the high cost and complexity of the analytical instrumentation, and the frequent maintenance requirements precludes practical use in continuous (no down time) process monitoring. Typical analytical separation chemistries for ^{99}Tc are effective only for the pertechnetate species. Therefore, in order to enable total ^{99}Tc analysis via radiochemical measurement, oxidation chemistries and procedures must be developed to convert all Tc to pertechnetate. Radiochemical analysis of ^{99}Tc entails sample oxidation, ^{99}Tc(VII) separation (typically using ion exchange or solvent extraction), and liquid scintillation or gas proportional counting. Manual radiochemical procedures are tedious, time consuming, and not practical for process control applications. Nevertheless, a wet radiochemical analysis approach may become practical for process monitoring applications if it is successfully automated.

A fully automated radiochemical analyzer system has been developed for rapid analysis of total ^{99}Tc in aged LAW streams. The automated radiochemical analyzer instrument executes fluid handling steps required to perform acidification of the caustic sample, microwave-assisted sample oxidation using peroxidisulfate oxidant, separation of ^{99}Tc(VII) from radioactive interferences using an anion exchange column, and delivery of the separated pertechnetate to a flow-through scintillation detector. In this manner, an automated radiochemical measurement is performed in a single functional unit. The instrument design incorporates advanced digital fluid handling techniques using multiple zero dead volume digital syringe pumps and multiple valves for sample and reagent delivery. Comprehensive multithreaded control software was developed to enable fully automated asynchronous operation of the instrument components, as well as data processing, storage, and display.

The automated sample treatment protocol begins with sample acidification followed by the first digestion step. This initial treatment ensures removal of the nitrites, which are abundant in LAW matrixes and interfere with the subsequent oxidation. In addition, initial heating promotes rapid dissolution of the $Al(OH)_3$ precipitate which forms during acidification of the caustic LAW matrix. Nitric acid concentration and volume are selected to ensure complete dissolution of the $Al(OH)_3$ species upon heating, while maintaining relatively high pertechnetate uptake values on the anion exchange sorbent material during sample loading. The initial sample acidification procedure is followed by the second digestion treatment using sodium peroxidisulfate as the oxidizing reagent to convert the reduced Tc species to pertechnetate.

Pertechnetate separation was accomplished using macroreticular strongly basic anion exchange resin (AGMP-1, Biorad). Compared to the extraction chromatographic chemistry using the Aliquat-336-based TEVA-resin (Eichrom Technologies) which exhibits similar uptake characteristics, solid-phase anion exchange material offers much longer column life under elevated backpressure conditions. Pertechnetate elution kinetics of the extraction chromatographic material was superior to that of solid-phase anion exchange sorbent of similar particle size (20-50 µm). Nevertheless, rapid pertechnetate elution from the anion exchange column was possible using strong nitric acid solution when the eluent flow direction through the column was reversed. The pertechnetate separation approach based on anion exchange in nitric acid media offers adequate separation selectivity for the determination of ^{99}Tc in aged nuclear waste matrixes. It enables reliable separation from major radioactive constituents of aged waste (e.g. $^{90}Sr/^{90}Y$, ^{137}Cs). However, a comprehensive column wash sequence using 0.2mol/L nitric acid, 1mol/L sodium hydroxide, 0.2mol/L nitric-0.5mol/L oxalic acid, and 2mol/L nitric acid was required prior to pertechnetate elution to ensure reliable removal of interfering anionic radionuclides (e.g., isotopes of Sn, Sb, and Ru).

A flow-through scintillation detector equipped with a lithium glass solid scintillator flow cell was used for detection of the eluted $^{99}Tc(VII)$. The glass scintillator enabled absolute detection efficiency of ~55% and exhibited excellent stability in 8mol/L nitric acid media used for pertechnetate elution.

In order to enable reliable ^{99}Tc quantification in varying sample matrixes and remote instrument calibration, an automated standard addition technique was implemented as a part of the analytical protocol. The standard addition approach is based on the introduction of an aliquot of a ^{99}Tc standard solution during the sample acidification step. The ^{99}Tc standard is in a nitric acid solution of the same concentration used for sample acidification. To perform the standard addition measurement, the Tc monitor instrument automatically substitutes a given volume of the nitric acid for the ^{99}Tc standard solution during the sample acidification step of the automated analysis procedure. The volume of the standard is selected by the software to yield a 3 fold higher signal relative to the analysis of an unspiked sample. The total effective analytical efficiency (product of the recovery efficiency and the detection efficiency) is calculated based on the

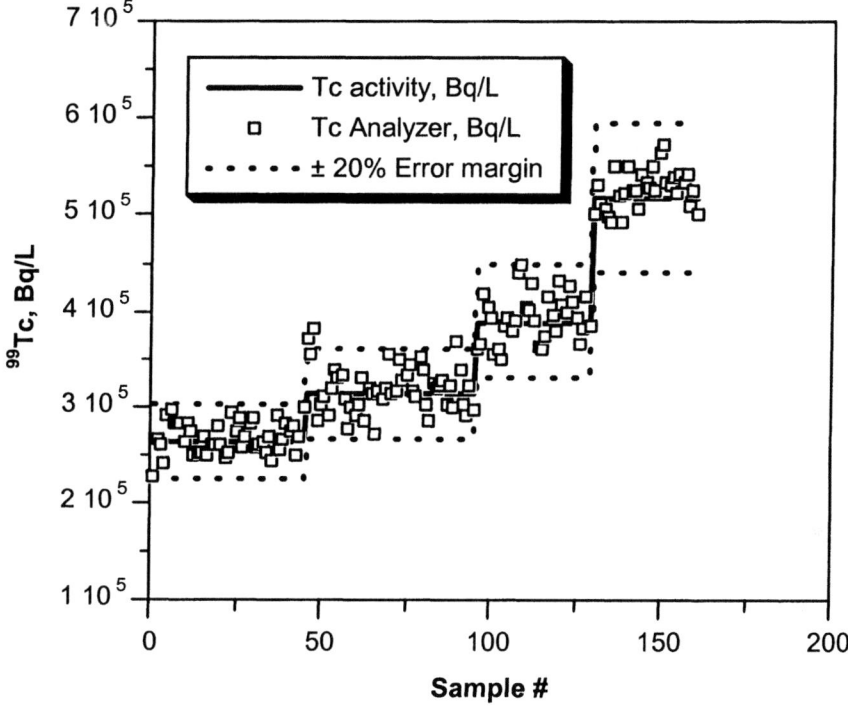

Figure 10. Results of the automated total ^{99}Tc analysis in LAW matrix with varying ^{99}Tc content performed over extended period of operation. Analysis frequency is 4 samples per hour. Automated standard addition was performed with every forth sample and used for quantification.

difference in analytical response obtained for the analysis of the spiked and unspiked samples. This approach provides a reliable method for remote, matrix matched instrument calibration. Automated standard addition can be periodically used during routine Tc monitor operation to enable verification of the instrument performance and correction for varying analysis conditions. The automated radiochemical analysis procedure is rapid, with a total analysis time of 12.5 minutes per sample. The total analysis time for the standard addition measurement was 22 minutes (including analysis of both unspiked and spiked samples).

The analyzer instrument was successfully tested against various LAW samples from the US DOE Hanford site, including those with high organic content. The automated radiochemical analyzer instrument provided accurate total ^{99}Tc quantification as verified by independent sample analysis using ICP-

MS. Figure 10 shows results of the total ^{99}Tc analysis in the LAW matrix under automated operation over a period of several days.

Automated Separations for ICP MS Analysis of Actinides

Compared to radiometric detection techniques, ICP MS can potentially offer lower limits of detection for long-lived radioisotopes, shorter analysis times, and simpler analytical procedures. The ICP MS technique is particularly well suited for analysis of the actinides. However, actinide measurements using low-resolution quadrupole ICP MS instruments with direct sample introduction are subject to isobaric interferences (e.g., ^{241}Am/^{241}Pu, ^{238}U/^{238}Pu, ^{244}Cm/^{244}Pu), molecular interferences (e.g., ^{239}Pu/^{238}U^{1}H), and spectral interferences (e.g., ^{237}Np/^{238}U).*(19)* Chemical separations are therefore necessary for unambiguous quantification of the individual actinide isotopes. In addition, chemical separations are advantageous to improve analysis sensitivity and detection limits by enabling analyte preconcentration and removal of the sample matrix.

On-line high performance ion chromatography (HPIC) has been used to overcome interferences in the ICP MS analysis of individual actinide isotopes in such sample matrixes as irradiated nuclear fuels.*(19-22)* The HPIC separation format features excellent separation efficiency attained through the use of high performance separation columns. Nevertheless, HPIC requires the use of expensive columns with low capacity and the separations must be carried at high pressures. Consequently, HPIC separations may not be ideally suited for routine analytical applications and development of more straightforward on-line separation methods would be advantageous.

A sequential injection (SI) separation system has been developed for separation and analysis of Am, Pu, and Np isotopes using on-line inductively coupled plasma mass spectrometry (ICP MS) detection.*(3)* On-line actinide separations were carried out using an actinide-specific extraction chromatographic material, TRU-resin (Eichrom Technologies, Inc.). This sorbent uses octyl(phenyl)-N, N-diisobutylcarbamoylmethylphosphine oxide (CMPO) in tri-n-butyl phosphate (TBP) as the stationary organic phase which has high affinity and selectivity for actinides in oxidation states III, IV and VI in nitric acid media. Group and individual actinide separations using TRU resin where developed and optimized for compatibility with the sample introduction requirement of the on-line ICP MS detector and to address isobaric, molecular, and spectral interferences encountered in the analysis of Am, Pu and Np isotopes.

The separation begins with loading a sample in nitric acid onto the column and then using additional acid to remove common matrix constituents. Trivalent and tetravalent

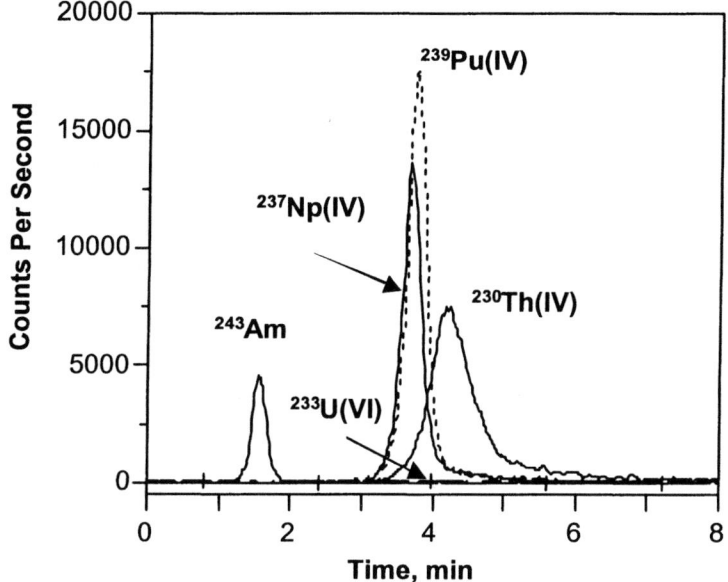

Figure 11. ICP MS detector signals illustrating sequential actinide elution sequence using 2mol/L HCl/1mol/L HC-/0.07mol/L oxalic acid eluent. Elution flow rate was 1 mL/min

actinides were sequentially eluted using a combination of HCl and HCl-oxalic acid eluents. Uranium remains on the column and can be subsequently eluted using ammonium bioxalate as a complexant. The sequential actinide group separation sequence is shown in Figure 11. In addition, we developed a separation sequence in which Np(IV) and Pu(IV) species could be selectively and rapidly eluted using 0.5mol/L nitric acid-0.05 mol/L oxalic acid eluent, while trivalent actinides, thorium, and uranium remained on the column.

Because analytes can be preconcentrated during sample load, this separation approach is well suited for trace analysis of Np and Pu in various nuclear and environmental samples. In the separation sequences described above, uranium separation factors were limited by the formation of U(IV) species, which exhibited separation behaviors similar to that of Pu(IV) and Np(IV). Nevertheless, separation factors as high as 3.0×10^5 could be achieved at the 100 ppm U concentration levels using Fe(II) sulfamate as the Pu reductant. The TRU-resin columns exhibited limited column life and column replacement was necessary for each measurement. Automated column-switching and a renewable column technique were recently developed for the automated replacement of either the separation column or sorbent bed, respectively. *(23)* On-line separation sequences were successfully used for accurate determination of actinide isotopes in aged tank and vitrified nuclear waste samples.

Discussion

Classical radiochemical analysis principles continue to play an important role in modern analytical radiochemistry. As noted in a recent article by Bickel et. al entitled "Radiochemistry: inconvenient but indispensable", "Radiochemistry has always been and still is a crucial tool in the field of radionuclide determination, ... particularly in the case of alpha and beta emitters."(24) Radiochemical analysis principles combined with the recent developments in selective separation chemistries, radiation detection, and fluidic instrumentation make automation of radiochemical analyses feasible.

An intimate integration of separation chemistries with radiation detectors can yield radionuclide selective sensor concepts for *in-situ* measurements of beta- and alpha- emitters as required for long-term monitoring and stewardship applications. Practical radionuclide sensor concepts using selective sorption chemistries combined with scintillation and diode detection have been developed and demonstrated for the analysis of key beta emitters (^{99}Tc, ^{90}Sr) and alpha emitters (actinides). Modern fluidic techniques and instrumentation can be used to develop automated radionuclide analyzers in which sample preparation, separation, and radiometric detection steps are integrated in a single functional instrument capable of rapid, unattended radiochemical measurements for process control applications. This has been exemplified in the development of the total ^{99}Tc analyzer instrument based on automated radiochemical analysis procedure. Automated radiochemical analyzers and sensors demonstrate that radiochemical analyses beyond the analytical laboratory is both possible and practical.

Acknowledgements

This research was supported by the Environmental Management Science Program of the Office of Environmental Management, U.S. Department of Energy. The Pacific Northwest National Laboratory is a multiprogram national laboratory operated for the U.S. Department of Energy by Battelle Memorial Institute.

References

1. Grate, J. W.; Egorov, O. B. *Anal. Chem.* **1998**, *70*, 779A-788A.
2. Choppin, G. R. *Anal. Sciences* **1995**, *11*, 143-147.
3. Egorov, O. B.; O'Hara, M. J.; Farmer, O. T., III; Grate, J. W. *Analyst* **2001**, *126*, 1594-1601.
4. Garcia Alonso, J. I. *Anal. Chim. Acta* **1995**, *312*, 57-78.
5. Ross, R. R.; Noyce, J. R.; Lardy, M. M. *Radioact. Radiochem.* **1993**, *4*, 24-37.
6. Smith, M. R.; Wyse, E. J.; Koppenaal, D. W. *J. Radioanal. Nucl. Chem.* **1992**, *160*, 341-354.
7. Horwitz, E. P.; Dietz, M. L.; Diamond, H.; LaRosa, J. J.; Fairman, W. D. *Anal. Chim. Acta* **1990**, *238*, 263-271.
8. Horwitz, E. P.; Chiarizia, R.; Dietz, M. L. *React. Funct. Polym.* **1997**, *33*, 25-36.
9. Kim, G.; Burnett, W.; Horwitz, E. P. *Anal. Chem.* **2000**, *72*, 4882-4887.
10. Burnett, W. C.; Corbett, D. R.; Schultz, M.; Horwitz, E. P.; Chiariza, R.; Dietz, M.; Thakkar, A.; Fern, M. *J. Radioanal. Nucl. Chem.* **1997**, *226*, 121-127.
11. Marion, J. B.; Young, F. C. *Nuclear Reaction Analysis. Graphs and Tables*; John Wiley & Sons, Inc.: New York; 1968.
12. Egorov, O.; O'Hara, M. J.; Grate, J. W., "Radionuclide selective sensors for water monitoring: ^{99}Tc(VII) detection in Hanford groundwater", In *Spectrum 2002;* American Nuclear Society: La Grange Park, Ill, 2002; 928-931.
13. Janata, J.; Bezegh, A., *Anal. Chem.* **1988**, *60*, 62R
14. Egorov, O. B.; Fiskum, S. K.; O'Hara, M. J.; Grate, J. W. *Anal. Chem.* **1999**, *71*, 5420-5429.
15. DeVol, T. A.; Egorov, O. B.; Roane, J. E.; Paulenova, A.; Grate, J. W. *J. Radioanal. Nucl. Chem.* **2001**, *249*, 181-189.
16. L'Annunziata, M., Ed. *Handbook of Radioactivity Analysis*; Academic Press: San Diego; 1998.
17. Krapivin, M. I.; Lebedev, I. A.; Myasoedov, B. F.; Yudina, V. G.; Yakobson, A. A.; Frenkel, V. Y. *Radiokhimia* **1979**, *21*, 321-323.
18. Feist, I.; Vdolecek, K.; Konecny, C. *Radiochem. Radioanal. Lett.* **1978**, *36*, 101-106.
19. Garcia Alonso, J. I.; Sena, R.; Arbore, P.; Betti, M.; Koch, L. *J. Anal. Atom. Spectrom.* **1995**, *10*, 381-393.
20. Rollin, S.; Kopatjtic, Z.; Wernli, B.; Magyar, B. *J. Chromat. A.* **1996**, 139-149.

21. Barrero Moreno, J. M.; Betti, M.; Garcia Alonso, J. I. *J. Anal. Atom. Spectrom.*, **1997**, *12*, 381-387.
22. Betti, M. *J. Chromat. A* **1997**, 369-379.
23. Egorov, O.; O'Hara, M. J.; Grate, J. W.; Ruzicka, *J. Anal. Chem.* **1999**, *71*, 345-352.
24. Bickel, M.; Holmes, L.; Janzon, C.; Koulouris, G.; Pilvo, R.; Slowikowski, B.; Hill, C. *Appl. Rad. Isot.* **2000**, *53*, 5-11.

Chapter 18

Rapid Actinide Column Extraction Methods for Bioassay Samples

S. L. Maxwell, III and D. J. Fauth

Westinghouse Savannah River Company, Aiken, SC 29808

Significant improvements have been made in extraction chromatography methods for bioassay samples in the last decade. Both single column and tandem column methods have been utilized in laboratories to analyze bioassay samples. A new, rapid separation method to assay actinides in urine samples has been developed at the Westinghouse Savannah River Site (SRS) that illustrates one of the newest developments that has been made in this field. This method combines some of the features of both single column and tandem column methods, utilizing two highly selective resins stacked to form a single column. The new method separates plutonium, neptunium, uranium, americium and strontium-90 from urine samples with high chemical recovery and excellent removal of matrix interferences such as thorium. Fast flow rates are achieved by using small particle size resin cartridges and a vacuum box separation system that will separate up to 24 samples at a time. The method uses calcium phosphate precipitation and stacked TEVA Resin® and TRU Resin® cartridges to separate and purify the actinides. Plutonium and neptunium are separated on TEVA Resin®, while uranium and americium are simultaneously retained and separated on TRU Resin®. Plutonium-236 tracer is used to allow simultaneous separation and measurement of both plutonium and neptunium using TEVA Resin®. Strontium-90 can also be separated on Sr Resin® by evaporating and redissolving load and rinse solutions collected from the TEVA/TRU column and separating strontium on Sr-Resin. This unique approach can be used with urine samples because iron is not present at significant levels in urine and plutonium reduction is accomplished without adding iron (II) to the sample. The advantage of this approach is that actinides can be loaded onto two separate resins in a single load step. This method offers significant advantages when a large number of actinides are analyzed, but is just one of the several different extraction chromatography methods now available for bioassay samples.

Introduction

There have been significant advances in last five to ten years in radiochemical separations, with broad application in a wide range of labs. These improvements in column extraction chromatography have advanced analytical technology in process labs, bioassay labs, and environmental labs. Though these labs analyze sample types that are often very different, all have certain commonalties. Despite their differences, sample preparation is essential in all these labs to preconcentrate analytes and remove matrix interferences prior to assay.

Column extraction chromatography has become very popular over the last decade for analytical separations. It offers several advantages over large column ion exchange and liquid –liquid solvent extraction. Because extractant-coated resins are often more selective than ion exchange, these new methods are usually simpler than older ion exchange techniques. In addition, column extraction methods typically generate less liquid waste, can be employed using lower acid strengths and do not create hazardous organic solvent waste. Many of the new extraction chromatography resins available were developed at Argonne National Laboratory and are now marketed by Eichrom Technologies, Inc. (Darien, IL). The PG Research Foundation , directed by Dr. E. Philip Horwitz, continues to fund research into new separation techniques. For example, a new extraction chromatography method for environmental samples was recently funded by the PG Research Foundation (*1*). Dr. Horwitz was group leader of the Chemical Separations Group at Argonne National Laboratory for many years and has made significant contributions to the field of separation science for both process and analytical applications.

Recently, the use of vacuum boxes and smaller particle size resin cartridges has become increasingly popular to reduce separation times. These boxes provide flow rates five times faster than gravity flow methods. This approach also allows the operator to apply increased vacuum to any "stubborn" columns that do not flow as fast as others in the batch. With gravity flow, one or more slow columns can significantly increase the time it takes to process the entire batch. The smaller particle size resin used with vacuum is 50 to 100 micron size. A typical particle size used with gravity-flow column methods is 100-150 microns. The smaller particle resin typically provides better resolution of elution bands than larger particle resin. The Westinghouse Savannah River Site Central Laboratory began using vacuum boxes in the 1980's to speed up ion exchange separations and later began applying this approach to extraction chromatography resins from Eichrom Technologies. The advent of prepacked resin cartridges produced by Eichrom Technologies has made vacuum box separations even more popular. (*2*)

In the 1990's, there was a need to upgrade radiochemistry methods at the Savannah River Site. The Savannah River Site Central Analytical Laboratory (CLAB) replaced a wide range of solvent extraction methods in CLAB used for the previous twenty to thirty years for actinide separation techniques. This eliminated mixed waste problems caused by the use of solvents such as hexane and thenolytrifluoroacetone (TTA)-xylene. New tandem methods for process and waste analyses were developed at the Savannah River Site and implemented using rapid column extraction chromatography for a wide range of process analyses (*3,4,5*).

In 1994 the SRS Bioassay lab also began to upgrade methods analytical methods used in that laboratory. Previous ion exchange methods generated large volumes of acid waste and all too often resulted in inconsistent tracer recoveries. An increase of requests at SRS to measure more than one actinide in urine samples prompted the need to consolidate actinide analysis into a single sequential method. New methods were successfully implemented to improve the SRS actinide method for urine.

In addition, there was a need to upgrade the method for fecal samples at SRS, where total dissolution is difficult, yet very important. Fecal analysis of actinides is often an important factor in establishing total dose received from a radiological incident. Analytical methods have been reported that achieve total dissolution and analyze actinides in fecal samples. Though effective, they sometimes require extensive preparation steps to separate the actinides. (*6*) A new SRS fecal method recovers actinides from large fecal samples in a small volume of nitric acid with minimal phosphate present. This solution can be easily loaded onto small extraction chromatography columns for rapid separation and analysis. This method provides total sample dissolution, high recovery of actinides and excellent purification of plutonium and americium for measurement by alpha-particle spectrometry (7) The new method uses Diphonix Resin® to extract actinides from fecal samples that have been ashed and redissolved in dilute hydrochloric acid-hydrofluoric acid and hydrochloric acid-boric acid solutions. Diphonix Resin®, a resin with geminally-substituted diphosphonic acid groups chemically-bonded to a styrene-divinylbenzene matrix, was developed by Argonne National Laboratory and the University of Tennessee. (8, 9,10) Diphonix exhibits a high affinity for actinide ions in the tri-, tetra and hexavalent oxidation states. Diphonix Resin has also been used in soil methods to preconcentrate actinides (11,12,13). Although the new SRS urine method does not require Diphonix Resin to preconcentrate actinides, it does use a similar extraction chromatography approach.

The new SRS urine method takes advantage of new resin cartridge technology that separates plutonium, neptunium, americium and uranium using a single column. This method will be described in greater detail to illustrate one of the newest developments in extraction chromatography, the stacking of resin cartridges to form a single column. The SRS urine method builds on earlier work in the field, which began with single column work and moved to anion exchange coupled with extraction chromatography. The newest approach is a stacked cartridge method that offers the selectivity of two different extractants in a single column.

Column Extraction Method Improvements for Urine

Monitoring of actinides in urine is an important analysis in the nuclear industry. A variety of methods have been reported in the literature for measurements of actinides in urine. Early solid phase extraction methods typically used one resin for actinide separations in urine samples. Horwitz et al. separated actinides from urine on a resin containing octyl (phenyl)-N,N-diisobutylcarbamolymethylphosphine oxide (CMPO) and tributylphosphate (TBP) extractant, now marketed as TRU Resin ® by Eichrom Technologies, Inc. (Darien, IL). (14) TRU Resin retains hexavalent, tetravalent and trivalent actinides in the presence of nitric acid. Actinides were stripped as a group from the resin together using 0.1 mol/L ammonium bioxalate. An attempt was also made to elute the actinides sequentially using 2 mol/L nitric acid, 2 mol/L hydrochloric acid and 2 mol/L hydrochloric acid containing the reductant hydroquinone (0.1mol/L). In reference 14, bands showed overlap due to tailing and did not exhibit adequate resolution of the various actinides. In a subsequent paper by Horwitz et al., the elution behavior of actinides and the impact of matrix interferences on TRU Resin separations were thoroughly examined (*15*). A reasonably good separation of five actinides was achieved, but the paper concluded the separation of this many actinides on one resin was probably too tedious and difficult to execute for lab analytical work. The paper recommended using TRU Resin in combination with other resins such as strongly basic anion resin. The use of TRU Resin instead of older solvent extraction or ion exchange techniques was a significant step forward, but it was difficult to separate all the actinides with adequate resolution on a single resin. Harduin et al. used TRU Resin to separate actinides from urine samples and elute them in a single fraction. (*16*) This work used peak deconvolution software to attempt to separate a few of the overlapping actinide peaks found in the alpha spectrum; however, the need to isolate actinides in separate fractions was noted. The separation of strontium from urine samples was also enhanced by the development of Sr Resin (Eichrom Technologies, Inc., Darien, IL), which contains the 4,4'(5')-di-t-butylcyclohexano 18-crown-6 (crown ether) extractant selective for strontium. (*17,18*). This simple separation was a significant improvement over older ion exchange, solvent extraction and precipitation techniques.

Coupling Extraction Chromatography with Anion Exchange

Although there have been improvements in single column separations since this early work was reported, recent column extraction methods have usually employed tandem columns to better separate actinide fractions for analysis. Alvarez and Navarro described a method that used AG X-2 anion resin (Bio-Rad, Inc.) to separate plutonium and TRU Resin to separate americium from urine (*19*). Sr Resin was used to further purify strontium for analysis by beta counting. The load and rinse solution from the anion resin containing americium and strontium had to be evaporated to dryness and redissolved in 2 mol/L nitric acid to separate americium on TRU Resin. The load and rinse solutions from TRU Resin containing strontium had to be collected and evaporated to dryness to readjust the acidity to 8 mol/L nitric acid for separation of strontium on Sr Resin. Typical chemical recoveries reported were 83% for plutonium, 81% for americium and 86% for strontium. Coupling anion exchange with TRU Resin allowed adequate resolution of individual actinides, but still generated a relatively large acid waste volume. Sr Resin effectively collected strontium after the load and rinse solutions from AG-X-2 TRU Resin were evaporated and redissolved in strong nitric acid. The evaporation steps also require a significant amount of time.

New Options

Additional extraction chromatography options for a variety of types of samples have been developed. Horwitz et al. also described the elution behavior of a new extraction chromatography resin for uranium separations that was coated with diamylamylphosphonate extractant (20). Eichrom Technologies, Inc. (Darien, IL) markets this resin as UTEVA Resin®. UTEVA Resin retains hexavalent and tetravalent actinides in the presence of nitric acid. Langston et al. used UTEVA Resin and TRU Resin in tandem to separate uranium, neptunium and thorium from urine (21). In this method, uranium was separated on UTEVA Resin, while neptunium and thorium were separated together on TRU Resin. Uranium chemical recoveries varied from 73-89%, neptunium recoveries were 59-74% and thorium recoveries were 82-97%. This method, however, did not separate other actinides from each other, eluting them from TRU Resin as a group. As noted earlier, there are potential interferences when several actinides are eluted together. Anil Thakker of Eichrom Technologies recently reported a method using UTEVA Resin and TRU Resin to measure uranium, plutonium and americium in urine. (22) Thakker demonstrated UTEVA and TRU Resin cartridges could be successfully applied to urine samples while using a vacuum box manifold. The tracer recoveries were good and the decontamination from urine matrix interferences was effective. The UTEVA Resin and TRU Resin tandem approach has been used in many laboratories for a wide variety of matrices. In this method, uranium and thorium are separated on UTEVA Resin, while plutonium and americium are separated on TRU Resin. Plutonium must be reduced with a reductant or combination of reductants such as ferrous sulfamate and ascorbic acid so that Pu will pass through UTEVA Resin as Pu (III) and be retained on TRU Resin. Uranium is eluted from UTEVA Resin and plutonium and americium are eluted separately from TRU Resin. Stacking the UTEVA and TRU Resin cartridges saves time. The load and rinse solution from UTEVA Resin does not have to be collected and then loaded to TRU Resin; both columns are loaded simultaneously.

In the SRS bioassay laboratory, there was a need to separate and measure not only plutonium, uranium and americium, but neptunium-237 as well. The UTEVA plus TRU Resin method is not applicable because it separates plutonium, americium and uranium. Neptunium and thorium are collected together in a discarded rinse from UTEVA Resin. The new method would also need effective thorium-228 removal, excellent tracer recoveries and actinide and strontium analysis that involves a single sample preparation step.

SRS Stacked Cartridge Approach

A new stacked column method was developed to meet those needs, utilizing two selective resins stacked to form a single column (23). TEVA Resin and TRU Resin cartridges are used to perform the separations, stacked and used with a vacuum box as shown in Figure 1. The new method separates plutonium, neptunium, uranium, americium and strontium-90 with high chemical recovery and excellent thorium removal. Plutonium and neptunium are separated on TEVA Resin, while uranium and americium are simultaneously retained and separated on a TRU Resin cartridge. Strontium is collected from the load and rinse solution from a single column, evaporated, redissolved and separated on Sr Resin. This unique approach can be used with urine samples because iron is not present at significant levels in these samples. Plutonium reduction is accomplished without adding iron (II) to the sample.

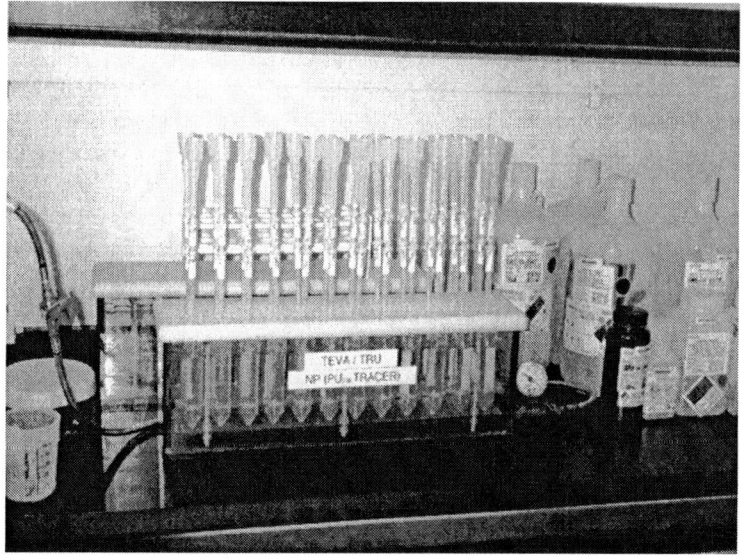

Figure 1. Stacked TEVA Plus TRU Resin Column on Vacuum Box

The advantage of this approach is that actinides, including neptunium, can be loaded onto two separate resins in a single load step. Labor costs were reduced significantly and rework has been reduced by 50%. Vacuum boxes and cartridges have allowed much faster analyses than before. TEVA Resin, which contains Aliquat^{TM} 336 extractant, has quaternary amine groups that retain tetravalent actinides. This resin has similar behavior to anion resin with quaternary amine groups such as Dowex 1x8 Resin (Dow Chemical Co.), but can be used with lower nitric acid concentrations. TEVA Resin also shows less uranium tailing than anion resin (24). By using two extraction chromatography resins instead of coupling anion exchange with TRU Resin, less acid waste is generated. An additional advantage is that plutonium and neptunium can be separated using nitric acid in the 3 mol/L range instead of 7-8 mol/L range. This allows easier reduction of reduction of neptunium to Np (IV) and plutonium to Pu (III) prior to oxidation to Pu (IV). UTEVA -TRU Resin cartridges can also be stacked and used with vacuum boxes. Both methods successfully measure actinides, however, the TEVA Resin plus TRU resin approach allows separation and analysis of neptunium. The UTEVA-TRU Resin method does not. The TEVA-TRU Resin method allows plutonium and neptunium to be separated and collected in the same fraction. This is very important because it allows neptunium is to be traced with Pu-236 tracer. Though not measured in urine at SRS, thorium can also be collected in the hydrochloric acid rinse fraction from TEVA Resin for assay. One additional advantage of this method is that when only plutonium and neptunium analysis is required, the TRU Resin cartridge is not needed.

For urine analysis where neptunium is not required, the UTEVA-TRU Resin stacked column is an equally viable approach. Uranium is, however, retained more strongly on TRU Resin than on UTEVA Resin, especially from a high phosphate urine matrix. The TEVA - TRU Resin stacked column method has been applied to other matrices, including soil and water samples. A barium sulfate precipitation has been used that removes iron, eliminating any interference from iron in soil on the stacked TEVA Resin plus TRU Resin column (25). This stacked TEVA plus TRU method where plutonium and neptunium are separated on TEVA Resin requires no iron to be present in the sample load solution. In contrast, the UTEVA –TRU Resin method is not limited in this way because iron is reduced to iron (II), which does not interfere on TRU Resin. The TEVA -TRU Resin method can, however, also be used in a tandem way without stacking the cartridges when iron is present, as in fecal samples. (26) In this case, the load and rinse solutions from TEVA Resin are collected, iron is reduced to non-interfering iron (II) using sulfamic acid and ascorbic acid, and this solution is loaded onto TRU Resin.

SRS Urine Method Protocol and Testing

The following protocol was followed in the TEVA - TRU Resin stacked column method. Urine samples were acidified with nitric acid and allowed to stand for two hours. The appropriate tracers (Pu-236, U-232, and Am-243) were added to 500 mL aliquots of urine sample.

Two drops of 1-octanol and 1 mL of 3 mol/L calcium nitrate were added to each sample. Samples were heated on low heat for 1.5 hours and cooled to room temperature. After cooling, 5 mL of 3 mol/L ammonium hydrogen phosphate was added to each sample and the sample was stirred. The samples were adjusted to pH 9 or above with ammonium hydroxide and the precipitate was allowed to settle for at least one hour. The precipitate and supernate were centrifuged at 3000 rpm for 35 minutes. After decanting the supernate, the precipitate was dissolved in approximately 20 mL of concentrated nitric acid and ashed to dryness on a hot plate at approximately 170-200C. The samples were ashed with 30 wt% hydrogen peroxide several times and then ashed with a mixture of nitric acid and hydrogen peroxide until the residual salts were white.

The evaporated-resin digest was redissolved in the appropriate acid solution for subsequent-column separations. In this work the residues were redissolved in approximately 6 mL of 6 mol/L nitric acid and 6 mL of 2.5 mol/L aluminum nitrate (previously scrubbed by passing through UTEVA Resin to remove traces of uranium). The solution was warmed slightly to ensure complete redissolution. The final solution contains approximately 12 to 15 mL of 2.5 mol/L nitric acid-1.25 mol/L aluminum nitrate. Relatively high aluminum nitrate concentrations are used to complex phosphate in urine samples and minimize phosphate interference on actinide retention.

Two-mL TEVA Resin cartridges and two-mL TRU Resin cartridge were stacked to isolate the actinides of interest. The vacuum box system (Eichrom Technologies, Inc.) used can handle up to 24 samples at a time. The TRU cartridge was placed below the TEVA column by luer connection. Plutonium and neptunium were retained on TEVA Resin and americium and uranium on TRU Resin. Ferric ions can interfere with americium retention on TRU Resin. Since there are no significant levels of iron in urine, the TRU cartridge can be used in a stacked column with TEVA Resin if the valence adjustment used does not require iron.

The valence of plutonium and neptunium was adjusted to Pu (IV) and Np (IV) by adding 0.5 mL of 1.5 mol/L sulfamic acid and 2 mLs of 1.5 mol/L ascorbic acid, waiting 3 minutes, and adding 2 mL of 4 mol/L sodium nitrite. The kinetics of the oxidation of Np (IV) to Np (V) by the nitrite ion are extremely slow at room temperature, so neptunium stays in the extractable Np (IV) valence state under these conditions. (27) After the valence adjustment, the sample solution was loaded onto the stacked TEVA plus TRU column. The TEVA and TRU column was rinsed with 20 mL of 3 mol/L nitric acid to remove matrix components. After the rinsing with nitric acid, the TRU cartridge was removed. To remove thorium from the TEVA column, 3 mLs of 9 mol/L hydrochloric acid and 25 to 30 mL of 8 mol/L hydrochloric acid were added. The plutonium and neptunium were stripped from TEVA Resin with 30 mLs of 0.1 mol/L hydrochloric acid-0.05 mol/L hydrofluoric acid –0.1 mol/L ammonium iodide. Recent work has indicated that plutonium stripping from TEVA resin is improved by using rongalite (sodium formaldehyde sulfoxylate) or titanium chloride as a reductant instead of iodide (*28*). Four mL of 0.02 mol/L sulfuric acid and approximately 3 mLs of 15.7 mol/L nitric acid were added to each sample and the sample solution was evaporated. Rongalite is compatible with electrodeposition while titanium is not. Titanium (III) chloride may be used instead of rongalite or ammonium iodide if cerium fluoride microprecipitation is utilized to prepare the alpha mounts. Both are strong reductants and work well in stripping plutonium from TEVA Resin.

Initially, a second-column separation using 1 mL of TEVA Resin was employed to ensure complete removal of all traces of Th-228. It was found that if the U-232 tracer were scrubbed to remove Th-228 daughter that a second TEVA column was not required. To scrub the U-232 tracer, it was adjusted to 4 mol/L nitric acid and passed through a TEVA Resin cartridge to remove Th-228. The Th-228 was retained on TEVA Resin and the U-232 passed through the resin. The final solution activity was validated versus a traceable uranium standard using alpha spectrometry. The "scrubbing" is performed every 12-18 months.

The americium was stripped from each TRU cartridge using 15 mL of 4 mol/L hydrochloric acid. The uranium was stripped using 20 mL of ammonium bioxalate.

To prepare for electrodeposition, solutions were evaporated, wet-ashed using 15.7 mol/L nitric acid and 30 wt% hydrogen peroxide, redissolved in a sodium bisulfate matrix and electroplated for 2.5 hours using 0.5 amp current. Cerium fluoride microprecipitation was performed using 50 micrograms of cerium in the presence of hydrofluoric acid and filtration and mounting on Gelman 25 mm filters. Solutions prepared for cerium fluoride precipitation did not have to be evaporated prior to filtration. By using only 50 micrograms of cerium, peak width was only slightly larger than peaks from electrodeposited mounts.

For strontium analysis, load and rinse solutions were collected from the TEVA –TRU Resin stacked column, evaporated on a hot plate, and redissolved in 15 mL of 6 mol/L nitric acid. Each solution was loaded on to a 2 mL Sr Resin cartridge. The column was rinsed with 15 mL of 8 mol/L nitric acid and stripped with 10 mL of 0.05 mol/L nitric acid. The strip solutions were evaporated on annealed planchets. The planchets were cooled and counted for 20 minutes using a gas proportional counter. Sr-90 spikes (3.42 Bq) were added to blank urine samples to perform Sr-90 yield corrections.

In an initial test performed using TEVA Resin only, ferrous sulfate and ascorbic acid were used to adjust the plutonium valence to Pu (III) and sodium nitrite was used to adjust the plutonium valence to Pu (IV) (*29*). The average Pu-242 tracer recovery when cerium fluoride precipitation is used was 102%

(±7%@1s). When samples were analyzed using ammonium iodide in the TEVA strip solution with electrodeposition, an average tracer recovery of 79% (±6.2%@1s) was obtained. The lower efficiency of electroplating for these samples may be explained by traces of fluoride that were not completely removed, despite multiple ashing steps with nitric acid and hydrogen peroxide and the addition of 4 mLs of 0.02 mol/L sulfuric acid to enhance fluoride volatilization. There may also have been traces of plutonium that were not removed from TEVA Resin.

The accuracy achieved using the TEVA-TRU Resin method on urine samples spiked with plutonium, neptunium, americium, uranium and strontium was tested (29). The urine samples contained Pu-238 in the range 2.42E-3 to 8.25E-2 Bq/L (N=12), Pu-239 in the range 3.67E-4 to 6.03E-2 Bq/L (N=12), and Am-241 in the range 9.17E-3 to 5.83E-2 Bq/L (N=7). The samples contained U-234 in the range 3.28E-3 to 3.40E-2 Bq/L (N=4), U-238 in the range 2.45E-3 to 5.12E-2 Bq/L (N=4) and Sr-90 in the range 0.163 to 6.96 Bq/L (N=6).

The average bias for Pu-238 and Pu-239 was -14.7% and $+12.4\%$ respectively. The average bias for Am-241 measurements was -3.4%. For uranium, the average bias for U-234 and U-238 was +7.8% and +1.5% respectively. The Sr-90 recoveries used to perform yield corrections averaged 90.2%. The Sr-90 bias averaged –4.8% for the spiked samples.

The average bias results are well within the Department of Energy Laboratory Accreditation program (DOELAP) bias criteria of –25% to +50%. The average blank values for each radionuclide shown were sufficiently low to be acceptable for SRS bioassay needs.

Figure 2 shows tracer recoveries when rongalite was used in the TEVA stripping solution for a batch of twenty urine samples (30). The average plutonium tracer recovery using rongalite in the strip solution and electroplating was 99%. Since rongalite is a stronger reducing agent than ammonium iodide, the stripping of plutonium from TEVA Resin is more effective. In addition, rongalite decomposes into sulfate during the ashing steps and this likely enhances the removal of fluoride ions, which can interfere with electroplating. The amount of bisulfate added in the electrodeposition procedure was reduced to allow for the amount of sulfate resulting from the rongalite.

Figure 3 shows tracer recoveries using TEVA Resin to analyze 500 mL urine samples with Pu-236 tracer (7.10E-3 Bq) added to all samples (31). Ten of the twenty samples were spiked with known amounts of Pu-239 (1.43E-2 Bq/L), Np-237 (8.92E-3 Bq/L) and Pu-238 (1.83E-3 Bq/L). In this test using stacked TEVA and TRU resin cartridges, sulfamic acid and ascorbic acid were used to adjust the plutonium valence to Pu (III) and sodium nitrite was used to adjust the plutonium valence to Pu (IV). Alpha mounts were prepared using electroplating. The average measured values were as follows: Pu-239 (1.41E-2 Bq/L), Np-237

(8.47E-3 Bq/L) and Pu-238 (2.38E-3 Bq/L). The average bias for Pu-238 and Pu-239 was +30% and +1.0% respectively. The average bias for Np-237 was +5.0%. The larger bias for Pu-238 may have related to the very low level of Pu-238 standard added. The typical uncertainty for plutonium using this method is ± 16% (2s) at 1.6E-2 Bq/L and ± 30% (2s) at 1.6E-3 Bq/L. For neptunium, the uncertainty is approximately ± 30% (2s) at 8.3E-3 Bq/L..

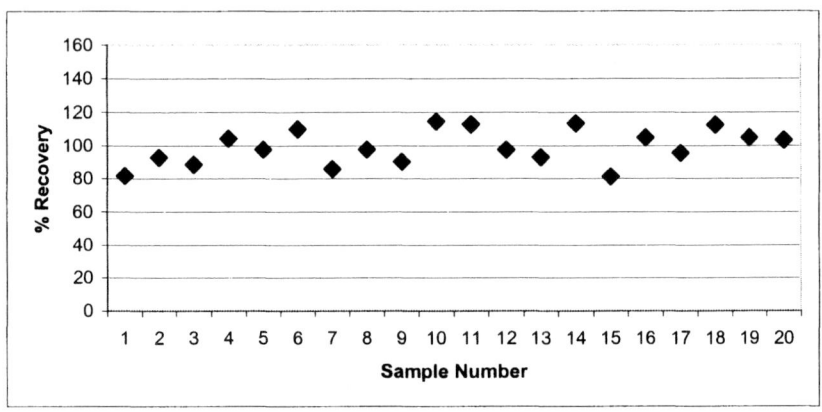

Figure 2. Pu-236 Tracer Recoveries on TEVA Resin

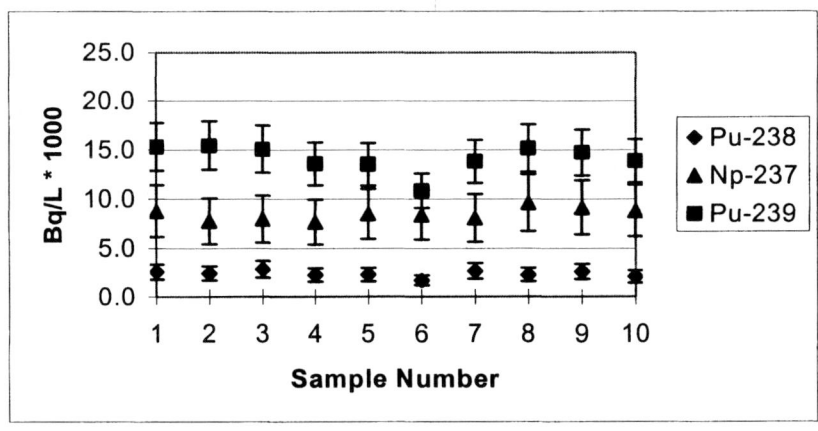

Known values:
Pu-239 =1.43E-2 Bq/L
Np-237=8.92E-3 Bq/L
Pu-238 =1.83E-3 Bq/L

Avg. measured:
Pu-239 =1.41E-3 Bq/L
Np-237=8.47E-3 Bq/L
Pu-238 =2.38E-3 Bq/L

Figure 3. Pu and Np Results on Spiked Urine Samples

In the same testing, americium and uranium were separated on TRU Resin. Am-243 tracer (2.58E-2 Bq) and U-232 tracer (9.23E-3 Bq) were added. The average tracer Am-243 recovery was 96.9% (±21.2% @2s) and the average U-232 recovery was 84.7% (±25.8% @2s) when samples were electroplated.

Conclusion

There has been a significant improvement in actinide separations for bioassay samples over the last decade, beginning with the groundbreaking work at Argonne National Laboratory to develop new extraction chromatographic resins. These resins are now available through marketing by Eichrom Technologies. Ongoing research in the field by many laboratories, including research supported by the PG Research Foundation, has continued to advance extraction chromatography. The vacuum box stacked cartridge approach, pioneered for radiochemical applications at the Westinghouse Savannah River Site Laboratory, has made these column extraction methods faster and less time-consuming. Tandem separations can now be combined into a single, rapid load step using this approach. There are a variety of viable extraction chromatography methods available to the bioassay laboratory that improve chemical recoveries, generate less waste and remove interferences. The evolution of single column extraction methods to tandem column methods has greatly improved bioassay analyses. There is better resolution of alpha spectrometry peaks and interferences are eliminated. Tandem separations can now be combined into a single, rapid load step. The SRS stacked cartridge method is a step forward in the radiochemical analysis of actinides in urine samples. Loading actinides (Pu, Np, U, Am and Th) onto two highly selective resins in a single step is a significant advance. The TEVA -TRU Resin stacked column method separates not only plutonium, uranium, and americium, but neptunium and thorium as well. Strontium can also be analyzed by collecting the load and rinse solution from the stacked column, evaporating the solution and performing a strontium separation on Sr Resin. This method provides high chemical recovery for actinides and strontium, excellent cleanup from interferences such as Th-228, and rapid column flow rates to save time. The UTEVA - TRU Resin method also provides an effective means of separating actinides and strontium from urine and fecal samples using extraction chromatography. This method can also be used with vacuum box flow rates and a stacked cartridge approach. With either method, actinides are loaded in a single step, chemical recoveries are high, flow rates are rapid, matrix interferences are eliminated and waste volumes are minimized. Continued advances in extraction chromatography in the last 10 to 15 years have greatly improved the capabilities of laboratories to analyze bioassay samples.

There are a variety of options available to allow today's bioassay laboratory to customize extraction chromatography separations to meet its specific needs.

Acknowledgements

This work was performed under the auspices of the Department of Energy, DOE Contract No. DE-AC09-96SR18500. The authors also wish to acknowledge A. Harper Shull, statistician at the Westinghouse Savannah River Co., for his statistical evaluation of the data.

References

1. Burnett, W.C.; Horwitz, E. P., Kim, G. *Anal. Chem.* **2000**, *72*, 4882-4887.
2. Eichrom Technologies, Inc. Newsletter, 7, Issue 1, November 2000, URL http://www.eichrom.com/analytical/radio/newsletters/news/news.htm.
3. Horwitz, E. Philip; Dietz, M.L.; Chiarizia, R., Diamond; H., Maxwell, S.L.; Nelson, M.R. *Anal. Chim. Acta* **1995**, *310*, 63-78.
4. Maxwell III, S.L. *Radioactivity & Radiochemistry* **1997**, *8*/4, 36-44.
5. Maxwell III, S.L.; Satkowski, J. *Radioact. & Radiochem.* **2001**, *12*/2, 12-20.
6. Sill, D.S.; Bohrer, S. E. *Radioact. & Radiochem.* **2000**, *11*/3, 28-34.
7. Maxwell III, S.L.; Fauth, D.J. *J. Radioanal. Nucl. Chem.* **2001**, *250*, 117-121.
8. Horwitz, E.P.; Chiarizia, R.; Diamond, H.; Gatrone, R.C.; Alexandratos, S.D.; Trochimczuk, A.Q.; Crick, E.W. *Solvent Extraction Ion Exchange* **1993**, *11*, 943-966.
9. Chiarizia, R.; Horwitz, E.P.; Gatrone, R.C.; Alexandratos, S.D.; Trochimczuk, A.Q.; Crick, E.W. *Solvent Extraction Ion Exchange* **1993**, *11*, 967-985.
10. Chiarizia, R.; Horwitz, E.P.; Alexandratos, S.D.; Gula, M. J. *Sep. Sc. Techn.* **1997**, *32,* 1-35.
11. Smith, L.L.; Crain, J.S.; Yaeger, J.S.; Horwitz, E.P; Diamond, H.; Chiarizia, R. *J. Radioanal. Nucl. Chem.* **1995**, *194*, 151-156.
12. Burnett, W.C.; Horwitz, E. P.; Kim, G. *Anal. Chem.* **2000**, *72*, 4882-4887.
13. Maxwell, S.L.; Nichols, S.T. *Radioact. & Radiochem.* **2000**, *11*/4., 1-9.
14. Horwitz, E. P.; Dietz, M.L.; LaRosa J. J.; Fairman, W. D.; Nelson, D. M. *Anal. Chim. Acta* **1990**, *238*, 263-271.
15. Horwitz, E. P.; Chiarizia, R.; Dietz, M.L.; Diamond, H.; Nelson, D. M. *Anal. Chim. Acta* **1993**, *281*, 361-372.
16. Harduin, J.C.; Peleau, R.; Piechowski, J. *Radioprotection* **1993**, *28*, 291-304.

17. Horwitz, E. P.; Dietz, M.L.; Fisher, D. E. *Ana.l Chem.* **1991**, *631*, 522-525.
18. Dietz, M.L.; Horwitz, E. P.; Nelson, D.M.; Wahlgren, M. *Health Physics* **1991**, *61*, 871-877.
19. Alvarez, A; Navarro, N. *Applied Radiation and Isotopes* **1996**, *47*/9&10, 869-873.
20. Horwitz, E. P.; Dietz, M.L.; Chiarizia, R.; Diamond, H.; Essling, A.M.; Graczyk, D. *Anal. Chim. Acta* **1992**, *266*, 25-37.
21. Langston, R; Nguyen., S; Simpson, T. *Proceedings of the 41st[h] Annual Conference on Bioassay, Analytical and Environmental Radiochemistry,* Boston, MA, 1995.
22. Thakker, A.H. *J. Radioanal. Nucl. Chem.* **2001**, *248*, 453-456.
23. Maxwell III, S.L.; Fauth, D.J. *Radioact. & Radiochem.* **2000**, *11*/3, 28-24.
24. Maxwell III, S.L. *Radioact. & Radiochem.* **1997**, *8*/4, 36-44.
25. Ivey, W. *Proceedings of the 48th Annual Radiobioassay and Radiochemical Measurements Conference,* 2002.
26. Maxwell III, S.L.; Fauth, D.J. *J. Radioanal. Nucl. Chem.* **2001**, *250*, 117-121.
27. Maxwell III, S.L. *Radioact. & Radiochem.* **1997**, *8*/4, 36-44.
28. Maxwell, S.L.; D.J. Fauth, Westinghouse Savannah River Co., unpublished
29. Maxwell III, S.L.; Fauth, D.J. *Radioact. & Radiochem.* **2000**, *11*/3, 28-24.
30. Maxwell, S.L.; D.J. Fauth, Westinghouse Savannah River Co., unpublished.
31. Maxwell, S.L.; D.J. Fauth, Westinghouse Savannah River Co., unpublished.

Chapter 19

Determination of Isotopic Thorium in Biological Samples by Combined Alpha Spectrometry and Neutron Activation Analysis

S. E. Glover

Isotope and Nuclear Chemistry, Los Alamos National Laboratory, Los Alamos, NM 87545

Thorium is a naturally occurring element for whom all isotopes are radioactive. Many of these isotopes are alpha emitting radionuclides, some of which have limits for inhalation lower than plutonium in current regulations. Neutron activation analysis can provide for the low-level determination of ^{232}Th but can not determine other isotopes of dosimetric importance. Biological and environmental samples often have large quantities of materials which activate strongly, limiting the capabilities of instrumental neutron activiation analysis. This paper will discuss the application of a combined technique using alpha spectrometry and radiochemical neutron activiation analysis for the determination of isotopic thorium.

A number of radiometric and non-radiometric methods have been used for the determination of ^{232}Th in biological and environmental samples. These include alpha spectrometry(1,2), gamma ray spectrometry (3), instrumental neutron activation analysis (INAA) (3), radiochemical neutron activation analysis (RNAA) (4-9), pre-concentration neutron activation analysis (PCNAA) (10), absorption spectroscopy (6), and inductively coupled plasma mass spectrometry (ICP-MS) (11-14). However, only alpha spectrometry following radiochemical separation allows the determination of the thorium isotopes with the greatest dosimetric impact for biological samples, ^{228}Th, ^{230}Th, and ^{232}Th. Alpha spectrometry also allows the use of a tracer (e.g. ^{229}Th or ^{234}Th) for chemical yield determination.[1] While alpha spectrometry offers good radiometric detection limits (2.8x10^{-4} Bq/sample) (2), the very long half-life of ^{232}Th (4.5x10^9 years) makes the mass detection limit fairly high (~70 ng) compared to isotopes with shorter half-lives.

Studies of human populations by a number of investigators have provided some information on the distribution of thorium in the human body as the result of long-term chronic intake from environmental sources of thorium (i.e. non-occupational intake). All of these studies indicate that the skeleton is the major deposition site for thorium for persons exposed principally through inhalation as well as for persons exposed to environmental sources of thorium (presumably, mostly through inhalation). However, these studies on human exposure were based on the measurement of only a few samples from each body and in many cases the analytical results were of low precision because of the very low activities of thorium isotopes in human tissues from normal individuals.

Neutron activation analysis has long been used for the determination of ^{232}Th via the (n,γ) reaction and subsequent beta decay of the short lived ^{233}Th ($t_{1/2}$=22.3 min)[15] product to ^{233}Pa ($t_{1/2}$=27.0 days) (15). ^{232}Th has large (n,γ) cross-sections (σ_γ=7.37 b, I=85 b) (16) and ^{233}Pa is determined by measuring the 300 (6.2%), 312 (36%) or 340 (4.2%) keV gamma rays(15). In radiochemical neutron activation analysis (RNAA) for the determination of ^{232}Th, ^{233}Pa is separated from matrix elements following neutron activation to minimize interferences and reduce the gamma-ray background. While this technique is capable of much lower detection limits for ^{232}Th compared to alpha spectrometry, RNAA is not suitable for isotopic thorium analysis.

Neutron activation analysis is a multielement technique capable of low detection limits, good precision, and in some applications, can have essentially a zero blank. However, in some instances, it is helpful to use various pre-concentration methods to allow the use of larger samples or to remove specific interferences. These techniques, known as pre-concentration neutron activation analysis, PCNAA, have many of the same advantages of NAA but may have contribution to the blank by reagents and environmental factors which must be assessed. This paper will present one such work for the determination of isotopic thorium combining alpha spectrometry, PCNAA, and PCRNAA.

Experimental

Preparation of reagents and irradiation vials

A thorium standard (^{232}Th) for neutron activation analysis was prepared by dissolving Th(NO$_3$)$_4$•xH$_2$O (Johnson, Matthey, & Co.) in 1 M HNO$_3$ and the isotopic concentration (^{228}Th, ^{230}Th, ^{232}Th) determined by alpha spectrometry. The ^{229}Th tracer used for the determination of isotopic thorium was prepared from the National Institute for Standards and Technology (NIST) Standard Reference Material (SRM 4328A). ^{231}Pa was prepared from Amersham Certified Reference Materia PNP100101. Working solutions were prepared by volumetric dilution of a known weight of reference material to an appropriate working concentration (~0.2 Bq/mL).

All reagents (HNO$_3$, HCl) were trace metal grade (Fisher Scientific). De-ionized water was used for the preparation of all solutions and was prepared to 18 MΩ using distilled water in a Nanopure™ system.

Flip-top polyethylene vials (Fisher scientific) were soaked for 24 hr in 40% HNO$_3$ (v/v), rinsed with de-ionized water 2-3 times, and then soaked for 24 hr in de-ionized water. The vials were drained, soaked in acetone for 24 hr, and then dried in a laminar flow hood.

Electrodeposition disks were prepared from 0.25 mm thick sheets of 99.7% pure vanadium (Aldrich) machine punched to 5/8" diameter planchets. Disks were rinsed with acetone prior to use.

Sample preparation

Human tissue samples were dried at 110°C, ashed to 450°C and then wet ashed with HNO$_3$ and H$_2$O$_2$ (*17*). Residues from certain tissues (e.g. lung and lymph nodes) were treated with HF to dissolve silicates. Samples were then dissolved in 8 M HCl and an aliquot selected for analysis. Quality control samples consisting of an aliquot of diluted ^{232}Th standard (from the Th(NO$_3$)$_4$ stock solution) as well as a reagent blank were analyzed with each set of samples. The radiochemical recovery was determined by adding approximately 0.08 Bq (5 dpm) of ^{229}Th tracer to each sample aliquot.

Radiochemical separation of thorium from samples

The sample, quality control samples, and blanks were taken to dryness on a hot plate prior to separation and wet ashed repeatedly with concentrated HNO_3 to ensure that the ^{229}Th tracer and thorium in the sample were in the same chemical form (i.e. Th(IV)) and thoroughly distributed within the sample. The aliquot, containing ^{229}Th tracer, was then dissolved in 8 M HNO_3 and passed over 12 cm^3 of AG 1x8 (Cl^- form) that had been previously conditioned with 5 column volumes of 8 M HNO_3. The columns were then rinsed with 5 column volumes of 8 M HNO_3 and the thorium eluted with 5 column volumes of 9 M HCl (to prevent co-elution of ^{239}Pu in the sample). Two mL of 0.36 M $NaHSO_4$ were added to the eluent to prevent radiochemical losses during subsequent electrodeposition steps. The eluent was then taken to dryness and wet ashed with concentrated HNO_3 and in some cases concentrated H_2SO_4 to destroy any organic material that was present following separation.

Samples were electroplated according to the method of Glover et al. (*18*). Samples were dissolved in 5 mL of 0.75 M H_2SO_4, several drops of thymol blue indicator added, and then transferred into electrodeposition cells followed by two subsequent 3 mL rinses of 0.75 M H_2SO_4. The pH of the sample solution was adjusted to 1.5-2 using concentrated NH_3. One important difference was the use of 99.7% pure vanadium planchets rather than stainless steel planchets typically used for electrodeposition.

Determination of thorium by alpha spectrometry

After electrodeposition the ^{228}Th, ^{230}Th, ^{232}Th, and the ^{229}Th tracer in each sample were determined by alpha spectrometry in a Canberra Alpha Analyst system equipped with 450 mm^2 detectors calibrated over the range of 3.5 to 7 MeV in 1024 channels. Samples were counted on the second shelf (approximately 0.5 mm source-to-detector distance) which yielded approximately a 20% efficiency (counts/α emission). Detectors were energy calibrated using secondary sources of approximately 1 Bq each of ^{234}U, ^{238}U, ^{239}Pu, and ^{241}Am. The detectors were efficiency calibrated using secondary sources containing approximately 15 Bq of ^{242}Pu. These secondary sources were calibrated using NIST SRM 4906L, a ^{238}Pu point source, at the greatest source-to-detector geometry (~4 cm) to minimize geometry differences between the point source and the 5/8" planchet used for sample preparation. Background counts for each detector were of 300,000 seconds duration and samples were counted for 100,000 to 300,000 seconds. The chemical yield of the separation and electrodeposition steps was obtained by the ratio of the net counts of ^{229}Th versus the expected count rate of the decay corrected tracer.

Determination of ^{232}Th by PCRNAA (ion exchange method)

Samples were irradiated at the 1 MW TRIGA III fueled research reactor located at Washington State University for 6 hours with a thermal neutron flux of 6.5×10^{13} cm^{-2}s^{-1}. The samples were allowed to decay for approximately 12 hours in the pool to allow the short lived activation products to decay.

The vanadium planchets were dissolved using 10 mL of 8 M HNO$_3$/0.025 M HF spiked with approximately 0.2 Bq of ^{231}Pa in a covered polyethylene beaker (Azlon™) suitable for heating to 130 °C. The addition of HF is required to keep the Pa in solution as it will rapidly hydrolyze. Also, this ensures the Pa was in chemical equilibrium with the ^{231}Pa tracer which was also in 8 M HNO$_3$/0.025 M HF. Following completion of this exothermic reaction, the beaker was heated at 90 °C for 15 minutes to insure completion of the reaction. 90 mL of 9 M HCl were then added to the sample and allowed to cool to room temperature.

The ion exchange columns were prepared using a 10 mL Fast Rad™ polyethylene column (Environmental Express LTD) with 200 mL plastic reservoir containing 10 mL of Biorad AG 1x8 resin, 100-200 mesh. The columns were washed with 5 column volumes of 0.5 M HCl to remove all actinides and pre-conditioned with 5 column volumes of 9 M HCl prior to addition of the samples. Glass components were not used for any step in these procedures due to the presense of HF in the samples.

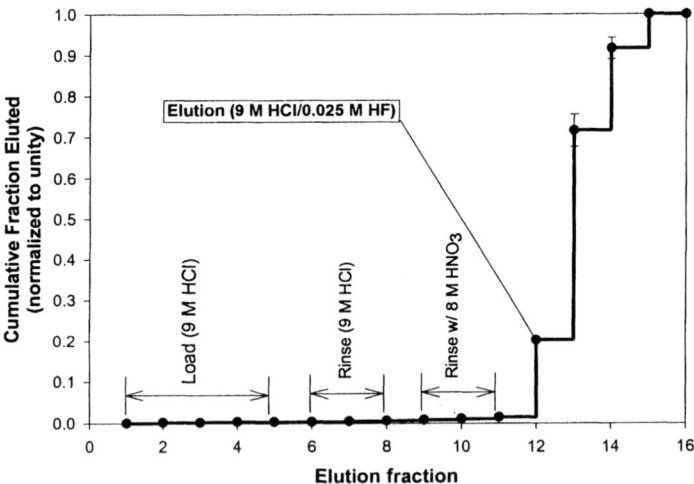

Figure 1. Elution profile of protactinium from anion exchange column method

Immediately prior to addition to the column, 2 mL of 0.5 M Al(NO$_3$)$_3$ was added to the sample to bind the flouride, the sample stirred thoroughly, and then added to the column. The beaker was rinsed three times with 9 M HCl and these rinses were also added to the column. Each step was allowed to pass completely through the column prior to addition of the next wash step. The column was washed with 5 column volumes of 9 M HCl, then rinsed twice with 2.5 column volumes of 8 H HNO_3. The Pa was then eluted with 10 column volumes of 9 M HCl/0.025 M HF into a polyethylene beaker. (Figure 1)

Electrodeposition of Pa

It was quickly determined that a modification of the electrodeposition method would be required for completion of this work due to the HF used in the elution of Pa. This method and its evaluation will be discussed in detail elsewhere. A brief description of the method involves adding 1 mL of 9 M H_2SO_4 to the eluent (in a plastic beaker capable of heating to 130 °C) and evaporating to a constant volume at 90 °C (H_2SO_4 does not evaporate at this temperature) and then following the previously described method with a pH of 2, an electrodeposition time of 1.5 hours, at a constant current of 0.75 amps. Samples were then counted by alpha spectrometry for 100,000 seconds in the previously described system.

Results and discussion

Alpha spectrometry of alpha emitting thorium isotopes

The alpha spectrometric determination of isotopic thorium in biogogical samples has been reported in several publications. Both ^{234}Th and ^{229}Th have been used as radiochemical tracers. Alpha emitting ^{229}Th has a number of properties which make it superior to ^{234}Th for tracer recover including a 7340 y half-life and availability as an SRM from NIST. Disadvantages of using ^{229}Th, however, include recoil contamination of the detector with short-lived, alpha-emitting progeny (^{225}Ac and its daughters) which can interfere with ^{228}Th measurements at high activities. Also, low probability (~0.3%) ^{229}Th alpha emissions are present in the ^{230}Th region of interest which must be accounted for to preclude bias for low activity measurements. A cross-over value of (5±1) x 10^{-3} counts ^{230}Th/^{229}Th was experimentally determined for the reported method and has been used to correct the counts of ^{230}Th.

Electrodeposited thorium sources must have good alpha spectrometric resolution (<30 keV FWHM) because of the proximity of the ^{229}Th tracer alpha peaks to ^{230}Th. The reliability of the measurement of ^{230}Th and ^{232}Th by alpha spectrometry was evaluated by measuring QC samples spiked over a range of activities. Excellent agreement between the measured value and the expected value was obtained for activities greater than the MDA of the method.

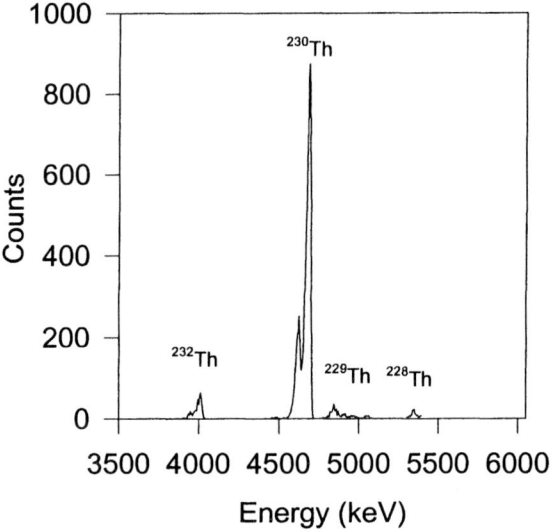

Figure 2:Thorium alpha spectra

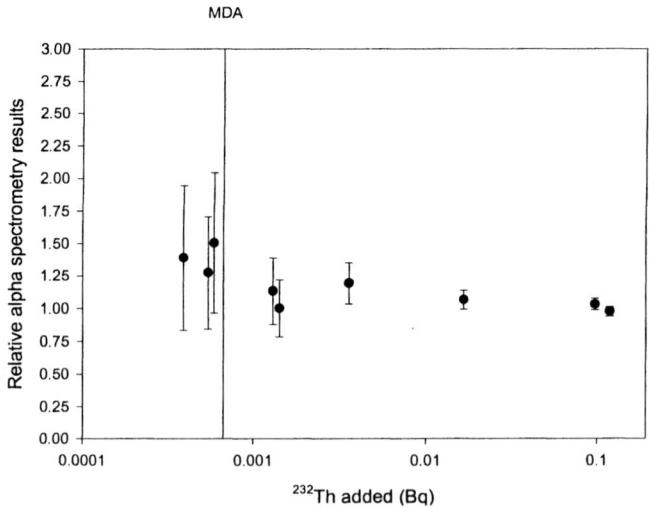

Figure 3: Plot of relative alpha spectrometry results (normalized to unity) versus expected values.

Figure 2 shows an alpha spectrum for a sample containing ^{232}Th, ^{230}Th, ^{228}Th and isotopic tracer ^{229}Th. The tracer allows for the conversion of the number of counts to activity because there is no change in detector efficiency over the energy range of interest. The number of counts in each region of interest is determined. The background spectrum for each detector was previously obtained and the same region of interest used for each isotope in the sample used to determine the appropriate background. Figure 3 shows the results of the determination of various activities of ^{232}Th which clearly shows that the method is limited to approximately 10^{-3} Bq, and 10^{-2} Bq for good precision measurements.

Determination of ^{232}Th by PCNAA and PCRNAA

Selection of an appropriate substrate was critical to the success of this method. Typically stainless steel or platinum planchets are used for alpha spectrometry. Both are unsuitable for subsequent determination of ^{232}Th by NAA because of the large activity due to (n,γ) reaction products (e.g. ^{59}Fe, ^{51}Cr, ^{191,197}Pt) which increases the background under the ^{233}Pa gamma ray peaks. Neutron irradiation of vanadium planchets, however, produces predominantly ^{52}V ($t_{1/2}$=3.76 min) from the ^{51}V(n,γ)^{52}V reaction plus small amounts of ^{47}Sc and ^{48}Sc from fast neutron (n,α) reaction on ^{50}V and ^{51}V, respectively. The radionuclides induced by the irradiation of the vanadium planchet did not interfere in the determination of ^{233}Pa but do contribute to the background under the ^{233}Pa γ-rays and make it necessary to optimize counting conditions. The signal to noise ratio was optimized after 9-10 days decay.

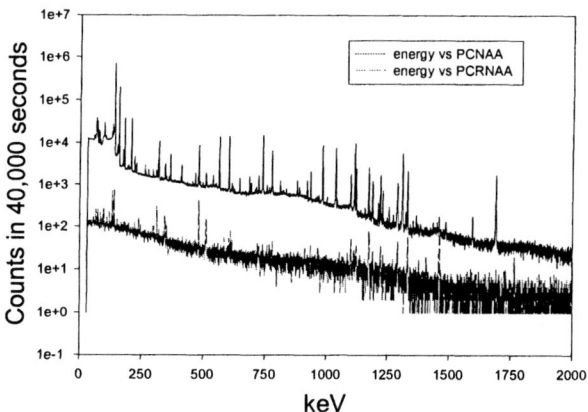

Figure 4: Spectra for the determination of ^{232}Th by PCNAA and PCRNAA using a high resolution HPGe detector following a 10 day cooling period.

The gamma ray spectrum of an irradiated vanadium planchet electroplated with ^{232}Th showing three main gamma ray emissions (300, 312, 340 keV) from ^{233}Pa is shown in Figure 4. The spectrum of a radiochemical blank carried through the method (tracer, separation, electrodeposition, alpha spectrometry, and neutron activation) is shown in Figure 4 was found to be free of peaks interfering with the determination of ^{233}Pa. Gamma-ray peaks resulting from impurities or nuclear reactions in the vanadium disk include the 320 keV peak from ^{51}Cr and the very minor 308 and 316 keV gamma ray peaks from ^{182}Ta.

The precision and accuracy of the PCNAA determination of ^{232}Th was measured for a series of ^{232}Th quality control standards and the results are shown in Figure 5. The data demonstrate that the method generates accurate and precise results for ^{232}Th determination for activities well below those achievable by alpha spectrometry. The fixed geometry of the planchets allows for highly reproducible positioning of sources and standards. Figure 5 further shows that there is a small, but not negligible, blank.

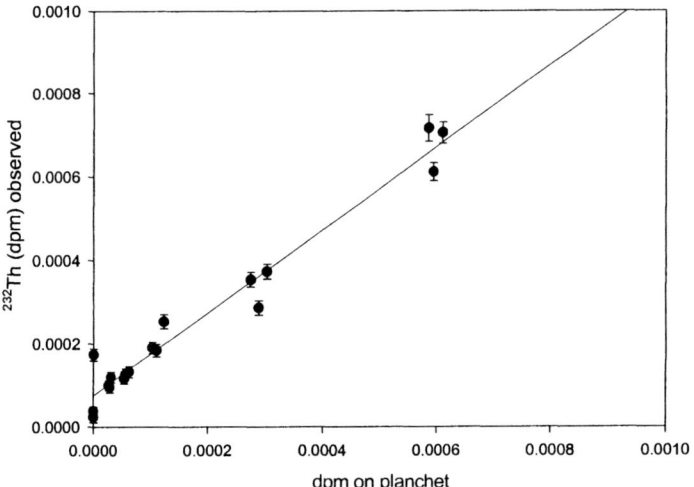

Figure 5: Graph of ^{232}Th determined by PCNAA versus expected (no correction for the blank).

PCNAA offers a very large increase in sensitivity for the determination of ^{232}Th compared to alpha spectrometry and this is evident from the improved detection limits and precision for low activity samples. A sensitivity enhancement of approximately 83,000 versus the alpha spectrometric determination of ^{232}Th was obtained.

The detection limit of ^{232}Th by this method varies with the count time of the sample. A detection limit in tissue samples of approximately 5×10^{-6} Bq/sample (3×10^{-4} dpm) was obtained using the stated parameters, approximately $^{1}/_{50}$th the detection limit of ^{232}Th by alpha spectrometry. Detection limits were calculated based on ANSI NI3.30 criteria.[19] Unlike many methods, the PCNAA method can be used to analyze very large samples. The activity concentration limit (Bq/g) depends primarily on the amount of sample available for analysis because matrix interferences are removed before neutron activation. The overall detection limits (MDA) exhibited by the PCNAA alpha spectrometry method for isotopic thorium compare very well with other reported methods (see Table 1).

Determination of ^{232}Th PCRNAA

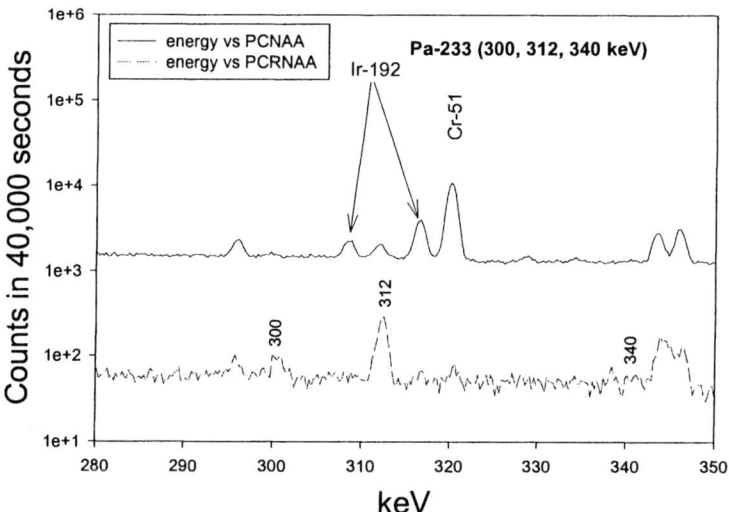

Figure 6: Comparison of PCNAA and PCRNAA in the 312 keV region used for the determination of ^{233}Pa showing the reduction in background and removal of spectral interferences.

Ion exchange chromatography was found to provide a very credible reduction in background interferences and consistently high recoveries (95 ± 5 %). Protactinium was found to be efficiently retained using the procedure previously described (Figure 1). Figure 6 shows an example of a sample counted after irradiation but prior to radiochemical separation of Pa for 80,000 seconds on a 48% HPGe detector in a low background shield containing 2 µBq of ^{232}Th with a higher than typical ^{192}Ir content (notice the 308 and 316 keV peaks) which surround the 312 keV peak used to determine the ^{233}Pa. Figure 6 further shows the results of this same sample following RNAA by ion exchange chromatography as previously described counted for the same time and duration. The ^{192}Ir content was reduced by factor of 30 for this example, and is typically much higher, in many cases with only trace levels of ^{192}Ir remaining. Background for the 312 keV region was reduced by a factor of 10-15 for the samples, resulting in a significant improvement in both the detection limit and the ability to determine ^{232}Th in the blank. The effective detection limit for the method using the conditions as stated is 3.5 x10^{-7} Bq for ^{232}Th and it is capable of 3-5% precision at levels above the limit of quantification. The method showed itself to provide reliable results for the determination of ^{232}Th (Figure 7).

Table I: Comparison of detection limits for important thorium isotopes

Isotope	*Detection limit (Bq)*		
	Alpha Spec	*PCNAA*	*PCRNAA*
^{228}Th	6x10^{-4}	-	-
^{230}Th	2.8x10^{-4}	-	-
^{232}Th	2.8x10^{-4}	5x10^{-6}	3.5x10^{-7}

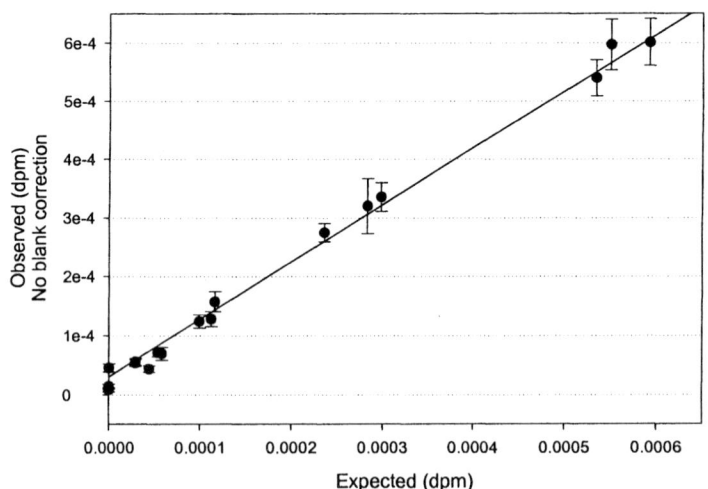

Figure 7: Graph of ^{232}Th determined by PCRNAA versus expected (no correction for the blank)

References

1. Wrenn, M.E.; Singh, N.P.; Cohen, N.; Ibrahim, S.A.; Saccomanno, G. NUREG/CR-1227 US Nuclear Regulatory Commission. Washington, D.C., 1981.
2. Environmental Measurements Laboratory Procedures Manual: HASL-300. 27th edition. U.S. Dept. of Energy, New York, 1992.
3. Kitamura, K.; Inazawa, Y.; Moritimo, T.; Sato, K; Higuchi, H.; Imai, K.; Watari, K.. *J. Radioanal. Nuclear Chem* **1997**, *217*, 175-180.
4. Clifton, R.J.; Farrow, M.; Hamilton, E.L. *Ann. Occ. Hyg.* **1971**, *14*, 303-310.
5. Sunta, C.M.; Dang, H.S.; Jaiswal, D.D. *J. Radioanal. Nuclear Chem.* **1987**, *1*, 149-154.
6. Lucas, Jr., H.F.; Edgington, D.N.; Markun, F. *Health Physics* **1970**, *19*, 739-742.
7. Edgington, D.N. *Int. J. Applied Rad. Isot.* **1967**, *18*, 11-18.
8. Picer, M.; Strohal, P. *Anal. Chim. Acta* **1968**, *40*, 131-136.
9. Jaiswal, D.D.; Dang, H.S.; Sunta, C.M. *J. Radioanal. Nuclear Chem* **1985**, *88/2*, 225-229.
10. Glover, S.E., Filby, R.H., Clark, S.B. *J. of Radioanal. Nuclear Chem.* **1998**, *234*, 201-208.
11. Crain, J.S.; Smith, L.L.; Yaeger, J.S.; Alvarado, J.A. *J. Radional. Nuclear Chem* **1995**, *194*, 133-139.
12. Crain, J.S. *Spectroscopy* **1996**, *11*, 31-36.
13. Crain, J.S.; Mikesell, B.L.. *Appl. Spectroscopy* **1992**, *46*, 1498-1503.
14. Terry, K.W.; Hewson, G.S.; Meuner, G. *Health Physics* **1995**, *68*, 105-111.
15. *Table of Radioactive Isotopes*. Browne, E.; Firestone, R.B.; Editor Shirley, V.S. John Wiley & Sons, New York, 1986.
16. Walker, F.W.; Parrington, F.R.; Feiner; F. *Nuclides and Isotopes. 14th edition*. GE Nuclear, San Jose, CA. 1989.
17. McInroy, J.F.; Boyd, H.A.; Eutsler, B.C.; Romero, D. *Health Physics* **1985**, *49*, 587-592.
18. Glover, S.E., Filby, R.H., Clark, S.B. *J. of Radioanal. Nuclear Chem.* **1998**, *234*, 213-218.
19. *Performance Criteria for Radiobioassay; Standard N13.30*. American National Standards Institute. 1996.

Chapter 20

Applications of Radioanalytical Chemistry to Alzheimer's Disease

J. D. Robertson[1] and M. A. Lovell[2,3]

[1]Department of Chemistry and Missouri University Research Reactor, University of Missouri, Columbia, MO 65211
[2]Sanders-Brown Center on Aging, University of Kentucky, Lexington, KY 40536
[3]Department of Chemistry, University of Kentucky, Lexington, KY 40506

Two examples of the use of radioanalytical methods to probe key questions about the pathogenesis of Alzheimer's disease (AD) are presented. In the first, micro-beam proton-induced X-ray emission analysis was used to investigate imbalances of Zn in senile plaques (SP) and neurofibrillary tangles (NFT) in AD in light of observations that this metal can accelerate aggregation of amyloid beta peptide *in vitro*. In the second, accelerator mass spectrometry and the ^{14}C bomb pulse were used to determine the average "age" of NFT and SP in the AD brain to help define the time course and defining factors involved in the formation of these pathological features.

Alzheimer's disease (AD), an age associated dementing disorder, is the most common form of adult onset dementia and is the fourth leading cause of death in the United States (*1*). A community-based study suggests that 4 million persons in the US have AD (*2*) and that, with the aging of society, nearly 9 million individuals could be affected by the year 2040 (*3*). Clinically, AD is characterized by a gradually progressive decline in cognitive function and pathologically by neurofibrillary tangle (NFT), senile plaque (SP) and neuropil thread formation, and neuron and synapse loss, particularly in the hippocampus, amygdala, entorhinal cortex, neocortex and nucleus basalis of Meynert. The major histopathologic features of AD are senile plaques (SP), composed primarily of amyloid-β (Aβ) peptide and neurofibrillary tangles (NFT), composed of paired helical filaments containing hyperphosphorylated tau. These pathologic lesions markers remain the key criteria for the neuropathologic diagnosis of AD (*4*). The major barrier to treating and/or preventing AD is a lack of understanding of the etiology and pathogenesis of neuron degeneration. Numerous etiologic/pathogenic mechanisms have been suggested for the cause of AD including: genetic defects, metabolic defects, oxidative stress, endogenous toxins, mitochondrial defects, latent or slow virus, trace element toxicity or some combination of the above. In this work, we review our use of radiochemical methods to investigate the potential role of trace metals in AD and the time course involved in the formation of NFT and SP.

Trace Metals and AD

A decade ago, the association of metals and AD conjured up thoughts of tossing out aluminum cookware and deodorant and not drinking beverages from aluminum cans. The hypothesis that exogenous aluminum (Al) plays a role in the development of AD now receives little attention, in part, because of radioanalytical studies of Al in AD brain. Ehmann and Markesbery used instrumental neutron activation analysis (INAA), along with graphite furnace atomic absorption spectrometry (GFAAS) and laser microprobe mass spectrometry (LMMS) to study possible Al imbalances in brains of AD patients compared with age-matched controls. While slight elevations of Al were observed in small, bulk AD brain samples using GFAAS and in some AD brain cellular/subcellular components of the hippocampus using LMMS. INAA analyses did not detect an Al imbalance in AD brain when much larger samples were taken from a wider range of brain regions (*5*). Moreover, Watt and Grime et al. demonstrated through a series of micro-particle induced x-ray emission (μPIXE) measurements that Al was not, as proposed, enhanced in SP and NFT of AD brain (*6*).

While several other elements have been implicated in AD, including Hg (*7*) and Si (*8*), Zn has become the focus of considerable interest. Zinc is a crucial component in over 300 enzymes including Cu/Zn superoxide dismutase, and in various eukaryotic transcription factors. The possible connection of Zn to AD was first proposed by Burnett who suggested that Zn deficiencies lead to dementia and that Zn supplements might delay or prevent the onset of dementia (9).

The brain contains three Zn pools: 1) a membrane bound metalloprotein; 2) a vesicular pool localized in nerve terminal synaptic vesicles; and 3) an ionic pool of free or loosely bound ions in the cytoplasm (10). Of these sources, the vesicular pool, which is easily chelated, is thought to be the most important because it is released with glutamate during neurotransmission and may reach concentrations of 300 µM in the synapse. Unless these gradients of Zn are immediately sequestered they could potentially induce neurodegeneration. Although Zn is essential for normal function of the brain and may act in a protective role at low concentration, experimental studies have shown high (100 and 1000 µM) concentrations of Zn are toxic to neurons *in vitro* (11, 12) and *in vivo* (13, 14).

In 1994, Bush and co-workers reported that Zn at low physiological concentrations induced Aβ aggregation *in vitro* (15,16). These reports, and the INAA studies that showed significant Zn elevations at the "bulk" level in AD hippocampus and amygdala (7,17), led us to investigate the association between Zn and SP in AD brain using the radioanalytical technique micro-PIXE. The measurement is challenging because SP are only 20 to 40 microns in diameter and because Zn occurs at trace (10 to 100 µg/g) levels in the plaques and surrounding tissue.

The technique of using an accelerated particle beam for x-ray emission analysis was first introduced at the Lund Institute of Technology in 1970 (18). PIXE, like other elemental analysis x-ray spectroscopic techniques, utilizes the x-rays that are emitted from the atoms in a sample when that sample is exposed to an excitation source. The energies of the resultant x-rays are characteristic of the elements from which they are emitted and the number of x-rays of a given energy is proportional to the mass of that corresponding element in the sample. The use of a proton beam as an excitation source offers several advantages over other x-ray techniques including a higher rate of data accumulation across the entire periodic table and better overall sensitivities, especially for the lower atomic number elements. As illustrated in Figure 1, the better sensitivity of PIXE in comparison to electron excitation is due to a lower bremsstrahlung background (see *19*). The fractional mass sensitivities achieved with PIXE are typically on the order of 1 µg per gram for samples whose total target masses range between 10 mg for standard systems to 1 pg for a proton microprobe. Of course, the chief disadvantage of PIXE is the reason that it is considered a radioanalytical technique; it requires the use of a particle accelerator.

The results of the micro-PIXE analysis of ten SP and surrounding neuropil from 9 AD subjects and neuropil from 5 age-matched control subjects are presented in Table I. A detailed description of the micro-PIXE measurements can be found in reference (20). Comparison of overall SP Zn levels with AD neuropil using the Mann-Whitney U-test (non-normal distributions) demonstrated a significant elevation of Zn ($p<0.05$) in SP compared with surrounding neuropil. Likewise, comparison of AD and control neuropil indicated a significant elevation of Zn ($p<0.05$) in AD.

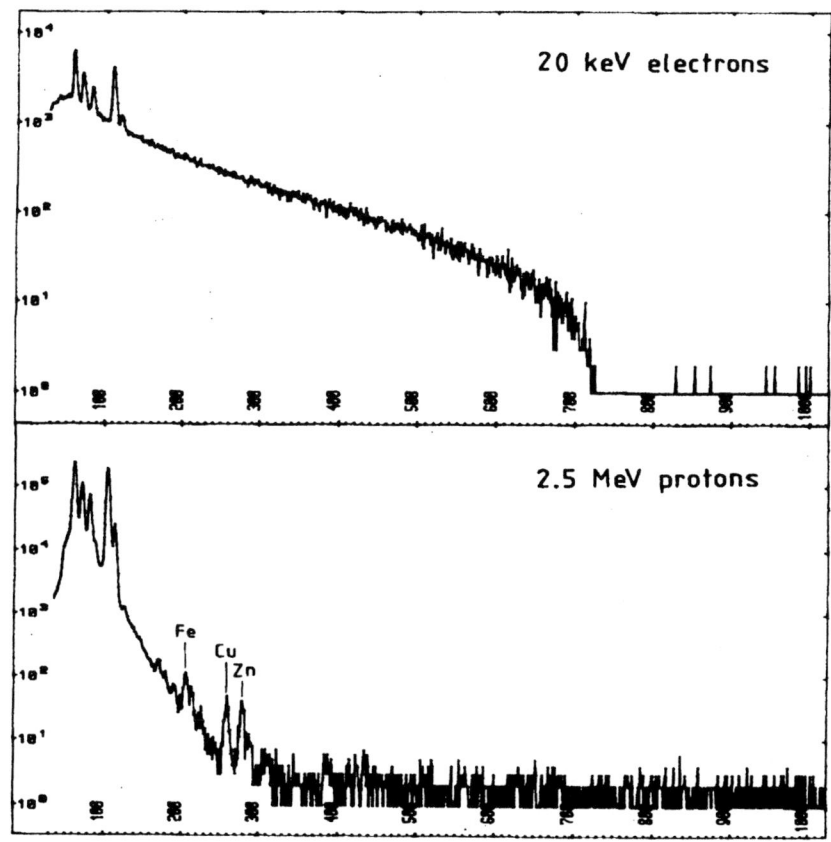

Figure 1. *The X-ray spectra of brain tissue taken with an electron microprobe and a proton microprobe. From PIXE: A Novel Technique for Elemental Analysis, S.A.E. Johansson and J.L. Campbell, 1988, John Wiley and Sons Limited. Reproduced with permission.*

AD is, most likely, a disease of multiple, interrelated, etiologic/pathogenic factors and it is naïve to conclude from our measurements that the disease results from a simple excess of a trace-element like zinc. On the other hand, a possible role for the disruption of Zn homeostasis in AD is inferred from several experimental models and the effects of Zn on a number of protein/peptide systems that are known to be altered in AD. For example, amyloid precursor protein (APP)

Table I. Average (n=10) and standard deviation of the zinc concentrations (µg/g) in SP in AD subjects and in neuropil in AD and control subjects.

Subject	AD SP	Neuropil	Subject	Control Neuropil
1AD	38 ± 8	23 ± 2	1C	38 ± 20
2AD	36 ± 19	18 ± 3	2C	23 ± 4
3AD	146 ± 49	112 ± 34	3C	30 ± 10
4AD	54 ± 22	29 ± 6	4C	28 ± 5
5AD	58 ± 23	54 ± 6	5C	24 ± 2
6AD	36 ± 10	37 ± 10		
7AD	104 ± 50	82 ± 8		
8AD	140 ± 97	72 ± 59		
9AD	163 ± 108	88 ± 13		

contains a ligand binding site for Zn and at concentrations less than 50 µM, Zn inhibits degradation of APP and the proteolytic cleavage of Aβ (21) leading to the generation of the toxic Aβ fragment and perhaps the excess levels of Aβ observed in AD. Of considerable note is the *in vivo* evidence recently reported by Lee et al. (22) that altered Zn transporter expression and the synaptic Zn released during neurotransmission is responsible for amyloid deposition in transgenic mice expressing cerebral amyloid plaque pathology.

Time Course of NFT and SP in AD

There is a paucity of knowledge about the factors and time course involved in the formation of NFT and SP. Because no current diagnostic procedure can detect the presence of these pathologic features in living subjects, it is unclear whether Aβ deposition occurs initially and influences neuron degeneration and NFT formation, or if NFT occur first followed by Aβ deposition, or they occur concomitantly. Although recent studies suggest that NFT formation is a primary event (23,24) these autopsy studies only represent an endpoint measure and do not provide definite temporal information. The temporal relationship between these pathologic features and cognitive impairment is also unclear. Current evidence indicates that NFT in particular may exist in aged prospectively evaluated nondemented subjects and formation of NFT and SP may begin even before age 30 (25). Studies of the relationship between NFT and SP densities and clinical parameters have been contradictory with nearly equal numbers of reports supporting a positive correlation between severity of cognitive impairment and degree of SP or NFT accumulation as those that do not (26)

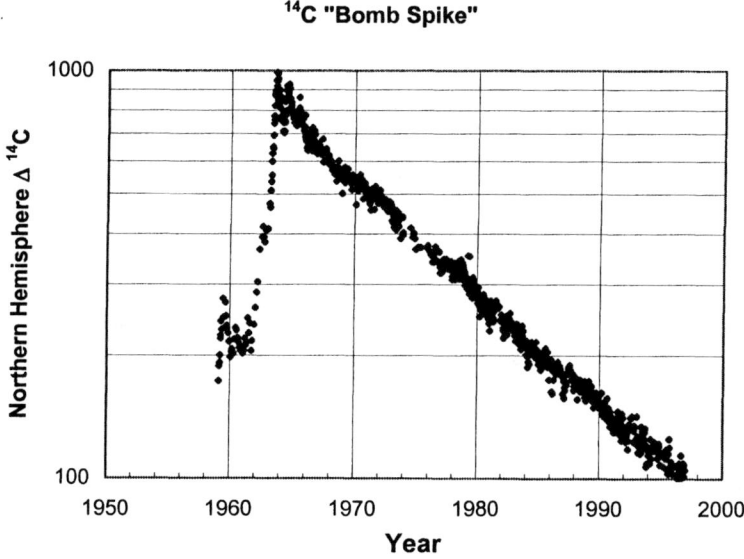

Figure 2. Northern hemisphere environmental 14C levels before and after above ground nuclear testing in per mil units.

Carbon-14 is produced in the atmosphere through cosmic ray interaction with nitrogen. Atmospheric testing of nuclear weapons in the early 1960s doubled the natural concentration of this radioisotope as shown in Figure 2 (adapted from (27,28,29). After the peak in 1963, ^{14}C levels began decreasing exponentially to the contemporary level of 110% that of 1950 (Figure 2) and should reach prebomb levels within 20 years. The observed decrease is due to sequestration of CO_2 in environmental reservoirs and the introduction of ^{14}C depleted CO_2 from burning fossil fuels. As previously demonstrated in analyses of gallstones (30), the elevation in ^{14}C above the contemporary level can be used to date biological samples in which carbon does not turn over. The ^{14}C content of any tissue is a function of dietary ^{14}C levels and the residence time of carbon in that tissue. Several studies of ^{14}C in human tissues during the late 1960s and early 1970s demonstrated incorporation of bomb-pulse ^{14}C into a variety of organs with variable turnover rates (31, 32). The slowest turnover rates are observed in collagen and bone, whereas brain exhibits the fastest. Because NFT and SP are essentially insoluble, we hypothesized that they turn over carbon at a slow rate and that the ^{14}C levels in these structures would be indicative of their average age of formation.

The results of the use of accelerator mass spectrometry (AMS) and the measurement of ^{14}C levels in isolated SP and NFT to determine the average age of formation of isolated SP and NFT fractions in bulk brain samples of six AD subjects are given in Table II. Details of the experiment are presented in reference (26). Although a heterogeneous process of formation is indicated, what is clear from this work is that these two pathologic features, once formed, have a much slower carbon turnover rate than normal brain tissue and are not in a formation/enzymatic degradation equilibrium.

Table II. Average Year of Formation of NFT and SP Isolated from AD Bulk Brain Samples

Subject	NFT	SP
1B	1988.4 ± 0.7	1984.0 ± 0.6
2B	1989.0 ± 0.7	1991.2 ± 0.8
3B	1981.3 ± 0.6	1984.6 ± 0.6
4B	1978.4 ± 0.8	--
5B	--	1977 ± 1
6B	1988.0 ± 0.6	1987.9 ± 0.6

To verify that aged brain free of NFT and SP actively turns over carbon, samples of cerebellum were analyzed from 5 prospectively studied, normal control subjects ages 52, 63, 74, 81, and 92, all autopsied in 1995. Results of the analyses of cerebellum for the 5 subjects indicate that the individuals turned over carbon at equal rates without regard to age suggesting that ages of SP and NFT fractions measured for AD subjects were not skewed due to the age of the subject. The fact that the average ages of cerebellum samples from subjects of different ages are consistent with the year of death minus 1.4 indicates a rapid turnover rate for carbon in human brain without significant decline due to aging. The lag time of 1.4 years is attributed to the time between photosynthesis in plants and consumption of the "atmospheric" carbon by man.

The small number of subjects in our study and the inherent subject-to-subject variability associated with NFT and SP formation prevents drawing definitive conclusions regarding NFT and SP formation at this time. Because the current measurements are based on bulk samples, the represent an average of NFT and SP isolated from the combined brain regions. Despite being an average measure of the year of formation of NFT and SP, this is the first, and currently only, study that quantitatively assigns a year of formation based on a physical component of the pathologic lesions that can be related to a well defined physical quantity such as atmospheric bomb-pulse ^{14}C.

References

1. Katzman R. *Archives Neurol* 1976,*33* 217.
2. Evans, D.A.; Funkenstein, H.H.; Albert, M.S.; Scherr, P.A.; Cook, N.R.; Chown, M.J.; Herbert, L.E.; Hennekens, C.H.; Taylor, J.D. *JAMA* **1989**, *262,* 2551.
3. National Institiute on Aging. Progress report on Alzheimer's disease. NIH Publications # 95-3994, Washington, DC US Government Printing Office, 1995.
4. National Institute on Aging and Reagan Institute Working Group on Diagnostic Criteria for the Neuropathological Assessment of Alzheimer's disease. *Neurobiol. Aging* **18(S4)**, S1 (1997).
5. Ehmann, W. D.; Markesbery, W. R.   Sanders-Brown Center Aging, University Kentucky, Lexington, KY, USA.   Life Chemistry Reports (1994), 11(1), 11-28.
6. Makjanic, J.; Watt, F. *Nucl. Instr. Methds.* **1999**, *B150,* 167-172.
7. Ehmann, W.D.; Markesbery, W.R.; Alauddin, M.; Hossain; T.M. and Burbaker E.H. *Neurotoxicology* **1986** *7,* 197-206.
8. Candy, J.M.; Klinowski, J.; Perry, R.H.; et al. *Lancet* **1986**, *1,* 354-357.
9. Burnet F.M. *Lancet* **1986**, *1* 186.
10. Freund, W.D.; Reddig, S. *Brain Res.* **1994**, *654* 257.
11. D. W. Choi, D.W.; Yokoyama, M.; Koh, J. *Neuroscience* **1988**, *24,* 67.
12. Duncan, M.W.; Marini, A.M.; Watters, R.; Kopin, I.J.; Markey, S.P. *J. Neurosci.* **1992**, *12* 1523.
13. M. I. Chuah, M.I.; Tennet, R.; Jacobs, I. *J. Neurosci. Res.***1995**, *42,* 470.
14. Cuajungco, M.P.; Lees, G.J. *NeuroReport* **1996**, *7,* 1301.
15. Bush, A.I.; Pettingell, W.H.; Multhap, G.; Paradis, J.D.; Vonsattel, J.P. *Science* **1994**, *265,* 1464.
16. Bush,A.I.; Moir, R.D.; Rosenkranz, K.M.; Tanzi, R.E. *Science* **1995**, *268,* 1921.
17. Wenstrup, D.; Ehmann, W.D.; Markesbery, W.R. *Brain Res.* **1990**, *533,*125-131.
18. Johansson, T.B.; Akselsson, K.R.; Johansson, S.A.E. *Nucl. Instr. Meth. Phys. Res.* **1970**, *84,* 141.
19. Johansson, S.A.E.; Campbell, J.L. *PIXE A Novel Technique for Elemental Analysis*, Wiley: New York, NY, 1988.

20. Lovell, M.A.; Robertson, J.D.; Teesdale, W.J.; Campbell, J.L.; Markesbery, W.R. *J. Neurol. Sci.* **1998**, *158*, 47.
21. Bush, A.I.; Multhaup, G.; Moir, R.D.; Williamson, T.G.; Small, D.H.; Rumble, B.; Pollwein, P.; Beyreuther, K.; Masters, C.L. *J. Biol. Chem.* **1993**, *268*, 16109.
22. Lee, Joo-Yong, Cole, T.B., Palmiter, R.D., Suh S.W., Koh J.Y. *PNAS* **2002**, *99(11)*, 7705-7710.
23. Delacourte, A.; David, J.P.; Seargeant, N.; Buee, L.; Wattez, A.; Vermersch, P.; Ghozali, F.; Fallet-Bianco, C.; Pasquier, F.; Lebert, G.; Petit, H.; Di Manza, C. *Neurology* **1999**, *52*, 1158-1165.
24. Price, J.L.; Morris, J.C. *Ann Neurol* **1999**, *45*, 358-368.
25. Braak, H.; Braak, E. *Acta Neuropathol* **1991**, *82*, 239-259.
26. Lovell, M.A.; Robertson, J.D.; Buchholz, B.A.; Xie, C.; Markesbery, W.R. *Neurobiol Aging* **2002**, *23*, 179-186.
27. Http://cdiac.esd.ornl.gov/epubs/ndp/ndp/ndp057/ndp057.htm.
28. Http://cdiac.esd.ornl.gov/trends/co2/contents.htm
29. Levin, I.; Kromer, B.; Schoch-Fischer, H. *Radiocarbon* **1985**, *27*, 1-19.
30. Druffel, E.M.; Mok, H.Y.I. *Radiocarbon* **1983**, *25*, 629-636.
31. Broecker, W.S.; Schulert, A.; Olson, E.A. *Science* **1959**, *130*, 331-332.
32. Harkness, D.D.; Walton, A. *Nature* **1972**, *240*, 302-303.

Chapter 21

Recent Developments in Neutron Activation Analysis: 1997–2002

S. Landsberger

Nuclear Engineering Teaching Laboratory, University of Texas, Pickle Research Campus, R–9000, Austin, TX 78712

A review of the recent developments and applications of neutron activation analysis in the past five years is presented in this chapter. Topics covered include reference materials, archaeology, geochemistry, engineering materials, environmental, plants and crops, medical and clinical studies, facilities improvements, and software and instrumentation development. Neutron activation analysis is now more widely used than ever before in a plethora of investigations by many worldwide laboratories and the participating research groups.

Introduction

It has been two thirds of a century since Hevesey discovered neutron activation analysis (NAA) in 1936 and developed its application for elemental analysis. During the next two decades NAA was at the forefront of using gamma-ray spectroscopy in conjunction with the continually updated sodium iodide detectors and modern electronics for a variety of studies, including those of cold war national significance. But it was not until President Eisenhower's initiative of Atoms for Peace and the emergence of the International Atomic

Energy Agency (IAEA), both in 1956, that many new areas for nuclear applications using research reactors as a primary source for radioisotope production were constructed. A multitude of papers began to appear followed by national, regional and international conferences. By the 1960's, NAA was unquestionably the prime method for non-destructive and multielemental analysis. But by the middle 1970's, a lot of competition to NAA had arrived. Multielemental techniques such as particle induced x-ray emission (PIXE) and inductively coupled plasma - atomic emission spectroscopy (ICP-AES) and then inductively coupled plasma - mass spectrometry (ICP-MS), were serious alternatives to NAA. During the 1980's, there were many dire predictions that NAA would die a peaceful death and many of the users would turn to ICP methods. However, it became clear that both ICP-AES and ICP-MS methods had problems in sample dissolution and many signal interferences. In fact many new types of samples required a significant amount of research time to eliminate or reduces these problems. In the 1990's, there was a resurgence of interest in the applications of NAA. Many groups utilizing the methods of NAA formed research teams with other non-nuclear scientists to study a multitude of environmental, geological, archaeological, materials, and medical problems. These types of research interactions were made easier by the use of fast desktop computers for spectrum analysis, statistical evaluations and modeling of the results. Private industry also produced better resolution detectors, larger detectors to enable faster analytical turn around times, loss free counting systems, and very user friendly spectrum analysis packages. Many types of automation came into being, both for sample irradiation and sample gamma ray counting. Methods such as Compton suppression and epithermal NAA, with their unique capabilities, are now routinely used in laboratories worldwide. Equally important is the enormous effort put on quality assurance and quality control programs on the NAA product.

A review of the recent developments of mainly reactor based NAA in the past five years is presented. Instead of performing an exhaustive literature research on the multitude of uses in the many journals, the author has decided to concentrate mainly, but not exclusively, on the Journal of Radioanalytical and Nuclear Chemistry to obtain information. These references well represent the large number applications and developments of NAA. None of the theory of NAA or that of gamma ray spectroscopy is presented. Theses topics are well covered in other books, review articles, and references therein (*1-5*).

Reference Materials

Many national and international agencies continue to employ neutron activation analysis methods as one of the very few non-destructive methods for elemental determination, both for certified and informational elemental concentrations. These materials are made available to laboratories to verify the methodologies used in quality control. The National Institute of Standards and Technology and the US Geological Survey extensively use NAA in their many certification programs. In 1996 (6) an in depth intercomparison of the first IAEA airborne particulate matter was made for fine and course particles. Later on the use of NAA, PIXE and x-ray fluorescence (XRF) methods have were used test homogeneity of new IAEA reference air filters (7). NAA was used for quality control of biological reference materials in Canada (8), while in China it was used rare-earth elemental determination (9). In Poland homogeneity studies on various environmental and geological reference materials were done for microanalytical techniques; sample weights of 10 mg or less (10). Further studies in Portugal provided data on sixteen geochemical samples (11), while the preparation of four Czech reference soils was done using NAA (12). Groups in the USA (13), China (14) and Japan (15, 16) have effectively used NAA in the analysis of geological and soil reference materials. Korean rice has been used for the certification of various trace and toxic heavy metals (17), while the IAEA has used NAA for several elements in a Japanese reference diet material (18). A description of the collection and characterization of air particulate matter for use in reference materials has been reported (19). The IAEA through its member states continues its vigorous program in presenting updated databases of natural reference materials, many of which have been analyzed by NAA (20), while an IAEA internet database of natural matrix reference materials has been reported (21, 22). These examples are only a fraction of the many reference materials that benefited by the use of NAA in the certification process.

Archaeology

One area in NAA that has maintained its position in analytical chemistry and that continues to grow is in archaeological specimens. Glascock (23) effectively reviewed the status of NAA in archaeology and geochemistry including its ability to be used in source identification and pattern recognition on a wide variety of samples. NAA's unique ability for multielemental determination of small amounts of material for rare earth and other trace elements makes it ideal for studies involving many samples to be used in statistical analyses. An archaeometric investigation of clay sources and archaeological artifacts in Roman Pavia, led researchers to conclude that most of the ceramic production was produced locally (24). Application of NAA to art objects in Polish fine arts academies and museums led to the understanding of materials used in the various processes, as well as approximating their places of origin (*25*). Equally successful were the chemical characterizations of Brazilian ceramics (*26, 27*). Non-destructive analysis of European and North American trade beads (*28, 29*) gave insight to possible trade routes between these two continents. Further studies included obsidian samples from Central Europe and the Aegean regions, (*30*), pumice from the Aegean region (*31, 32*) and Eastern Mediterranean regions (*33*), medieval ceramics from the Rhineland (*34*), obsidians from Mexican quarries (*35*), Hungary (*36*), Armenia (*37*) and Anatolia (*38*). Compositional studies of ancient copper (*39*), ancient Romanian coins (*40*) and gold objects (*41*) have been undertaken. Other studies have included analysis of medieval French sculptures (*42*) for source identification and provenance studies of pottery fragments from Egypt (*43*). Mean concentrations of the popular Ohio Red Clay reference material have been published for data from several laboratories (*44*).

Food and Diet

Many groups in the world have investigated the major and trace elemental composition of specific foods and various diets. Simulated Iranian mixed diets for essential and toxic elements were investigated for daily intakes to the population (*45*), while similar studies were conducted for nutritional studies for adult Koreans (*46*). Zinc and selenium soil deficiencies has led to the study of these two essential elements in diet samples in selected Turkish populations (*47*), while the intake of fifteen essential elements in diets in university groups (*48*) and pre-school children, healthy adults, and elderly people (*49*) and pre-school children in nurseries (*50*) in Brazil were evaluated. Other determinations of trace elements in Indian vegetarian diets (*51*), other Indian diets (*52*) and in Canada (*53, 54*) were evaluated. Specific popular food items have also been studied for nutritional and possibly toxic effects. These include cheese, eggs, fish, fowl and meats in Canada (*55*), a popular oriental herb in China (*56*), chewing gum in Pakistan (*57*), betel nuts and associated chewing materials among Asian groups in Great Britain (*58*), dried desserts in Greece (*59*), areca nuts and limes in Taiwan (*60*), iodine in various Czech foodstuffs (*61*), salt samples from Ghana, (*62*) algal products in Italy (*63*), and mercury content in Malaysian seafood (*64*). The effect of the addition of rare-earth elements on the increase of rice yield was investigated in Vietnam (*65*). Multielemental analysis of agroindustrial by-products on animal feed were determined in Brazil (*66*), while the transfer of trace elements from food packaging materials into food was also determined (*67, 68*).

Engineering Materials

Many newly fabricated materials need ultra sensitive analytical methods to determine the impurities for quality control. NAA has been used on a variety of engineering materials primarily for chemical characterization. These studies include trace impurities in nickel-based alloys (*69*), manganese in high purity iron (*70*), radiochemical activation analysis of high purity aluminum (*71*) and impurities in gallium oxide (*72*). Certification of ion implanted arsenic in silicon (*73*), determination of copper and iron in silicon wafers (*74*), characterization of trace elements in silicon single crystals (*75*) and silicon wafers (*76, 77*) have all greatly benefited from NAA. The determination of trace elements of lithious ceramic materials to be used as tritogenic breeders for future fusion reactors was done using NAA (*78*), while measurements of phosphorous in metals (*79*) and refractory metals (*80*) were done by radiochemical and ion exchange methods, respectively. The precision analysis of Nb-Ti ratios in a superconducting alloy was determined by a NAA-PIXE combined technique (*81*). A good overview of the unique NAA capabilities for the high purity materials industry has been presented (*82*).

Environmental Monitoring

One of the most popular applications of NAA is that of environmental monitoring. With the exception of lead, almost all environmentally important heavy metals can be determined. Advances in Compton suppression methods and epithermal NAA have replaced many laborious radiochemical methods particularly for air filters and various biological materials, and to a smaller extent for contaminated soils. Steinnes gave a good overview of neutron activation techniques in environmental studies (*83*), while IAEA health-related monitoring programs and assessment of airborne particulate matter and biological studies have been well outlined (*84, 85*). An intercomparison and determination of environmental standards by NAA has been presented (*86*). This section is subdivided into air pollution monitoring, biomonitoring, solid waste and wastewater monitoring.

Air Pollution

NAA of air samples has been in existence for more than three decades and it continues to be a popular method of choice in many different countries. Large amounts of NAA data are available from such investigations to perform source-receptor modeling in conjunction with wind trajectories. Comprehensive overviews on IAEA projects on air pollution (*87*) and nuclear techniques to air particulate matter studies (*88*) have been presented. Air pollution monitoring in urban and suburban areas of Korea (*89, 90*), Bangkok (*91*), Sri Lanka (*92*), Beijing (*93*), Italy (*94, 95*), Toronto (*96*), Czech Republic (*97*), Lisbon (*98*), Istanbul (*99*), Vietnam (*100*) and the USA (*101*) have been carried out. Size distribution in sub micron particulate matter was investigated in the Great Smokey Mountain National Park in the USA (*102*), while the effect of traffic emissions on roadside plants was examined in Saudi Arabia (*103*). Indoor pollutants were characterized in dwellings in Mexico (*104*) using NAA and XRF methods. Regional studies of aerosol composition in Southern Africa (*105*) and over the Great Lakes (*106, 107*) have been reported. Elemental characterization of aerosols in remote areas of the Arctic (*108*) and remote and industrial areas of Siberia have also been described (*109*).

Biomonitoring

Evaluation of air pollution via analysis of biomonitors has greatly increased in the past fifteen years, notably in Europe, and has gained more popularity in other parts of the world. Typically, lichen native to a country is chosen as the receptor. Careful consideration must be given to control studies and any inherent blank values in the lichens themselves. However, biomonitoring has been a very effective way of estimating the deposition of heavy metals onto land surface. Biomonitoring of air pollution has been exemplified by various IAEA programs (*110*). Studies in Portugal (*111-114*) and Brazil (*115-117*) have effectively used several types of lichens over very wide land grids to determine elemental concentrations resulting from air pollution. Studies of air pollution in Beijing's major industrial areas have been done with biomonitors (*118, 119*). Analytical and statistical elemental composition of lichen, including the use of enrichment factors has been made in Argentina (*120*). Analysis of tree rings in Brazil (*121*) and Germany (*122*) and tree bark (*123*) have also been studied to reconstruct historical pollution events.

Solid Waste

Solid waste samples can either be analyzed on the matrix itself (e.g. fly ash, sludge, etc) or on the receptor site (soil or sediment). A case study was presented yielding information on possibilities for NAA research in pre and post radiochemical analysis, exchange properties of waste particles and groundwater, and leaching dynamics of solid waste (*124*). An overview of multielemental analysis of solid wastes and leachates using NAA and ICP-AES methods was presented (*125*). Elemental characterization of coal, sawdust, fly ash and landfill waste samples (*126*) and estuarine sediments (*127*) have been surveyed in Great Britain, while solid waste from India (*128*), ash samples from a Canadian power plant (*129*) and mercury levels of sediment and soil samples in Brazil (*130*) were evaluated. Fingerprints of pollution migration in twenty-one Antarctic marine sediments were investigated for forty elements (*131*). Continued work in the analysis of phosphate fertilizers from Romania (*132*) and Libya (*133*) for potential heavy metal and radiological contamination has been explored. Analytical comparisons between NAA and ICP for background soils (*134*) and various environmental samples including soil, sediment, plant (*135*) and water were evaluated. More recently, NAA and XRF were employed to determine lead and copper contamination from industrial activities and a firing range (*136, 137*) and heavy metals in urban soils in the USA (*138*).

Wastewater

Most multielemental wastewater analyses are done ICP-AES or ICP-MS techniques. This is primarily due for the need to employ preconcentration methods for NAA and the excellent detection limits by ICP methods. However, NAA is still used in several studies. In Japan (*139*) copper in water samples was determined preceded by preconcentration on activated carbon, while preconcentration evaporation methods were used in Romanian water samples (*140*). Trace metals in Taiwan seawater and river water in China (*141*) were evaluated using magnesium oxide as a preconcentration agent (*142*). A combination of NAA and ICP-MS was used to determine sixty elements in river water in Brazil (*143*).

Biological Systems

Trace elemental analysis of biological systems, as an indication of environmental effect of heavy metals has been used for many years, and continues to be popular in NAA studies. Hair analysis still remains a popular technique to assess heavy metal exposure. NAA methods have been used in studies in Korea (*144*), Malaysia (*145*), Brazil (*146, 147*) and Romania (*148, 149*). Compositional baseline studies of hair have been reported in Iran (*150, 151*). Extractable organohalogens in shrimp in the north Atlantic were evaluated using NAA (*152*). Toxic and trace elements in algae for IAEA standard reference materials (*153*) and Ghanaian seaweeds (*154*) were reported and shown to have good potential to be biological indicators of environmental pollution. Several species of marine invertebrates were shown to be reliable indicators of environmental conditions (*155*). Studies reporting aluminum in fish (*156*), mercury in biological reference materials (*157*) and selenium in aquatic species have appeared (*158*). Elemental distribution of tobacco components and smoke (*159, 160*) provided good evidence of potential metal contamination of indoor air where smoking was prevalent. NAA has also been utilized to determine arsenic and antimony in bushes collected near a lead smelter in western Texas (*161*), while plants have been studied to assess heavy metal uptake of polluted soil in Russia (*162*).

Plants and Crops

Major elements in plants and crops such as calcium, magnesium, potassium, sodium, and essential trace elements such as iron, manganese, molybdenum, selenium and zinc can easily be determined by NAA. As well, rare-earth elements (REE), which are also readily determined, have shown to increase production rates. Hence, NAA is an ideal method to non-destructively determine the major, minor and trace elements in a variety of growth studies. Many potentially toxic elements such as, aluminum, antimony, arsenic, chromium, and mercury can also be determined. A study of the nutrient recycling in a plant-soil system was done using NAA (*163*), while the multielemental profile patterns during the lifecycle of a soybean was reported (*164, 165*). Studies in rare-earth content in mature and developing fern leaves (*166*), soil and pine needle (*167*) and natural plants (*168, 169*) have been reported. The interrelationship between REEs and calcium, and fern leaves (*170, 171*) has appeared. Studies have been reported on mineral extraction of legume and grass species (*172*), effects of fertilizer on mineral composition on absorption of major, minor and trace elements on supplemental feeding crops (*173, 174*), and the potential effect of microbial activity on the uptake of elements in wheat seedlings (*175*). Study of the relationship between iridium concentrations in hen eggshell and iridium-enriched feed by NAA was reported. This gave an indication of the ratio from the feed over the eggshell via the gastrointestinal pathway (*176*). Elemental analysis in cultured cells of tobacco and grape treated with potentially toxic levels of aluminum was studied (*177*). A review of radiochemical NAA for biological samples (*178*), stability of bromine and iodine in long term storage of freeze-dried biological materials (*179*) and losses of elements in plant samples under dry ashing conditions have been reported (*180*).

Medical and Clinical

Clinical and medical NAA studies, on human and animal subjects, have increased enormously in the past decade. Spyrou (*181*) has given an overview of the problems and applications of NAA in the biomedical area. There have been several studies on trace elements (*182-186*) and phosphorous (*187*) in human bone. Other studies have used NAA to determine trace elements in human brain including associations with Alzheimer's disease (*188-194*), skin (*195*), cancerous uterine (*196*) and in blood of hypersensitive subjects (*197*). Selenium studies in human liver (*198, 199*), in diet supplements for subjects with ulcerative colitis (*200*), in long-term human status (*201*), and possible association with cancer mortality rates have been reported (*202*). Other studies have included determination of elemental composition of breast milk (*203-205*), thyroid (*206*), breast tissue (*207*), lung (*208*), and chromium in DNA, RNA and protein fractions (*209*). The study of metallic ion transfer induced by orthopedic surgical tools or by metallic prostheses has also appeared (*210*). A review of *in-vivo* NAA for medical applications has been presented by Krishnan (*211*), while NAA methods for pharmaceutical analysis have appeared (*212, 213*). Evaluation of pharmaceuticals using *in-vivo* inelastic scattering and NAA has also been presented (*214*). Malnutrition studies of Bangladesh children were completed by determining trace element concentrations in hair in control and as well as malnourished subjects (*215*) have been reported, while determination of trace elements in nail clippings of healthy children and those with cystic fibrosis (*216*) were compared. NAA has been applied in various animal studies that include hair analysis of equines (*217*), trace elements in organs and tissues of zinc deficient mice (*218*), selenium in bovine samples (*219*), trace elements in porcine brain (*220*), and trace elements in several kinds of tissue in even-toed ungulates such pigs, cows and sheep (*221*). Distribution of samarium and ytterbium in rats measured by enriched stable isotope tracer technique was also presented (*222*).

Geochemistry

In the 1960's and onward, NAA gained its fame in geochemistry studies through its ability to determine many rare-earth elements in a variety of geological specimens. Even though ICP-AES and ICP-MS has gained popularity for geological trace element research, NAA still retains its market share, albeit somewhat smaller. The lack of need for any laborious dissolution methods and potential signal interferences from ICP methods remain the primary reasons that NAA is still the method of choice in geochemistry research in many worldwide laboratories. The determination and distribution of trace elements in suspended matter and bottom sediments in the Japan, China and Bering seas have been reported (*223*). Other NAA investigations include: trace elements in brine an deposits in salt lake (*224*), sediments (*225-233*), rock salts (*234*), marine fossil shells (*235*), natural fluorite (*236*), pegmatites (*237*), aluminiferous materials (*238*), carbonatites (*239*), halogens in rocks (*240*) iridium in geological specimens by radiochemical methods (*241, 242*), opals (*243*), and monazite (*244*). An overview of NAA as a geochemical tool (*245*) has been outlined. Other investigations have involved single crystal element analysis in rock forming materials (*246*), mercury isotopic ratios in stone meteorites (*247*), determination of hafnium and zirconium in geological materials (*248*), trace analysis of limestone (*249*), rhyolitic rocks (*250*), peat bog (*251*) thin sections of rocks by micro-coring (*248*), and in nickel and cobalt concentrates (*252*) NAA has also been used effectively to study elemental components in soils (*253-257*).

Facilities

Prompt Gamma NAA

Prompt gamma NAA utilizes the emission of gamma rays while the sample is being irradiated. Essentially it is an on-line method that is complimentary to conventional NAA. In particular, elements such as hydrogen, boron, carbon, nitrogen, phosphorous, sulfur, cadmium, mercury and some rare-earth elements can be detected. The disadvantage of the method is that only one sample at a time can be irradiated usually at a much lower neutron flux. Ultimately this means longer irradiation times. However, with the advent of cold neutrons (slower than the usual thermal neutrons used in conventional NAA) and focused beams, many special types of samples can be analyzed. Descriptions of PGNAA facilities in Korea (*258*), Budapest (*259*), USA (*260*), Switzerland (*261*) and Japan (*262*) have been reported. An overview of PGNAA fundamentals and applications has recently been presented (*263*). Descriptions of focused cold neutron beams (*264*) and capillary optics (*265*) have been presented for the NIST reactor in USA, while an account of guided neutron beams in Japan has been reported (*266*). Monte Carlo neutral particle (MCNP) computer codes have been used to estimate background interference (*267*), optimize instrument design for analysis of cement raw materials in Portugal (*268*), and calibrate and validate the determination of chlorine in soil (*269*). A new gamma-ray spectrum catalog and library (*270, 271*) have been published greatly facilitating many other users. Identification of photopeak profile of full energy and single and double escape peaks in PGNAA has been described (*272*), while coincidence and anticoincidence measurements with pulsed cold neutrons beams have bee evaluated (*273*). PGNAA has continued to be used successfully in wide variety of studies. Samples include: metals (*274*), environmental samples (*275, 276*), rats (*277*), meat homogenates (*278*), archaeological samples (*279, 280*), stainless steel (*281*), and silicon in meteorites samples (*282*) and stable ^{30}Si (*283*) as tracer in migration studies in geological materials. Determination of boron in volcanic rocks (*284*), reference materials (*285*) and animal plasma, urine feces, and body tissue (*286*) by PGNAA is still very convenient due to the very high of ^{10}B(n, α) ^{7}Li cross section yielding results in a short period of time with very good detection limits. New research in neutron resonance capture analysis has recently been reported (*287*).

NAA of Large Volume Samples

Perhaps the most significant evolution of NAA has been the development of kilogram size samples for elemental analysis. The Delft group in the Netherlands pioneered this work in the 1990's. At present there exists no other known method that can be used to analyze such vast amounts of material. The applications are many and extremely useful in environmental monitoring which usually relies on small amounts of material. The Delft group developed algorithms for neutron self-shielding and gamma ray self-attenuation (*288*). A good overview of large sample NAA has been presented (*289*), while the accepted use of inhomogeneous large samples was discussed (*290*). Other groups began to develop semi-empirical methods for detector absolute efficiency (*291*) and correction factors for large volume NAA (*292*). Studies of plastics (*293*) and mining waste materials (*294*) revealed the practical usage of large volume samples with very reliable results. Other large sample NAA facilities in Germany (*295*) and Austria (*296*) have now been developed, while neutrons from a Russian accelerator (*297*) have been used for analysis of oil, fuel, petrol and gaseous condensates in large volume samples.

Innovations, Upgrades and Characterizations

Several NAA facilities have been developed which take into account accelerator (*298*) and cyclotron neutrons (*299*), ^{252}Cf sources (*300-302*), and the use of reactor "shut downs" that still emit neutrons for short-lived isotope NNA (*303*). While these neutron fluxes may be low, they can effectively be used for larger samples or for concentrations of elements in the percent range. Progress of NAA work in the Atomic Institute in Vienna (*304*), and effective NAA usage in small (*305*) and medium sized reactors (*306*) has been reported. Work continues on parameterization of detector efficiency for NAA standardization (*307*), characterization of neutron flux gradients (*308*) and optimization of automated NAA (*309*). Epithermal neutron flux characterization of the TRIGA Mark III reactor in Mexico (*310*), alternative methods for thermal neutron flux measurements (*311*), corrections for non-ideal behavior of epithermal neutron spectrum (*312*), interferences in NAA (*313*), the importance of double neutron capture (*314*) the determination of threshold NAA reactions (*315*), development of an internal standard (*316*) and estimation of activity for intermittent irradiation (*317*) have been reported. Neutron self-shielding for hydrogenous samples (*318*) and gamma-ray attenuation for high-Z samples (*319*) has been described. Cyclic activation analysis with epithermal neutrons was successfully employed for the determination of fluorine (*320*).

One of the most significant changes in NAA in the past decade has been the routine use of Compton suppression methods to electronically reduce the detection limits of activated samples by using anti-coincidence methods. Tests of multielemental analysis of bone samples (*321*), Arctic aerosols (*322*), biological reference materials (*323*) and brain samples (*188*) have been conducted using Compton suppression methods. The capabilities of Compton suppression methods in multielemental NAA (*324*) and an analytical comparison with large and well-type detectors have been made (*325*).

Standardless NAA, more commonly known as the k_0-method, is now a well-accepted method to employ in multielemental analyses. The method originally developed two decades ago by De Corte (*326*) is based on being able to correctly characterize the irradiation position in any part of the reactor as well as the efficiency calibration of a detector in any geometrical position as not to need any standards. The k_0-method has proven to be a valuable tool and is used in many laboratories including: Belgium (*327-331*), Germany (*332, 333*), Canada and Jamaica (*334*), Brazil (*335-337*), Cuba (*338, 339*), Netherlands (*340, 341*), Japan (*342, 343*), Hungary and Slovenia (*344*), France (*345*), India (*346*) and Belarus (*347, 348*).

Software, Statistics and Total Quality Management

Major software innovations have meant easier data collection with more reliable results. Some of these areas include: coincidence summing effects (*349, 350*), automation and computerization (*351, 352*), maintaining high count rates with a loss free counter (*353-355*), high count rate corrections with zero deadtime correction (*356, 357*), and pulse generating systems for dead-time and pile-up correction (*358*). Sophisticated peak fitting routines for simple and complex spectra have also been published (*359, 360*). Reviews on NAA software (*361*) and performance control of gamma ray spectrometers have appeared (*362*).

Statistical evaluations of NAA and gamma ray spectrometry data continue to play an ever-important role. Overviews of detectors and detection limits in NAA (*363*) and analysis and uncertainty (*364*) have been presented. Detection and quantification (*365*), ISO guide to uncertainty (*366*), metrology for NAA (*367*), Poisson statistics in nuclear spectrometry (*368*), precision NAA determination of macrocontents (*369*) and the use of NAA for precise nuclear data (*370*), have all had important contributions to keep neutron activation analysis in the forefront of competitive technologies.

Finally, total quality management (TQM) for NAA, once considered just a side research effort, is now a major component of all laboratories. It is equally important for those facilities that process only several certified reference materials on a yearly basis to those who process from hundreds to thousands of samples per year. TQM for NAA was first purposed by Bode (*371-374*) whereupon the NAA facility was the first laboratory to be ISO certified. Under new and stricter accountability almost all NAA laboratories have varying amounts of quality assurance and quality control programs.

Conclusions

Rather than disappearing, NAA is now more widely used in research activities than ever before. Besides the plethora of research articles that have appeared in this review, there are many worldwide universities and government laboratories that perform NAA work for industry, other government laboratories, and other universities. Therefore, any educated guess of the total usage of NAA would be grossly underestimated, since many NAA projects to these organizations often do not lead to research articles. The single greatest threat to NAA is not competition from other chemical techniques, but rather the decommissioning of research reactors. While the number of research reactors continues to increase in the developing world, there has been a marked decrease in the North America and Western Europe. In the USA the Department of Energy has now begun to seriously support various research activities at University reactors, including NAA. Hopefully this support in the USA will allow neutron activation analysis to flourish and continue be a dynamic and ever-evolving analytical method.

References

1. De Soete; Gijbels, R; Hoste. J. *Neutron Activation Analysis*, Wiley-Interscience, 1972.
2. Heydorn. H. *Neutron Activation Analysis for Clinical Trace Element Research*, Volume I and II, CRC Press, 1984.
3. Alfassi, Z. B. *Neutorn Activaton Analysis* Volume I & II, CRC Press, 1990.
4. Parry, S. J. Activation Spectrometry in Chemical Analysis, John- Wiley, 1991.
5. Landsberger, S. *J. Trace Micro.Tech.* **1992,** 10, 1-41.
6. Landsberger, S.; Wu D.; Vermette, S.J., Cizek, W. *J. Radioanal. Nucl. Chem.* **1997,** 215, 117-127
7. Kucera, J; Parr, R. M.; Smodis B., Fajgelj, A; Mattiuzzi, M; Havranek. *J. Radioanal. Nucl. Chem.* **2000,** 244, 121-126.
8. Ihnat, M. *J. Radioanal. Nucl. Chem.* **2000,** 245, 65-72.
9. Weizhi, T.; Bangfa N.; Wang, P.; Huiling, N. *J. Radioanal. Nucl. Chem.* **2000,** 245, 51-56.
10. Dybczynski, R; Danko, B.; Polkowska-Motrenko, H. *J. Radioanal. Nucl. Chem.* **2000,** 245 97-104.
11. Gouveia, M. A.; Prudencio, M. I. *J. Radioanal. Nucl. Chem.* **2000,** 245, 105-108.
12. Kucera, J.; Sychra, V.; Horakova, J; Soukal. L. *J. Radioanal. Nucl. Chem.* **1997,** 215, 147-155.
13. Ila, P; Frey, F. A. *J. Radioanal. Nucl. Chem.* **2000,** 244, 599-602.
14. Bangfa, N.; Yu, Z.; G, He.;Wang, P.; Tian, W. *J. Radioanal. Nucl. Chem.* **1997,** 215, 77-79.
15. Sakamoto, K.; Aota, N.;Miyamoto,Y; Kosanda, S.; Oura, Y.; Igarashi, M.; Nakanishi, T. *J. Radioanal. Nucl. Chem.* **1997,** 215, 69-76.
16. Tsukada, M.; Sato, D.; Endo, K.; Yanaga, M.; Currie, L. A.; Glascock, M. D.; Ondov, J. M.; Han, M. *J. Radioanal. Nucl. Chem.* **2000,** 246, 463-466.
17. Chung, Y. S.; Chung, Y. J.; Cho, K. H., Lee, J. H. *J. Radioanal. Nucl. Chem.* **1997,** 215, 129-134.
18. Kawaruma, H.; Parr, R. M.; Dang, H. S.; Tian, W.; Barnes, R. M.; Iyengar, G. V., *J. Radioanal. Nucl. Chem.* **2000,** 245, 123-126.
19. Parr, R.; Smodis, B. *Bio. Trace Elem. Res.* **1999,** 71-72, 169-179.
20. Parr, R. M.; Kawamura, H.; Iyengar, G. V. *Bio. Trace Elem. Res.* **1999,** 71-72, 5-13.
21. Bleise, A. R.; Smodis, B.; Glavic-Cindro, D.; Parr, R. M. *Bio. Trace Elem. Res.* **1999,** 71-72, 47-53.
22. Bleise, A. R.; Smodis, B.; Glavic-Cindro, D.; Parr, R. M. *J. Radioanal. Nucl. Chem.* **2000,** 248, 205-209.
23. Glascock, M. D. *J. Radioanal. Nucl. Chem.* **2000,** 244, 537.

24. Meloni, S.; Oddone, M.; Genova, N.; Cairo, A. *J. Radioanal. Nucl. Chem.* **2000**, 244, 553-558.
25. Panczyk, E.; Ligeza, M.; Walis, L. *J. Radioanal. Nucl. Chem.* **2000**, 244, 543-551.
26. Munita, C. S.; Paiva, R. P.; Alves, M. A.; Momose, E. F.; Saiki, M. *J. Radioanal. Nucl. Chem.* **2000**, 244, 575-578.
27. Munita, C. S.; Paiva, R. P.; Alves, M. A.; de Oliveira, P. M. S.; Momose, E. F. *J. Radioanal. Nucl. Chem.* **2001**, 248, 93-96.
28. Hancock, R. G. V.; MeKechnie, J.; Aufreiter, S.; Karklins, K.; Kapches, M.; Sempowski, M.; Moreau, J. -F.; Kenyon, I. *J. Radioanal. Nucl. Chem.* **2000**, 244, 567-573.
29. Sempowski, M. L.; Nohe, A. W.; Moreau, J. -F.; Kenyon, I.; Karklins, K.; Aufreiter, S.; Hancock, R. G. V. *J. Radioanal. Nucl. Chem.* **2000**, 244, 559-566.
30. Kilikoglou, V.; Bassiakos, Y.; Doonan, R. C.; Stratis, J. *J. Radioanal. Nucl. Chem.* **1997**, 216, 87-93.
31. Peltz, C.; Schmid, P.; Bichler, M. *J. Radioanal. Nucl. Chem.* **1999**, 242, 361-377.
32. Peltz, C.; Bichler, M. *J. Radioanal. Nucl. Chem.* **2001**, 248, 81-87.
33. Bichler, M.; Egger, H.; Preisinger, A.; Ritter, D., Stastny, P. *J. Radioanal. Nucl. Chem.* **1997**, 224, 7-14.
34. Mommsen, H.; Hein, A.; Hahnel, E. *J. Radioanal. Nucl. Chem.* **1997**, 216, 247-252.
35. Jimenez-Reyes, M.; Tenorio, D.; Esparza-Lopez, J. R.; Cruz-Jimenez, R. L.; Mandujano, C.; Elizalde, S. *J. Radioanal. Nucl. Chem.* **2001**, 250, 465-471.
36. Oddone, M.; Marton, P.; Bigazzi, G.; Biro, K. T. *J. Radioanal. Nucl. Chem.* **1999**, 240, 147-153.
37. Oddone, M.; Bigazzi, G.; Keheyan, Y.; Meloni, S. *J. Radioanal. Nucl. Chem.* **2000**, 243, 673-682.
38. Oddone, M.; Yegingil, Z.; Bigazzi, G.; Ercan, T.; Ozdogan, M.; *J. Radioanal. Nucl. Chem.* **1997**, 224, 27-38.
39. Olariu, A.; Besliu, C.; Belc, M.; Popesuc, I. V.; Badica, T.; Lazarovici, Gh. *J. Radioanal. Nucl. Chem.* **2000**, 243, 599-605.
40. Cojocaru, V.; Serbanescu, D. *J. Radioanal. Nucl. Chem.* **1997**, 222, 15-20.
41. Olariu, A.; Constantinescu, M.; Constantinescu, O.; Badica, T.; Popescu, I. V.; Besliu, C.; Leahu, D. *J. Radioanal. Nucl. Chem.* **1999**, 240, 261-267.
42. Holmes, L. L.; Harbottle, G. *J. Radioanal. Nucl. Chem.* **2001**, 248, 75-79.
43. Beal, J. W.; Olmez, I. *J. Radioanal. Nucl. Chem.* **1997**, 221, 9-17.
44. Kuleff, I.; Djingova, R. *J. Radioanal. Nucl. Chem.* **1998**, 237, 3-6.
45. Gharib, A. G.; Aminpour, A. A.; Ahmadiniar, A. *J. Radioanal. Nucl. Chem.* **2001**, 249, 47-60.

46. Cho, S. Y.; Lee, J. K.; Kang, S. H.; Chung, Y. S.; Lee, J. Y. *J. Radioanal. Nucl. Chem.* **2001**, 249, 39-45.
47. Aras, N. K.; Nazli, A.; Zhang, W.; Chatt, A. *J. Radioanal. Nucl. Chem.* **2001**, 249, 33-37.
48. Maihara, V. A.; Favaro, D. I. T.; Silva, V. N.; Gonzaga, I. B.; Silva, V. L.; Cunha, I. I. L.; Vasconcellos, M. B. A.; Cozzolino, S. M. F. *J. Radioanal. Nucl. Chem.* **2001**, 249, 21-24.
49. Favaro, D. I. T.; Maihara, V. A.; Mafra, D.; Souza, S. A.; Vasconcellos, M. B. A.; Cordeiro, M. B. C.; Cozzolino, S. M. F. *J. Radioanal. Nucl. Chem.* **2000**, 244, 241-245.
50. Favaro, D. I. T.; Chicourel, E. L.; Maihara, V. A.; Zangrande, K. C.; Rodrigues, M. I.; Barra, L. G.; Vasconcellos, M. B. A.; Cozzolino, S. M. F. *J. Radioanal. Nucl. Chem.* **2001**, 249, 15-19.
51. Singh, V.; Garg, A. N. *J. Radioanal. Nucl. Chem.* **1997**, 217, 139-145.
52. Dang, H. S.; Jaiswal, D. D.; Nair, S. *J. Radioanal. Nucl. Chem.* **2001**, 249, 95-101.
53. Zikovsky, L.; Soliman, K. *J. Radioanal. Nucl. Chem.* **2001**, 247, 171-173.
54. Shi, Y.; Sullivan, E. E.; Holzbecher, J.; Chatt, A. *Bio. Trace Elem. Res.* **1999**, 71-72, 377-386.
55. Zikovsky, L.; Soliman, K. *J. Radioanal. Nucl. Chem.* **2002**, 251, 507-509.
56. Chen, C. Y. *J. Radioanal. Nucl. Chem.* **2002**, 252, 551-558.
57. Zaidi, J. H.; Arif, M.; Fatima, I.; Ahmad, S. Qureshi, I. H. *J. Radioanal. Nucl. Chem.* **2000**, 243, 683-688.
58. Ridge, C.; Akanle, O.; Spyrou, N. M. *J. Radioanal. Nucl. Chem.* **2001**, 249, 67-70.
59. Kyritsis, A.; Kanias, G. D.; Tzia, C. *J. Radioanal. Nucl. Chem.* **1997**, 217, 209-219.
60. Wei, Y. Y.; Chung, C. *J. Radioanal. Nucl. Chem.* **1997**, 217, 45-51.
61. Kucera, J.; Randa, Z.; Soukal, L. *J. Radioanal. Nucl. Chem.* **2001**, 249, 61-65.
62. Nyarko, B. J. B.; Serfor-Armah, Y.; Osae, S.; Akaho, E. H. K.; Anim-Sampong, S.; Maakuu. B. T. *J. Radioanal. Nucl. Chem.* **2002**, 251, 281-284.
63. Avino, P.; Carconi, P. L.; Lepore, L.; Moauro, A. *J. Radioanal. Nucl. Chem.* **2000**, 244, 247-252.
64. Rahman, S. A.; Wood, A. K.; Sarmani, S.; Majid, A. A. *J. Radioanal. Nucl. Chem.* **1997**, 217, 53-56.
65. Pham Thi Huynh, M.; Carrot, F.; Chu Pham Ngoc, S.; Dang Vu, M.; Revel, G. *J. Radioanal. Nucl. Chem.* **1997**, 217, 95-99.
66. Teruya, C. M.; Armelin, M. J. A.; Silva Filho, J. C.; Silva, A. G.; Saiki, M. *J. Radioanal. Nucl. Chem.* **2000**, 244, 237-240.

67. Thompson, D.; Parry, S. J.; Benzing, R. *J. Radioanal. Nucl. Chem.* **1997**, 217, 147-150.
68. Parry. S.J. *J. Radioanal. Nucl. Chem.* **2001**, 248, 143-147.
69. Zaidi, J. H.; Waheed, S.; Ahmed, S. *J. Radioanal. Nucl. Chem.* **1999**, 242, 259-263.
70. Tomura, K.; Tomuro, H. *J. Radioanal. Nucl. Chem.* **1999**, 242, 147-153.
71. Zaidi, J. H.; Arif, M.; Fatima, I.; Ahmed, S.; Qureshi, I. H. *J. Radioanal. Nucl. Chem.* **1999**, 241, 123-127.
72. Kim, N. B.; Woo, H. J.; Lee, K. Y.; Yoon, Y. Y.; Chun, S. K.; Park, K. S. *J. Radioanal. Nucl. Chem.* **2000**, 245, 37-40.
73. Greenberg, R. R.; Lindstrom, R. M.; Simons, D. S. *J. Radioanal. Nucl. Chem.* **2000**, 245, 57-63.
74. Shigematsu, T.; Polasek, M.; Yonezawa, H.; Katoh, M. *J. Radioanal. Nucl. Chem.* **1997**, 216, 237-240.
75. Takeuchi, T.; Nakano, Y.; Fukuda, T.; Hirai, I.; Osawa, A.; Toyokura, N. *J. Radioanal. Nucl. Chem.* **1997**, 216, 165-169.
76. Kim, N. B.; Choi, H. W.; Chun, S. K.; Cho, S. Y.; Woo, H. J.; Park, K. S. *J. Radioanal. Nucl. Chem.* **2001**, 248, 125-128.
77. Swanson, C. C.; Filo, A. J.; Lavine, J. P. *J. Radioanal. Nucl. Chem.* **2001**, 248, 69-74.
78. Moauro, A.; Carconi, P. L.; Casadio, S. *J. Radioanal. Nucl. Chem.* **1997**, 216, 171-178.
79. Paul, R. L. *J. Radioanal. Nucl. Chem.* **2000**, 245, 11-15.
80. Kim, N. B.; Park, K. S. *J. Radioanal. Nucl. Chem.* **2000**, 245, 41-45.
81. Kim, Y, S.; Kim, D. K.; Lee, K. Y.; Choi, H. W.; Yoon, Y. Y.; Shim, G.; Kim, N, B. *J. Radioanal. Nucl. Chem.* **1997**, 216, 137-141.
82. Herrera, R. S.; Denison, J. R.; Spate, V. L.; Gudino, A. M.; Baskett, C. K.; Dubman, I. M.; Mason, M. M.; Bohl, E. E.; Williams, A.; Nichols, T.; Glascock, M. D.; Morris, J. S. *J. Radioanal. Nucl. Chem.* **2001**, 248, 39-44.
83. Steinnes, E. *J. Radioanal. Nucl. Chem.* **2000**, 243, 235-239.
84. Parr, R. M.; Smodis, B. *Bio. Trace Elem. Res.* **1999**, 71-72, 169-179.
85. Parr, R. M. *J. Radioanal. Nucl. Chem.* **2000**, 244, 17-21.
86. Chung, Y. S.; Jeong, E. S.; Cho, S. Y. *J. Radioanal. Nucl. Chem.* **1997**, 217, 71-76.
87. Smodis, B.; Bleise, A. *J. Radioanal. Nucl. Chem.* **2000**, 244, 97-102.
88. Zeisler, R.; Haselberger, N.; Makarewicz, M.; Ogris, R.; Parr, R. M.; Stone, S. F.; Valkovic, O.; Valkovic, V.; Wehrstein, E. *J. Radioanal. Nucl. Chem.* **2000**, 217, 5-10.
89. Chung, Y. S.; Moon, J. H.; Chung, Y. J.; Cho, S. Y.; Kang, S. H. *J. Radioanal. Nucl. Chem.* **1999**, 240, 79-94.
90. Chung, Y. S.; Chung, Y. J.; Jeong, E. S.; Cho, S. Y. *J. Radioanal. Nucl. Chem.* **1997**, 217, 83-89.

91. Nouchpramool, A.; Sumitra, T.; Leenanuphunt, V. *Bio. Trace Elem. Res.* **1999**, 71-72, 181-187.
92. Seneviratne, M. C. S.; Mahawatte, P.; Fernando, R. K. S.; Hewamanna, R.; Sumithrarachchi, C. *Bio. Trace Elem. Res.* **1999**, 71-72, 189-194.
93. Wang, X. *Bio. Trace Elem. Res.* **1999**, 71-72, 203-208.
94. Gallorini, M.; Rizzio, E.; Birattari, C.; Bonardi, M.; Groppi, F. *Bio. Trace Elem. Res.* **1999**, 71-72, 209-222.
95. Rizzio, E.; Bergamaschi, G.; Profumo, A.; Gallorini, M. *J. Radioanal. Nucl. Chem.* **2001**, 248, 21-28.
96. Jervis, R. E.; Tan, P. V.; Evans. G. J. *Bio. Trace Elem. Res.* **1999**, 71-72, 223-232.
97. Kucera, J.; Havranek, V.; Smolik, J.; Schwarz. J.; Vesely, V.; Kugler, J.; Sykorova, I.; Santroch, J. *Bio. Trace Elem. Res.* **1999**, 71-72, 233-245.
98. Reis, M. A.; Freitas, M. C.; Alves, L. C.; Marques, A. P.; Costa, C. *Bio. Trace Elem. Res.* **1999**, 71-72, 273-280.
99. Ozben, C.; Belin, B.; Guven H. *J. Radioanal. Nucl. Chem.* **1998**, 238, 101-104.
100. Binh, N. T.; Truong, Y.; Ngo, N. T.; Sieu, L. N.; Hien, P. D. *J. Radioanal. Nucl. Chem.* **2000**, 244, 103-107.
101. Heller-Zeisler, S. F.; Borgoul, P. V.; Moore, R. R.; Simoliar, M.; Suarez, A. E.; Ondov, J. M. *J. Radioanal. Nucl. Chem.* **2000**, 244, 93-96.
102. Gone, J. K.; Olmze, I.; Ames, M. R. *J. Radioanal. Nucl. Chem.* **2000**, 244, 133-139.
103. Altaf, W. J. *J. Radioanal. Nucl. Chem.* **1997**, 217, 91-94.
104. Martinez, T.; Lartigue, J.; Navarrete, M.; Avila, P.; Lopez, C.; Cabrera, L.; Vilchis, V. *J. Radioanal. Nucl. Chem.* **1997**, 216, 37-39.
105. Salma, I.; Maenhaut, W.; Annegarn, H. J.; Andreae, M. O.; Meixner, F. X.; Garstang, M. *J. Radioanal. Nucl. Chem.* **1997**, 216, 143-148.
106. Biegalski, S. R.; Landsberger, S.; Hoff, R. *J. Air and Waste Manag.* **1998**, 48, 227.237,
107. Landsberger, S.; Biegalski, S. R. *Bio. Trace Elem. Res.* **1999**, 71-72, 247-256.
108. Landsberger, S.; Zhang, P.; Wu, D.; Chatt, A. *J. Radioanal. Nucl. Chem.* **1997**, 217, 11-15.
109. Peresedov, V. F. *J. Radioanal. Nucl. Chem.* **1997**, 224, 21-25.
110. Smodis, B.; Parr, R. M. *Bio. Trace Elem. Res.* **1999**, 71-72, 257-266.
111. Freitas, M. C.; Reis, M. A.; Marques, A. P.; Wolterbeek, H. Th. *J. Radioanal. Nucl. Chem.* **2000**, 244, 109-113.
112. Freitas, M. C.; Reis, M. A.; Marques, A. P.; Wolterbeek, H. Th. *J. Radioanal. Nucl. Chem.* **2001**, 249, 307-315.
113. Freitas, M. C.; Nobre, A. S. *J. Radioanal. Nucl. Chem.* **1997**, 217, 17-20.

114.Freitas, M. C.; Reis, M. A.; Alves, L. C.; Wolterbeek, H. Th.; Verburg, T.; Gouveia, M. A. *J. Radioanal. Nucl. Chem.* **1997**, 217, 21-30.
115.Figueiredo, A. M. G.; Saiki, M.; Ticianelli, R. B.; Domingos, M.; Alves, E. S.; Markert, B. *J. Radioanal. Nucl. Chem.* **2001**, 249, 391-395.
116.Coccaro, D. M. B.; Saiki, M.; Vasconcellos, M. B. A.; Marcelli, M. P. *J. Radioanal. Nucl. Chem.* **2000**, 244, 141-145.
117.Saiki, M.; Horimoto, L. K.; Vasconcellos, M. B. A.; Marcelli, M. P.; Coccaro, D. M. B. *J. Radioanal. Nucl. Chem.* **2001**, 249, 317-320.
118.Saiki, M.; Chaparro, C. G.; Vasconcellos, M. B. A.; Marcelli, M. P. *J. Radioanal. Nucl. Chem.* **1997**, 217, 111-115.
119.Ni, B.; Tian, W.; Nie, H.; Wang, P.; He, G. *Bio. Trace Elem. Res.* **1999**, 71-72, 267-272.
120.Calvelo, S.; Baccala, N.; Arribere, M. A.; Guevara, S. R.; Bubach, D. *J. Radioanal. Nucl. Chem.* **1997**, 222, 99-104.
121.Oliveira, H.; Fernandes, E. A. N.; Ferraz, E. S. B. *J. Radioanal. Nucl. Chem.* **1997**, 217, 125-129.
122.Wallner, G. *J. Radioanal. Nucl. Chem.* **1998**, 238, 149-153.
123.Pacheco, A. M. G.; Freitas, M. C.; Barros, L. I. C.; Figueira, R. *J. Radioanal. Nucl. Chem.* **2001**, 249, 327-331.
124.Jervis, R. E.; Rodgers, J. D.; Evans, G. J. *J. Radioanal. Nucl. Chem.* **1997**, 215, 15-21.
125.Landsberger, S.; Kaminski, M.; Basunia, M.; Iskander, F. Y. *J. Radioanal. Nucl. Chem.* **2000**, 244, 35-40.
126.Alamin, M. B.; Spyrou, N. M. *J. Radioanal. Nucl. Chem.* **1997**, 216, 41-45.
127.Randle, K.; Al-Jundi, J. *J. Radioanal. Nucl. Chem.* **2001**, 249, 361-367.
128.Garg, A. N.; Ramakrishna, V. V. S.; Singh, V.; Olaniya, M. S. *J. Radioanal. Nucl. Chem.* **1997**, 217, 31-37.
129.Tzonev, V.; Zikovsky, L. *J. Radioanal. Nucl. Chem.* **2001**, 250, 563-564.
130.Goncalves, C.; Favaro, D. I. T.; Melfi, A. J.; de Oliveira, S. M. B.; Vasconcellos, M. B. A.; Fostier, A. H.; Guimaraes, J. R. D.; Boulet, R.; Forti, M. C. *J. Radioanal. Nucl. Chem.* **2000**, 243, 789-796.
131.Waheed, S.; Ahmad, S.; Rahman, A.; Qureshi, I. H. *J. Radioanal. Nucl. Chem.* **2001**, 250, 97-107.
132.Pantelica, A. I.; Salagean, M. N.; Georgescu, I. I.; Pincovschi, E. T. *J. Radioanal. Nucl. Chem.* **1997**, 216, 261-264.
133.El-Ghawi, U.; Patzay, G.; Vajda, N.; Bodizs, D. *J. Radioanal. Nucl. Chem.* **1999**, 242, 693-701.
134.Lee, S. Y.; Watkins, D. R.; Jackson, B. L.; Schmoyer, R. L.; Lietzke, D. A.; Burgoa, B. B.; Branson, J. T.; Ammons, J. T. *J. Radioanal. Nucl. Chem.* **1997**, 217, 57-64.
135.Revel, G.; Ayrault, S. *J. Radioanal. Nucl. Chem.* **2000**, 244, 73-80.

136. Landsberger, S.; Iskander, F.; Basunia, S.; Barnes, D.; Kaminski, M. D. *Bio. Trace Elem. Res.* **1999**, 71-72, 387-396.
137. Basunia, S.; Landsberger, S. *J. Air and Waste Manag.* **2001**, 51, 174-185.
138. Kaminski, M. D.; Landsberger, S. *J. Air and Waste Manag.* **2000**, 60, 1667-1679.
139. Sakai, Y.; Tomura, T.; Ohshita, K.; Koshimizu, S. *J. Radioanal. Nucl. Chem.* **1998**, 230, 261-263.
140. Lucaciu, A.; Staicu, L.; Spiridon, S.; Scintee, N.; Arizan, D. *J. Radioanal. Nucl. Chem.* **1997**, 216, 29-31.
141. Yeh, S. J.; Lin, C. P.; Tsai, C. S.; Yang, H. T.; Ke, C. N. *J. Radioanal. Nucl. Chem.* **1997**, 216. 19-24.
142. Lo, J. M.; Lin, K. S.; Wei, J. C.; Lee, J. D. *J. Radioanal. Nucl. Chem.* **1997**, 216, 121-124.
143. Veado, M. A. R. V.; Pinte, G.; Oliveira, A. H.; Revel, G. *J. Radioanal. Nucl. Chem.* **1997**, 217, 101-106.
144. Cho, S. Y.; Awh, O. D.; Chung, Y. J.; Chung, Y. S. *J. Radioanal. Nucl. Chem.* **1997**, 217, 107-109.
145. Sarmani, S. B.; Hassan, R. B.; Abdullah, M. P.; Hamzah, A. *J. Radioanal. Nucl. Chem.* **1997**, 216, 25-27.
146. Vasconcellos, M. B. A.; Bode, P.; Paletti, G.; Catharino, M. G. M.; Ammerlaan, A. K.; Saiki, M.; Favaro, D. I. T.; Byrne, A. R.; Baruzzi, R.; Rodrigues, D. A. *J. Radioanal. Nucl. Chem.* **2000**, 244, 81-85.
147. Vasconcellos, M. B. A.; Bode, P.; Ammerlaan, A. K.; Saiki, M.; Paletti, G.; Catharino, M. G. M.; Favaro, D. I. T.; Baruzzi, R.; Rodrigues, D. A. *J. Radioanal. Nucl. Chem.* **2001**, 249, 491-494.
148. Georgescu, R.; Pantelica, A.; Salagean, M.; Cracinn, D.; Constantinescu, M.; Constantinescu, O.; Frangopol, P. T. *J. Radioanal. Nucl. Chem.* **1997**, 224, 147-150.
149. Georgescu, R.; Pantelica, A.; Craciun, D.; Grosescu, R. *J. Radioanal. Nucl. Chem.* **1998**, 231, 3-9.
150. Faghihian, H., Rahbarnia, H. *J. Radioanal. Nucl. Chem.* **2002**, 251, 427-430.
151. Tavakkoli, A.; Ahmadiniar, A.; Shirini, R. *J. Radioanal. Nucl. Chem.* **2000**, 243, 731-735.
152. Botttaro, C. S.; Kiceniuk, J. W.; Chatt, A. *Bio. Trace Elem. Res.* **1999**, 71-72, 149-166.
153. Chung, Y. S.; Moon, J. H.; Chung, Y. J.; Lee, K. Y.; Yoon, Y. Y. *J. Radioanal. Nucl. Chem.* **1999**, 240, 95-100.
154. Serfor-Armah, Y.; Nyarko, B. J. B.; Osae, E. K.; Carboo, D.; Seku, F. *J. Radioanal. Nucl. Chem.* **1999**, 242, 193-197.
155. Fukushima, M.; Tamate, H.; Nakano, Y. *J. Radioanal. Nucl. Chem.* **2000**, 244, 55-59.

156. Rietz, B.; Heydron, K.; Pritzl, G. *J. Radioanal. Nucl. Chem.* **1997**, 216, 113-116.
157. Dermelj, M.; Byren, A. R. *J. Radioanal. Nucl. Chem.* **1997,** 216, 13-18.
158. Yusof, A. M.; Misni, M.; Wood, A. K. H. *J. Radioanal. Nucl. Chem.* **1997**, 216, 59-63.
159. Wu, D.; Landsberger, S.; Larson, S. M. *J. Radioanal. Nucl. Chem.* **1997**, 217, 77-82.
160. Oliveira, H.; Fernandes, E. A. N.; Bacchi, M. A.; Sarries, G. A.; Tagliaferro, F. S. *J. Radioanal. Nucl. Chem.* **2000**, 244, 299-302.
161. Gardea-Torresdey, J.; Landsberger, S.; O'Kelly, D.; Tiemann, K. J.; Parsons, J. G. *J. Radioanal. Nucl. Chem.* **2001**, 250, 583-586.
162. Shtangeeva, I. V.; Vuorinen, A.; Rietz, B.; Christiansen, G.; Carlson, L. *J. Radioanal. Nucl. Chem.* **2001**, 249, 369-374.
163. Furukawa, J.; Kataoka, T.; Nakanishi, T. M. *J. Radioanal. Nucl. Chem.* **2000**, 244, 283-287.
164. Ueoka, S.; Furukawa, J.; Nakanishi, T. M. *J. Radioanal. Nucl. Chem.* **2001**, 249, 469-473.
165. Ueoka, S.; Furukawa, J.; Nakanishi, T. M. *J. Radioanal. Nucl. Chem.* **2001**, 249, 475-480.
166. Takada, J.; Nishimura, K.; Tanaka, Y.; Fujii, N.; Akaboshi, M. *J. Radioanal. Nucl. Chem.* **2002,** 251, 149-152.
167. Peresedov, V. F.; Gundorina, S. F.; Ostrovnaya, T. M. *J. Radioanal. Nucl. Chem.* **1997**, 219, 105-110.
168. Wang, Y. Q.; Sun, J. X.; Chen, H. M.; Guo, F. Q. *J. Radioanal. Nucl. Chem.* **1997,** 219, 99-103.
169. Liu, P. L.; Tian, J. L.; Jervis, R. E.; Li, Y. Q. *J. Radioanal. Nucl. Chem.* **1997,** 217, 39-43.
170. Takada, J.; Sumino, T.; Nishimura, K.; Tanaka, Y.; Kawamoto, K.; Akaboshi, M. *J. Radioanal. Nucl. Chem.* **1999**, 239, 609-612.
171. Ozaki, T.; Enomoto, S.; Minai, Y.; Ambe, S.; Ambe. F.; Tominaga, T. *J. Radioanal. Nucl. Chem.* **1997,** 217, 117-124.
172. Piasentin, R. M.; Armelin, M. J. A.; Primavesi, O.; Cruvinel, P. E. *J. Radioanal. Nucl. Chem.* **1998**, 238, 7-12.
173. Piasentin, R. M.; Armelin, M. J. A.; Primavesi, O.; Saiki, M. *J. Radioanal. Nucl. Chem.* **2001,** 249, 83-87.
174. Piasentin, R. M.; Armelin, M. J. A.; Primavesi, O.; Saiki, M. *J. Radioanal. Nucl. Chem.* **2000**, 244, 295-297.
175. Shtangeeva, I.; Heydorn, K.; Lissitskaya, T. *J. Radioanal. Nucl. Chem.* **2001**, 249, 375-380.
176. Gaochuang, Y.; Xueying, M.; Jinchun, W.; Yali, L.; Hong, O.; Zhaohui, Z.; Zhifang, C. *J. Radioanal. Nucl. Chem.* **2001,** 247, 567-570.

177. Tanoi, K.; Iikura, H.; Nakanishi, T. M. *J. Radioanal. Nucl. Chem.* **2001**, 249, 519-522.
178. de Goeij, J. J. M. *J. Radioanal. Nucl. Chem.* **2000**, 245, 5-9.
179. Zaichick, V.; Zaichick, S. *J. Radioanal. Nucl. Chem.* **2000**, 244, 279-282.
180. Koh, S.; Aoki, T.; Katayama, Y.; Takada, J. *J. Radioanal. Nucl. Chem.* **1999**, 239, 591-594.
181. Spyrou, N. M. *J. Radioanal. Nucl. Chem.* **1999**, 239, 59-70.
182. Aras, N. K.; Yilmaz, G.; Alkan, S.; Korkusuz, F. *J. Radioanal. Nucl. Chem.* **1999**, 239, 79-86.
183. Zaichick, V.; Dyatlov, A.; Zaichick, S. *J. Radioanal. Nucl. Chem.* **2000**, 244, 189-193.
184. Aras, N. K.; Yilmaz, G.; Korkusuz, F.; Olmez, I.; Sepici, B.; Eksioglu, F.; Bode, P. *J. Radioanal. Nucl. Chem.* **2000**, 244, 185-188.
185. El-Amri, F. A.; El-Kabroun, M. A. R. *J. Radioanal. Nucl. Chem.* **1997**, 217, 205-207.
186. Saiki, M.; Takata, M. K.; Kramarski, S.; Borelli, A. *Bio. Trace Elem. Res.* **1999**, 71-72, 41-46.
187. Porte, N.; Mauerhofer, E.; Denschlag, H. O. *J. Radioanal. Nucl. Chem.* **1997**, 220, 3-7.
188. Deibel, M. A.; Landsberger, S.; Wu, D.; Ehmann, W. D. *J. Radioanal. Nucl. Chem.* **1997**, 217, 153-155.
189. Zecca, L.; Tampellini, D.; Rizzio, E.; Giaveri, G.; Gallorini, M. *J. Radioanal. Nucl. Chem.* **2001**, 248, 129-131.
190. Panaya, A. E.; Spyrou, N. M.; Ubertalli, L. C.; Akanle, O. A.; Part, P. *J. Radioanal. Nucl. Chem.* **2000**, 244, 205-207.
191. Zecca, L.; Tampellini, D.; Costi, P.; Rizzio, E.; Giaveri, G.; Gallorini, M. *J. Radioanal. Nucl. Chem.* **2001**, 249, 449-454.
192. Panayi, A. E.; Spyrou, N. M.; Part, P. *J. Radioanal. Nucl. Chem.* **2001**, 249, 437-441.
193. Cutts, D. A.; Spyrou, N. M.; Maguire, R. P.; Leenders, K. L. *J. Radioanal. Nucl. Chem.* **2001**, 249, 455-460.
194. Cutts, D. A.; Maguire, R. P.; Stedman, J. D.; Leenders, K. L.; Spyrou, N. M. *Bio. Trace Elem. Res.* **1999**, 71-72, 541-549.
195. Hollands, R.; Spyrou, N. M.; Vijh, V.; Scales, J. T. *J. Radioanal. Nucl. Chem.* **1997**, 217, 185-187.
196. Zhong, H.; Tan, M.; Fu, Y.; Huang, J.; Tang, Z. *Bio. Trace Elem. Res.* **1999**, 71-72, 569-574.
197. Akanle, O. A.; Akintanmide, A.; Durosinmi, M. A.; Oluwole, A. F.; Spyrou, N. M. *Bio. Trace Elem. Res.* **1999**, 71-72, 611-616.
198. Chen, C.; Zhang, P.; Hou, X.; Chai, Z. *Bio. Trace Elem. Res.* **1999**, 71-72, 131-138.

199. Chen, C.; Lu, X.; Zhang, P.; Hou, X.; Chai, Z. *J. Radioanal. Nucl. Chem.* **2000**, 244, 199-203.
200. Stedman, J. D.; Spyrou, N. M.; Millar. A. D.; Altaf, W. J.; Akanle, O. A.; Rampton, D. S. *J. Radioanal. Nucl. Chem.* **1997**, 217, 189-191.
201. Baskett, C. K.; Spate, V. L.; Mason, M. M.; Nichols, T. A.; Williams, A.; Dubman, I. M.; Gudino, A.; Denison, J.; Morris, J. S. *J. Radioanal. Nucl. Chem.* **2001**, 249, 429-435.
202. Morris, J. S.; Rohan, T.; Soskolne, C. L.; Jain, M.; Horsman, T. L.; Spate, V. L.; Baskett, C. K.; Mason, M. M.; Nichols, T. A. *J. Radioanal. Nucl. Chem.* **2001**, 249, 421-427.
203. Akanle, O. A.; Balogun, F. A.; Owa, J. A.; Spyrou, N. M. *J. Radioanal. Nucl. Chem.* **2000**, 244, 231-235.
204. Sam, A. K.; Osman, M. M.; EL-Khangi, F, A. *J. Radioanal. Nucl. Chem.* **1998**, 231, 21-23.
205. Akanle, O. A.; Balogun, F. A.; Owa, J. A.; Spyrou, N. M. *J. Radioanal. Nucl. Chem.* **2001**, 249, 71-75.
206. Boulyga, S. F.; Petri, H.; Zhuk, I. V.; Kanash, N. V.; Malenchenko, A. F. *J. Radioanal. Nucl. Chem.* **1999**, 242, 335-340.
207. Ng, K. H.; Looi, L. M.; Bradley, D. A. *J. Radioanal. Nucl. Chem.* **1997**, 217, 193-199.
208. Altaf, W. J.; Spyrou, N. M. *J. Radioanal. Nucl. Chem.* **1997**, 217, 201-204.
209. Ding, W. J.; Qian, Q. F.; Hou, X. L.; Feng, W. Y.; Chai, Z. F.; Wangke *J. Radioanal. Nucl. Chem.* **2000**, 244, 259-262.
210. Oudadesse, H.; Irigaray, J. L.; Brun, V.; Zani, A.; Terver, S.; Vaneuville, G. *J. Radioanal. Nucl. Chem.* **2000**, 244, 195-198.
211. Krishnan, S.; Sturtrudge, W. C. *J. Radioanal. Nucl. Chem.* **1999**, 239, 71-77.
212. Capote Rodriguez, G.; Hernandez Rivero, A.; Moreno Bermudez, J.; Ribeiro Guevara, S.; Molina Insfran, J.; Perez Zayas, G. *J. Radioanal. Nucl. Chem.* **1997**, 223, 221-223.
213. Mosulishvili, L. M.; Kirkesali, Ye. I.; Belokobylsky, A. I.; Khizanishvili, A. I.; Frontasyeva, M. V.; Gundorina, S. F.; Oprea, C. D. *J. Radioanal. Nucl. Chem.* **2002**, 252, 15-20.
214. Kehayias, J. J. *J. Radioanal. Nucl. Chem.* **2000**, 244, 219-224.
215. Ali, M.; Khan, A. H.; Wahiduzzaman, A. K. M.; Malek, M. A. *J. Radioanal. Nucl. Chem.* **1997**, 219, 81-87.
216. Aguiar, A. R.; Saiki, M. *J. Radioanal. Nucl. Chem.* **2001**, 249, 413-416.
217. Armelin, M. J. A.; Avila, R. L.; Piasentin, R. M.; Saiki, M. *J. Radioanal. Nucl. Chem.* **2001**, 249, 417-419.
218. Yanaga, M.; Iwama, M.; Shinotsuka, K.; Takiguchi, K.; Noguchi, M.; Omori, T. *J. Radioanal. Nucl. Chem.* **2000**, 243, 661-667.

219. Resnizky, S.; Dallorso, M. E.; Pawlak, E. *Bio. Trace Elem. Res.* **1999**, 71-72, 343-347.
220. Panayi, A. E.; Spyrou, N. M.; Ubertalli, L. C.; White, M. A.; Part, P. *J. Radioanal. Nucl. Chem.* **1999**, 71-72, 529-540.
221. Fukushima, M.; Tamate, H.; Sato, S.; Terur, S.; Mitsugashira, T. *J. Radioanal. Nucl. Chem.* **1999**, 239, 595-599.
222. Li, F.; Wang, Y.; Zhang, Z.; Sun, J.; Xiao, H.; Chai, Z. *J. Radioanal. Nucl. Chem.* **2002**, 251, 437-441.
223. Kolesov, G. M.; Anikiev, V. V. *J. Radioanal. Nucl. Chem.* **1997**, 216, 299-308.
224. Yui, M.; Kikawada, Y.; Oi, T.; Honda, T.; Sun, D.; Shuai, K. *J. Radioanal. Nucl. Chem.* **1998**, 231, 83-86.
225. Kuno, A.; Sampei, K.; Matsuo, M.; Yonezawa, C.; Matsue, H.; Sawahate, H. *J. Radioanal. Nucl. Chem.* **1999**, 239, 587-590.
226. Dinescu, L. C.; Duliu, O. G.; Badea, M.; Mihailescu, N. G.; Vanghelie, I. M. *J. Radioanal. Nucl. Chem.* **1998**, 238, 75-81.
227. Yusof, A. M.; Akyil, S.; Wood, A. K. H. *J. Radioanal. Nucl. Chem.* **2001**, 249, 333-341.
228. Hossain, M. M. M.; Ohde, S. *J. Radioanal. Nucl. Chem.* **2002**, 251, 333-336.
229. Dinescu, L. C.; Duliu, O. G.; Andries, E. I. *J. Radioanal. Nucl. Chem.* **2000**, 244, 147-152.
230. Lara, L. B. L. S.; Fernandes, E. A. N.; Oliveira, H.; Bacchi, M. A.; Ferraz, E. S. B. *J. Radioanal. Nucl. Chem.* **1997**, 216, 279-284.
231. Minai, Y.; Tsunogai, U.; Takahashi, H.; Ishibashi, J.; Matsumoto, R.; Tominaga, T. *J. Radioanal. Nucl. Chem.* **1997**, 216, 265-277.
232. Georgescu, I. I.; Danis, A. S.; Mihai, A-S. G. *J. Radioanal. Nucl. Chem.* **1997**, 216, 253-260.
233. Molla, N. I.; Hossain, S. M.; Basunia, S.; Miah, R. U.; Rahman, M.; Sikder, D. H.; Chowdhury, M. I. *J. Radioanal. Nucl. Chem.* **1997**, 216, 213-215.
234. Yui, M.; Kikawada, Y.; Oi, T.; Honda, T.; Nozaki, T. *J. Radioanal. Nucl. Chem.* **1998**, 238, 3-6.
235. Ohde, S. *J. Radioanal. Nucl. Chem.* **1998**, 237, 51-54.
236. Balogun, F. A.; Tubosun, I. A.; Akanle, O.; Ojo, J. O.; Adesanmi, C. A.; Ajao, J. A.; Spyrou, N. M. *J. Radioanal. Nucl. Chem.* **1997**, 222, 35-38.
237. Pena Fortes, B.; Capote Rodriguez, G.; Riberiro Guevara, S.; Hernandez Rivero, A. T. *J. Radioanal. Nucl. Chem.* **1998**, 234, 175-177.
238. Tomuro, H.; Tomura, K. *J. Radioanal. Nucl. Chem.* **2001**, 250, 519-524.
239. Ohde, S.; Mataragio, J. P. *J. Radioanal. Nucl. Chem.* **1999**, 240, 325-328.
240. Ebihara, M.; Ozaki, H.; Kato, F.; Nakahara, H. *J. Radioanal. Nucl. Chem.* **1997**, 216, 107-112.

241. Dai, X.; Chai, Z.; Mao, X.; Ouyang, H. *J. Radioanal. Nucl. Chem.* **2000**, 245, 47-49.
242. Morcelli, C. P .R.; Figueiredo, A. M. G. *J. Radioanal. Nucl. Chem.* **2000**, 244, 619-621.
243. McOrist, G. D.; Smallwood, A. *J. Radioanal. Nucl. Chem.* **1997**, 233, 9-15.
244. Tsai, C. S.; Yeh, S. J. *J. Radioanal. Nucl. Chem.* **1997**, 216, 241-241.
245. Joron, J. L.; Treuil, M.; Raimbault, L. *J. Radioanal. Nucl. Chem.* **1997**, 216, 229-235.
246. Raimbault, L.; Peycelon, H.; Joron, J. L. *J. Radioanal. Nucl. Chem.* **1997**, 216, 221-228.
247. Thakur, A. N. *J. Radioanal. Nucl. Chem.* **1997**, 216, 151-59.
248. Lins, J. P.; Saiki, M. *J. Radioanal. Nucl. Chem.* **1997**, 216, 199-201.
249. Miyamoto, Y.; Sakamoto, K.; Mingqing, W. *J. Radioanal. Nucl. Chem.* **1997**, 216, 183-190.
250. Capote Rodriguez, G.; Pena Fortes, B.; Ribeiro Guevara, S.; Hernandez Rivero, A. T. *J. Radioanal. Nucl. Chem.* **1998**, 231, 171-174.
251. Efremova, T. T.; Efremov. S. P.; Koutzenogii, K. P.; Peresedov, V. F. *J. Radioanal. Nucl. Chem.* **1998**, 227, 31-36.
252. Lindstrom, D. J. *J. Radioanal. Nucl. Chem.* **2000**, 244, 511-515.
253. Capote Rodriguez, G.; Hernandez Rivero, A.; Moreno Bermudez, J.; Ribeiro Guevara, S.; Arrinere, M. A.; Molina Insfran, J.; Perez Zayas, G. *J. Radioanal. Nucl. Chem.* **1997**, 223, 217-220.
254. Earl-Goulet, J. R.; Mahaney, W. C.; Hancock, R. G. V.; Milner, M. W. *J. Radioanal. Nucl. Chem.* **1997**, 219, 717.
255. Genova, N.; Meloni, S.; Oddone, M.; Melis, P. *J. Radioanal. Nucl. Chem.* **2001**, 249, 355-360.
256. Capote Rodriguez, G.; Padilla Alvarze, R.; Perez Zayas, G.; Hernandez Rivero, A. T.; Lopez Reyes, M. C.; Ribeiro Guevear, S.; Molina Insfran, J. *J. Radioanal. Nucl. Chem.* **1998**, 237, 159-162.
257. Grant, C. N.; Lalor, G. C.; Vutchkov, M. K. *J. Radioanal. Nucl. Chem.* **1998**, 237, 109-112.
258. Byun, S. H.; Choi, H. D. *J. Radioanal. Nucl. Chem.* **2000**, 244, 413-416.
259. Molnar, G.; Belgya, T.; Dabolczi, L.; Fazekas, B.; Revay, Zs.; Veres, A.; Bikit, I.; Kiss, Z.; Ostor, J. *J. Radioanal. Nucl. Chem.* **1997**, 215, 111-115.
260. Paul, R. L.; Lindstrom, R. M.; Heald, A. E. *J. Radioanal. Nucl. Chem.* **1997**, 215, 63-68.
261. Crittin, M.; Kern, J.; Schenker, J.–L. *J. Radioanal. Nucl. Chem.* **2000**, 244, 399-404.
262. Yonezawa, C. *Bio. Trace Elem. Res.* **1999**, 71-72, 407-413.
263. Paul, R. L.; Lindstrom,R.M. *J. Radioanal. Nucl. Chem.* **2000**, 243, 181-189.
264. Chen-Mayer, H. H.; Mackey, E. A.; Paul, R. L.; Mildner, D. F. R. *J. Radioanal. Nucl. Chem.* **2000**, 244, 391-397.

265. Chen-Mayer, H. H.; Sharov, V. A.; Mildner, D. F. R.; Downing, R. G.; Paul, R. L.; Lindstrom, R. M.; Zeissler, C. J.; Xiao, Q. F. *J. Radioanal. Nucl. Chem.* **1997,** 215, 141-145.
266. Yonezawa, C.; Matsue, H. *J. Radioanal. Nucl. Chem.* **2000,** 244, 373-378.
267. Shypailo, R. J.; Ellis, K. J. *J. Radioanal. Nucl. Chem.* **2001,** 249, 407-412.
268. Oliveira, C.; Salgado, J.; Carvalho, F. G. *J. Radioanal. Nucl. Chem.* **1997,** 216, 191-198.
269. Howell, S. L.; Sigg, R. A.; Moore, F. S.; DeVol, T. A. *J. Radioanal. Nucl. Chem.* **2000,** 244, 173-178.
270. Revay, Zs.; Molnar, G. L.; Belgya, T.; Kasztovszky, Zs.; Firestone, R. B. *J. Radioanal. Nucl. Chem.* **2001,** 248, 395-399.
271. Revay, Zs.; Molnar, G. L.; Belgya, T.; Kasztovszky, Zs.; Firestone, R. B. *J. Radioanal. Nucl. Chem.* **2000,** 244, 383-389.
272. Kishikawa, T.;, Noguchi, S.; Yonezawa, C.; Matsue, H.; Nakamura, A.; Sawahata, H. *J. Radioanal. Nucl. Chem.* **1997,** 215, 211-217.
273. Zeisler, R.; Lamazw, G. P.; Chen-Mayer, H. H. *J. Radioanal. Nucl. Chem.* **2001,** 248, 35-38.
274. Kasztovszky, Zs.; Revay, Zs.; Belgya, T.; Molnar, G. L. *J. Radioanal. Nucl. Chem.* **2000,** 244, 379-382.
275. Yonezawa, C.; Matsue, H.; Hoshi, M. *J. Radioanal. Nucl. Chem.* **1997,** 215, 81-85.
276. Kuno, A.; Matsuo, M.; Takano, B.; Yonezawa, C.; Matsue, H.; Sawahate, H. *J. Radioanal. Nucl. Chem.* **1997,** 218, 169-176.
277. Oura, Y.; Enomoto, S.; Nakahara, H.; Matsue, H.; Yonezawa, C. *J. Radioanal. Nucl. Chem.* **2000,** 244, 311-315.
278. Anderson, D. L. *J. Radioanal. Nucl. Chem.* **2000,** 244, 225-229.
279. Nakahara, H.; Oura, Y.; Sueki, K.; Ebihara, M.; Sato, W.; Latif, Sk. A.; Tomizawa, T.; Enomoto, S.; Yonezawa, C.; Ito, Y. *J. Radioanal. Nucl. Chem.* **2000,** 244, 405-411.
280. Oura, Y.; Saito, A.; Sueki, K.; Nakahara, H.; Tomizawa, T.; Nishikawa, T.; Yonezawa, C.; Matsue, H.; Sawahata, W. *J. Radioanal. Nucl. Chem.* **1999,** 239, 581-585.
281. Rajput, M. U.; Ahmad, M.; Ahmad, W. *J. Radioanal. Nucl. Chem.* **2000,** 243, 719-722.
282. Latif, Sk. A.; Oura, Y.; Ebihara, M.; Kallemeyn, G. W.; Nakahara, H.; Yonezawa, C.; Matsue, T.; Sawahate, H. *J. Radioanal. Nucl. Chem.* **1999,** 239, 577-580.
283. Yoshikawa, H.; Yonezawa, C.; Kurosawa, T.; Hoshi, M.; Ishikawa, H. *J. Radioanal. Nucl. Chem.* **1997,** 215, 95-101.
284. Sano, T.; Fukuoka, T.; Hasenaka, T.; Yonezawa, C.; Matsue, H.; Sawahate, H. *J. Radioanal. Nucl. Chem.* **1999,** 239, 613-617.

285. Yonezawa, C.; Ruska, P. P., Matsue, H.; Magara, M.; Adachi, T. *J. Radioanal. Nucl. Chem.* **1999**, 239, 571-575.
286. Miyamoto, S.; Sutoh, M.; Shiomoto, A.; Yamazaki, S.; Nishimura, K.; Yonezawa, C.; matsue, H.; Hoshi, M. *J. Radioanal. Nucl. Chem.* **2000**, 244, 307-309.
287. Postma, H.; Blaauw, M.; Bode, P.; Mutti, P.; Corvi, F.; Siegler, P. *J. Radioanal. Nucl. Chem.* **2001**, 248, 115-120.
288. Overwater, R. M. W.; Bode, P.; de Goeij, J. J. M.; Hoogenboom, J. E. *Analy. Chem.* **1996**, 68, 341-348.
289. Bode, P.; Overwater, R. M. W.; De Goeij, J. J. M. *J. Radioanal. Nucl. Chem.* **1997**, 216, 5-11.
290. Lakmaker, O.; Blaauw, M. *J. Radioanal. Nucl. Chem.* **1997**, 216, 69-74.
291. Alamin, M. B.; Spyrou, N. M. *J. Radioanal. Nucl. Chem.* **1997**, 215, 205-209.
292. Shakir, N. S.; Jervis, R. E. *J. Radioanal. Nucl. Chem.* **2001**, 248, 61-68.
293. Hogewoning, S. J.; Bode, P. *J. Radioanal. Nucl. Chem.* **2000**, 244, 531-535.
294. Fernandes. E. A. N.; Bode, P. *J. Radioanal. Nucl. Chem.* **2000**, 244, 589-594.
295. Lin, X.; Henkelmann, R. *J. Radioanal. Nucl. Chem.* **2002**, 251, 197-204.
296. Gwozdz, R.; Grass, F. *J. Radioanal. Nucl. Chem.* **2000**, 244, 523-529.
297. Tsipenyuk, Y. M.; Firsov, V. I. *J. Radioanal. Nucl. Chem.* **1997**, 216, 47-50.
298. Jerde, E. A.; Glasgow, D. C.; Hastings, J. B. *J. Radioanal. Nucl. Chem.* **1999**, 242, 473-485.
299. Mukherjee, B. *J. Radioanal. Nucl. Chem.* **1998**, 231, 179-181.
300. Robinson, L.; Johnson, E.; Zhao, L. *J. Radioanal. Nucl. Chem.* **1998**, 238, 25-28.
301. Kim, W-S; Kim, H-S; Kin, J-Y.; Kim, J-H.; Kim, Y-U.; Lee, K-P. *J. Radioanal. Nucl. Chem.* **1997**, 216, 75-79.
302. DiPrete, D. P.; Peterson, S. F.; Sigg, R. A. *J. Radioanal. Nucl. Chem.* **2000**, 244, 343-347.
303. Jerde, E. A.; Glasgow, D. C. *J. Radioanal. Nucl. Chem.* **1999**, 242, 753-759.
304. Grass, F.; Buchtela, K.; Furch, T.; Gwozdz, R.; Hausch, E.; Hedrich, E.; Radev, R.; Westphal, G. P. *J. Radioanal. Nucl. Chem.* **2000**, 244, 333-338.
305. Das. H. A. *J. Radioanal. Nucl. Chem.* **1999**, 242, 587-593.
306. Vernetson, W. G. *J. Radioanal. Nucl. Chem.* **2000**, 244, 327-331.
307. Kennedy, G.; St-Pierre, J. *J. Radioanal. Nucl. Chem.* **1997**, 215, 235-239.
308. Matsushita, R.; Koyama, M.; Yamada, S.; Kobayashi, M.; Moriyama, H. *J. Radioanal. Nucl. Chem.* **1997**, 216, 95-99.
309. Papadopoulos, N. N.; Hatzakis, G. E.; Salevris, A. C.; Tsagas, N. F. *J. Radioanal. Nucl. Chem.* **1997**, 215, 103-110.

310. Diaz Rizo, O.; Herrera Peraza, E.; Lopez Reyes, M. C.; Alvarez Pellon I.; Manso Guevara, M. V.; Ixquiac Cabrera, M. *J. Radioanal. Nucl. Chem.* **1997,** 220, 95-97.
311. Osae, E. K.; Nyarko, B. J. B.; Serfor-Armah, Y.; Akaho, E. H. K. *J. Radioanal. Nucl. Chem.* **1998,** 238, 105-109.
312. Montoya R, E. H.; Cohen, I. M.; Mendoza Hidalgo, P.; Torres Chamorro, B.; Bedregal Salas, P. *J. Radioanal. Nucl. Chem.* **1999,** 240, 475-479.
313. Miyamoto, Y.; Haba, H.; Kajikawa, A.; Masumoto, K.; Nakanishi, T.; Sakamoto, K. *J. Radioanal. Nucl. Chem.* **1999,** 239, 165-175.
314. De Corte, F.; Steinnes, E.; de Neve, P.; Simonits, A. *J. Radioanal. Nucl. Chem.* **1997,** 215, 279-282.
315. Arribere, M. A.; Guevara, S. R.; Kestelman, A. J.; Cohen, I. M. *J. Radioanal. Nucl. Chem.* **1999,** 241, 25-31.
316. Byrne, A. R.; Dermelj, M. *J. Radioanal. Nucl. Chem.* **1997,** 223, 55.
317. Ahmad, K. *J. Radioanal. Nucl. Chem.* **1997,** 218, 71-75.
318. Blaauw, M.; Mackey, E. A. M. *J. Radioanal. Nucl. Chem.* **1997,** 216, 65-68.
319. Dodoo-Amoo, D.; Landsberger, S. *J. Radioanal. Nucl. Chem.* **2001,** 248, 327-332.
320. Parry, S. J.; Benzing, R.; Bolstad, K. L.; Steinnes, E. *J. Radioanal. Nucl. Chem.* **2000,** 244, 67-72.
321. Landsberger, S.; Basunia, M. S.; Ishander, F. *J. Radioanal. Nucl. Chem.* **2001,** 249, 303-305.
322. Porte, N.; Mauerhofer, E.; Denschlag, H. O. *J. Radioanal. Nucl. Chem.* **1997,** 224, 103-107.
323. Landsberger, S.; Wu, D. *Bio. Trace Elem. Res.* **1999,** 71-72, 453-461.
324. Lin, X.; Lierse, Ch.; Wahl, W. *J. Radioanal. Nucl. Chem.* **1997,** 215, 169-178.
325. Bode, P. *J. Radioanal. Nucl. Chem.* **1997,** 222, 127-132.
326. De Corte, F. *J. Radioanal. Nucl. Chem.* **2001,** 248, 13-20.
327. De Corte, F. *J. Radioanal. Nucl. Chem.* **2000,** 245, 157-161.
328. Van Lierde, S.; De Corte, F.; van Sluijs, R.; Bossus, D. *J. Radioanal. Nucl. Chem.* **2000,** 245, 179-184.
329. De Corte, F.; Van Lierde, S. *J. Radioanal. Nucl. Chem.* **2001,** 248, 103-107.
330. De Corte, F.; Van Lierde, S. *J. Radioanal. Nucl. Chem.* **2001,** 248, 97-101.
331. Pomme, S.; Hardeman, F.; Etxebarria, N.; Robouch, P. *J. Radioanal. Nucl. Chem.* **1997,** 215, 295-303.
332. Petri, H.; Kuppers, G. *J. Radioanal. Nucl. Chem.* **2000,** 245, 163-166.
333. Lin, X.; Li, X. *J. Radioanal. Nucl. Chem.* **1997,** 223, 47-53.
334. Kennedy, G.; St-Pierre, J.; Wang, K.; Zhang, Y.; Preston, J.; Grant, C.; Vutchkov, M. *J. Radioanal. Nucl. Chem.* **2000,** 245, 167-172.
335. de Menezes, M.; de Sabino, C.; Amaral, A. M.; Pereira Maia, E. C. *J. Radioanal. Nucl. Chem.* **2000,** 245, 173-178.

336. Bacchi, M. A.; Fernandes, E. A. N.; de Franca, E. J. *J. Radioanal. Nucl. Chem.* **2000**, 245, 209-213.
337. Bacchi, M. A.; Fernandes, E. A. N.; de Oliveira, H. *J. Radioanal. Nucl. Chem.* **2000**, 245, 217-222.
338. Herrera Peraza, E. F.; Diaz Rizo, O.; Cabrera, Montero, M. E.; Hernandez, A. T.; Contreras Folgar, R.; Izquiac Cabrera, M.; Hernandez Aguilar, O.; Manso Guevara, M. V.; Alvarez Pellon, I.; Lopez Dumenigo, R. A.; Padron Gonzalez, G.; Lopez Reyes, N. C. *J. Radioanal. Nucl. Chem.* **1999**, 240, 437-443.
339. Diaz Rizo, O.; Herrera Peraza, E. F.; Manso Guevara, M. V.; Alvarez Pellon, I.; Lopez Reyes, M. C.; Ixquiac Cabrera, M.; Montero Cabrera, M. E. *J. Radioanal. Nucl. Chem.* **1999**, 240, 445-450.
340. van Sluijs, R.; Bossus, D. A. W. *J. Radioanal. Nucl. Chem.* **1999**, 239, 601-603.
341. Blaauw, M. *J. Radioanal. Nucl. Chem.* **2000**, 245, 185-188.
342. Matsue, H.; Yonezawa, C. *J. Radioanal. Nucl. Chem.* **2000**, 245, 189-194.
343. Matsue, H.; Yonezawa, C. *J. Radioanal. Nucl. Chem.* **2001**, 249, 11-14.
344. De Corte, F.; De Wispelaere, A.; van Sluijs, R.; Bossus, D.; Simonits, A.; Kucera, J.; Frana, J.; Smodis, B.; Jacimovic, R. *J. Radioanal. Nucl. Chem.* **1997**, 215, 31-37.
345. Piccot, D.; Deschamps, C.; Delmas, R.; Revel, G. *J. Radioanal. Nucl. Chem.* **1997**, 215, 263-269.
346. Balaji, T.; Acharya, R. N.; Nair, A. G. C.; Reddy, A. V. R.; Rao, K. S.; Naidu, G. R. K.; Manohar, S. B. *J. Radioanal. Nucl. Chem.* **2000**, 243, 783-788.
347. Boulyga, S. F.; Zhuk, I. V.; Lomonosova, E. M.; Kieverz, M. K.; Denschlag, H. O.; Zauner, S.; Malenchenko, A. F.; Kanash, N. V.; Bazhanova, N. N. *J. Radioanal. Nucl. Chem.* **1997**, 222, 11-14.
348. Boulyga, E. G.; Boulyga, S. F. *J. Radioanal. Nucl. Chem.* **2000**, 246, 253-258.
349. Sima, O. *J. Radioanal. Nucl. Chem.* **2000**, 244, 669-673.
350. van Sluijs, R.; Bossus, D.; Blaauw, M.; Kennedy, G.; De Wispelaere, A.; van Lierde, S.; De Corte, F. *J. Radioanal. Nucl. Chem.* **2000**, 244, 675-680.
351. Ni, B.; Wang, P.; W.; Nie H.; Li, S.; Li, X.; Tian W,. *J. Radioanal. Nucl. Chem.* **2000**, 244, 665-668.
352. Westphal, G. P.; Lemmel, H.; Grass, F.; Gwozdz, R.; Jostl, K.; Schroder, P.; Hausch, E. *J. Radioanal. Nucl. Chem.* **2001**, 248, 53-60.
353. Zeisler, R. *J. Radioanal. Nucl. Chem.* **2000**, 244, 507-510.
354. Heydorn, K.; Damsgaard, E. *J. Radioanal. Nucl. Chem.* **1997**, 215, 157-160.
355. Kennedy, G.; Tye, P.; St-Pierre, J. *J. Radioanal. Nucl. Chem.* **2001**, 248, 339-343.

356. Upp, D. L.; Keyser, R. M.; Gedcke, D. A.; Twomey, T. R.; Bingham, R. D. *J. Radioanal. Nucl. Chem.* **2001**, 248, 377-383.
357. Blaauw, M.; Fleming, R. F.; Keyser. R. *J. Radioanal. Nucl. Chem.* **2001**, 248, 309-313.
358. Then, S. S.; Geurink, F. D. P.; Bode, P.; Lindstrom, R. M. *J. Radioanal. Nucl. Chem.* **1997**, 215, 249-252.
359. Fazekas, B.; Molnar, G.; Belgya, T.; Dabolczi, L.; Simonits, A. *J. Radioanal. Nucl. Chem.* **1997**, 215, 271-277.
360. Revay, Zs.; Belgya, T.; Ember, P. P.; Molnar, G. L. *J. Radioanal. Nucl. Chem.* **2001**, 248, 401-405.
361. Blaauw, M. *J. Radioanal. Nucl. Chem.* **2000**, 244, 661-664.
362. Ammerlaan, M. J. J.; Bode, P. *J. Radioanal. Nucl. Chem.* **1997**, 215, 253-256.
363. Bode, P. *J. Radioanal. Nucl. Chem.* **1997**, 222, 117-125.
364. Heydorn, K. *J. Radioanal. Nucl. Chem.* **2000**, 244, 7-15.
365. Currie, L. A. *J. Radioanal. Nucl. Chem.* **2000**, 245, 145-156.
366. Kucera, J.; Bode, P.; Stepanek, V. *J. Radioanal. Nucl. Chem.* **2000**, 245, 115-122.
367. Bode, P.; De Nadai Fernandes, E. A.; Greenberg, R. R. *J. Radioanal. Nucl. Chem.* **2000**, 245, 109-114.
368. Pomme, S.; Robouch, P.; Arana, G.; Eguskiza, M.; Maguregui, M. I. *J. Radioanal. Nucl. Chem.* **2000**, 244, 501-506.
369. Berger, A.; Goerner, W.; Haase, O. *J. Radioanal. Nucl. Chem.* **2000**, 244, 517-522.
370. Kafala, S. I.; MacMahon, T. D.; Borzakov, S. B *J. Radioanal. Nucl. Chem.* **1997**, 215, 193-204.
371. Bode, P. *J. Radioanal. Nucl. Chem.* **1997**, 215, 51-57.
372. Bode, P.; van Dijk, C. P. *J. Radioanal. Nucl. Chem.* **1997**, 215, 87-94.
373. Bode, P. *J. Radioanal. Nucl. Chem.* **2000**, 245, 127-132.
374. Bode, P. *J. Radioanal. Nucl. Chem.* **2000**, 245, 133-135

Author Index

Addleman, R. Shane, 246
Ashley, Kenneth R., 177
Ball, Jason R., 177
Bauer, Eve, 177
Bernard, Jonathan G., 177
Berning, Douglas E., 177
Boll, Rose A., 193
Borkowski, Marian, 140
Broda, R., 76
Cassette, P., 76
Cessna, J. T., 76
Chamberlin, Rebecca M., 177
Chiarizia, Renato, 140
Choppin, Gregory R., 12
Collé, R., 76
DeVol, T. A., 105
Dietz, Mark L., 161
Du, Miting, 193
Eaton, Gail F., 235
Egorov, Oleg B., 246
Fauth, D. J., 271
Fjeld, R. A., 105
Glasgow, D. C., 88
Glover, S. E., 286
Grate, Jay W., 246
Hafner, Ronald S., 218
Haines, Douglas K., 218
Hudson, G. B., 235
Husain, Liaquat, 218
Hutchinson, J. M. Robin, 38
Inn, Kenneth G. W., 38
Karam, Lisa, 38
Kronenberg, Andreas, 12
Kurosaki, Hiromu, 38
Landsberger, S., 307

Laue, Carola A., 2, 121
Leyba, J. D., 105
Lin, Zhichao, 38
Loveland, Walter D., 22
Lovell, M. A., 298
Martin, M. Z., 88
Martin, R. C., 88
Maxwell, S. L., III, 271
McMahon, Ciara, 38
Mirzadeh, Saed, 193
Moran, J. E., 235
Nash, Kenneth L., 2, 140
Nour, Svetlana, 38
O'Hara, Matthew J., 246
Otu, Emmanuel, 140
Outola, Iisa, 38
Parekh, Pravin P., 218
Paulenova, A., 105
Peretz, Fred, 193
Rabun, Robert L., 218
Rickert, Paul G., 140
Roane, J. E., 105
Robertson, J. David, 206, 298
Schroeder, Norman C., 177
Schultz, Michael, 38
Selvig, Linda, 38
Semkow, Thomas M., 218
Smith, David K., 121
Sylvester, Paul, 177
Vetter, K., 52
Williams, Philip G., 218
Wozniak, Gordon J., 218
Wu, Zhongyu, 38
Zimmerman, B. E., 76

Subject Index

A

α-Activity, α-contamination of soils, 129–132
Academic radiochemistry
 areas needing nuclear scientists, 17
 dismantling nuclear weapons, 13–14
 doctoral degrees in nuclear/radiochemistry by U.S. universities, 19f
 efficient separations, 16
 faculty and departments offering nuclear/radiochemistry graduate programs, 18f
 International Atomic Energy Agency (IAEA), 19–20
 international monitoring compliance of nonproliferation agreements, 14–15
 long term needs, 14–16
 National Science Foundation (NSF) policy, 18
 national survey of U.S. universities, 19–20
 nuclear power, 15–16
 Office of Education of Department of Energy (OECD), 18–19
 present situation, 13–14
 radiation dose effect research, 16
 radioactive waste, 15–16
 studies of personnel needs, 17–20
 study of nuclear phenomena, 12–13
Accelerated mass spectrometry (AMS), brain analysis of Alzheimer patients, 304
Actinides
 automated separation for, analysis of, 266–268
 Diphonix Resin® for extraction, 274
 elution on Dowex-50, 28f
 existence of α-emitters, 122
 fate and transport, 122–123
 fecal analysis, 273–274
 future work, 135
 monitoring in urine, 274–275
 need for chelating agents, 143
 Nevada Applied Ecology Group (NAEG) investigation results, 124t
 physics and chemistry, 35–36
 recycle, 153
 separations for bioassay samples, 283
 stacked cartridge approach, 277–278
 uptake by plant materials, 133
 vertical distribution in soil horizon, 134
 See also α-Autoradiography; Column extraction chromatography
Actinium, radiochemistry, 3
Actinium-225
 sales, 202f
 See also Thorium-229
Activated carbon, dual-irradiation particle-induced X-ray emission (PIXE), 210f
Activation analysis, radioanalytical technique, 32–33
Activity, total tritiated water (HTO), 229–230
Activity distribution pattern, detector plates, 132–133
Advanced Photon Source (APS), radiation damage testing, 100
Aerial contamination survey, α-emitter profiling, 133–134
Age dating

339

nuclear sciences, 9
See also Groundwater, age-dating
Airplanes
 tritium release, 225
 See also Tritium; World Trade Center (WTC)
Air pollution, neutron activation analysis (NAA), 311–312
Alaska, community building fire and tritium release, 225
Alchemy, middle ages, 2–3
Algorithms, gamma-ray tracking, 64
Alpha spectrometry
 alpha-emitting thorium isotopes, 291–293
 detection limits for thorium isotopes, 296t
 determination of thorium isotopes, 287
Aluminum, functionalized support for extraction chromatography, 171t
Alzheimer's disease (AD)
 accelerator mass spectrometry (AMS), 304
 average and standard deviation of Zn concentration in senile plaque (SP) and neuropil in AD subjects, 302t
 average year of formation of neurofibrillary tangles (NFT) and SP isolated from AD brain samples, 304t
 brain and Zn, 300
 ^{14}C content of tissue, 303–304
 characterization, 299
 disruption of Zn homeostasis, 301–302
 micro-particle induced X-ray emission (micro-PIXE) analysis of AD subjects, 300–301
 Northern hemisphere environmental ^{14}C levels, 303f
 time course of NFT and SP in AD, 302–305
 trace metals and AD, 299–302
 X-ray spectra of brain tissue, 301f
 zinc, 299-300

Am^{3+} extraction. *See* Radioanalytical chemistry
Analytical tools, nuclear sciences, 9–10
Anchored coatings, extraction chromatography (EXC) resin stabilization, 172
Angular resolution
 gamma-ray imaging, 68, 70
 geometrical efficiencies and gains as function of, 70f
Anion exchange
 coupling extraction chromatography with, 275
 elution profile of protactinium, 290f
 technetium concentration, 181
Anion exchange resin
 behavior of elements, 29
 elution of elements, 30f
Antimony, gunshot residue (GSR) analysis of jeans, 209, 212–213
Archaeology, neutron activation analysis (NAA), 309–310
Arsenic, determination in human hair, 206
Astatine, synthesis, 6
Atom
 Bohr model, 5
 Saturnian model, 5
Atomic mass, nucleus, 5
Authenticity, suspect gold coins, 213–216
Automation of radiochemical analysis
 analyzer system development, 263
 chemically selective diode detectors, 260–262
 chemical processing of low activity waste (LAW), 262–263
 Cherenkov detection approach, 258–260
 composite bed sensors, 253
 count rate after sensor equilibration, 255
 detector sensor and calibration curve of equilibration sensing approach for ^{99}Tc(VII), 256f

detector signals for reagentless
equilibration ^{99}Tc sensing, 258f
dual functionality sensor materials,
250–252
duplicate detector signals showing
flow-through operation of
chemically selective diode
detector, 262f
feasibility of ^{90}Sr sensing via ^{90}Y
Cherenkov radiation, 258–260
feasibility of chemically modified
diode detectors for in-situ
detection of alpha-emitters, 261f
flow-through scintillation detector,
264
functional requirements and general
approach, 248–249
ICP–MS detector signals for
sequential actinide elution
sequence, 267f
ICP–MS (inductively coupled plasma
mass spectrometry) for longer
lived radionuclides, 247
integration of separation chemistry
and radiation detectors, 268
loading and regeneration of Tc(VII)
selective composite bed sensor,
254f
pertechnetate separation, 264
pertechnetate uptake properties of
scintillating sensor material vs.
nitric acid concentration, 251f
preconcentrating minicolumn
sensors, 250–260
preferred methods for ^{99}Tc analysis,
263
quantification approaches, 254–257
reagentless ^{99}Tc(VII) sensing in
Hanford groundwater, 257–258
response of ^{99}Tc-sensor to injection
of sample solution containing
target analyte and unretained
species, 252f
sample treatment protocol, 263–264
schematic of radionuclide sensor
concept, 249f

separations for ICP–MS analysis of
actinides, 266–268
sequential injection (SI) separation
system, 266
standard addition approach, 264–265
stopped-flow counting interval, 255
^{99}Tc analysis in LAW matrix with
varying ^{99}Tc content, 265f
testing against LAW samples, 265–
266
total ^{99}Tc analyzer instrument, 262–
266
α-Autoradiography
actinide uptake by plant material,
133
activity distribution pattern, 132–133
advantages, 135
aerial contamination survey, 133–
134
CR-39 etching and evaluation
procedure, 126
CR-39 polymer of polyallyl diglycol
carbonate, 123
depths of α-emitting contamination,
131–132
design of in-situ soil exposure of CR-
39 detectors to α-emitters, 125–
126
details of Plutonium Valley, 128f
distribution of α-contamination in
soil, 129–130
experimental, 125–129
experimental parameter of in-situ
exposures of CR-39 detectors to
soils in Plutonium Valley, 129t
field exposure experiments, 128–129
field site–Plutonium Valley, Nevada
Test Site (NTS), 126–127
future work, 135
gunshot-like structures, 134
hot spots, 129–132
photographs of detector plates, 131f,
132f
points of concentrated α-activity,
129–132
rod-shaped structures, 134

schematic of exposure tool, 125f
schematic of radiographic process and photographic example images, 130f
solid state nuclear track detectors (SSNTD), 123
star-like structures, 134
technique of choice, 123
topographic map of NTS, 127f
types of hot spots structures, 134–135
vertical distribution in soil horizon, 134
Autoreduction, TcO_4^-, 189f

B

Barium, gunshot residue (GSR) analysis of jeans, 209, 212–213
Bathtub
  water leak into, 226
  World Trade Center complex, 219–220
  See also World Trade Center (WTC)
Biological systems, neutron activation analysis (NAA), 313–314
Biomentoring, neutron activation analysis (NAA), 312
Bleaching, Tc(IV)–EDTA complex, 182-183, 183f
Bohr, atom, 5
Bohrium, characterization, 36
Bomb pulse
  ^{14}C, 303, 305
  tritium, 236
Borehole logging, prompt gamma neutron activation analysis (PGNAA), 94–95
Bovine liver, particle-induced X-ray emission (PIXE), 211t
Brachytherapy, californium-252, 92–93
Brain
  Alzheimer's disease (AD) patients, 299

  Zn pools, 300
  See also Alzheimer's disease (AD)
Brass, naval, particle-induced X-ray emission (PIXE), 211t

C

Cadmium, functionalized support for extraction chromatography, 171t
Californium-252
  applications of ^{252}Cf neutron sources, 92–94
  brachytherapy, 92
  californium user facility (CUF) for neutron science, 99–101
  cancer therapy, 92–93
  comparison of neutron activation analysis (NAA) detection limits, 97–98
  CUF infrastructure and experimental capabilities, 99–100
  electroplates, 91
  fast-neutron-induced mutation of seeds, 101
  fissile material and nuclear waste characterization, 93–94
  fission-fragment emission, 92
  INAA (instrumental neutron activation analysis), 95–96
  INAA at CUF, 96
  INAA at high flux isotope reactor, 97–99
  isotopes, 91
  laser-induced breakdown spectroscopy (LIBS) with INAA, 98–99
  LIBS and prompt gamma NAA (PGNAA) detection of carbon in soils, 99
  neutron backscattering for land mine and contraband detection, 94
  neutron radiography, 93
  on-line process monitoring and borehole logging, 94–95
  PGNAA, 92, 94–95

properties and source forms, 91
radiation damage testing of
  permanent magnets, 100
radiation damage testing of
  semiconductor detectors, 100
source loan programs, 101–102
Californium user facility (CUF)
fast-neutron-induced mutation of
  seeds, 101
infrastructure and experimental
  capabilities, 99–100
instrumental neutron activation
  analysis (INAA) at CUF, 96
neutron science, 99–101
radiation damage testing of
  permanent magnets, 100
radiation damage testing of
  semiconductor detectors, 100
Cancer therapy, californium-252, 92–93
Carbamoylmethylphosphine oxide
  (CMPO), chelating agents, 144
Carbon-14
bomb pulse, 303, 305
content in tissues, 303–304
Northern hemisphere environmental
  levels, 303f
Carriers, precipitation, 25
Cartridges, column extraction
  chromatography, 273
Catalytic reduction, TcO$_4^-$, 187f
Cathode rays, charged particles, 4
Ceramics, medieval, neutron
  activation analysis (NAA), 310
Characterization, neutron activation
  analysis (NAA), 317–318
Charged particles, cathode rays, 4
Chelating agents
carbamoylmethylphosphine oxide
  (CMPO), 144
characterization of water-soluble,
  156–158
Chemical equilibrium, biphasic, 145
Chemically selective diode detectors
actinide selective organic extractants,
  260

duplicate detector signals showing
  flow-through operation, 262f
modern passivated ion implanted
  silicon (PIPS), 261
radionuclide sensing, 260
Chemical separation methods
detection techniques, 8
new elements, 7
Chemistry, 19th century, 3–4
Chemists, study of nuclear
  phenomena, 12–13
Cherenkov detection, feasibility of
  ^{90}Sr sensing via ^{90}Y, 258–260
Chernobyl, nuclear power accidents,
  15
Chromatography
radioisotopes, 110–111
separating radionuclides, 106
See also Column extraction
  chromatography; Extraction
  chromatography (EXC)
Chromium, functionalized support for
  extraction chromatography, 171t
CIEMAT (Centro de Investigaciones
  Energéticas Medioambientales y
  Tecnológicas)/NIST (National
  Institute of Standards and
  Technology)
efficiency tracing method, 77
See also Liquid scintillation (LS)
  counting
Clinical studies, neutron activation
  analysis (NAA), 315
Cold War
contamination of weapons sites,
  14
radiochemists and end of, 13
weapon production site
  decontamination, 14
Collimator-based gamma-ray imaging
  systems
efficiency values for conventional
  cameras, 57, 58f
illustration, 56f
limitations and trade-offs, 57
requirement of collimator, 55–56

See also Gamma-ray imaging systems
Collimator-less gamma-ray imaging systems
  advantage of double-sided strip detector (DSSD) geometry, 60–61
  algorithms, 64
  coincidence requirement, 61–62
  Compton camera, 59
  Compton scattering, 58–59
  Gamma-Ray Energy Tracking Array (GRETA), 62–63
  gamma-ray imaging detectors, 60–65
  gamma-ray tracking, 59–65
  high purity germanium (HPGe) detectors, 60
  high purity silicon (HPSi) detector, 61
  incidence of photons on HPSi detectors, 61–62
  incident gamma ray interacting by Compton scattering and absorption by photoelectric effect, 60*f*
  lithium-drifted silicon (Si(Li)) detector, 61
  nuclear physics experiments, 65–66
  potential advantage of Compton camera, 59
  schematic, 61*f*
  semi-conductor detectors, 65
  See also Gamma-ray imaging systems
Column extraction chromatography
  advantages, 272
  bioassay for plutonium, uranium, americium, and neptunium-237, 276
  coupling, with anion exchange, 275
  environmental sample method, 272–273
  monitoring actinides in urine, 274–275
  needing method for fecal analysis, 273–274
  new options, 276
  smaller particle size resin, 273
  SRS stacked cartridge approach, 227–228, 277–278
  TRU Resin®, 274–275
  uranium separations, 276
  UTEVA Resin®, 276
  vacuum boxes, 273
  See also Extraction chromatography (EXC)
Column washing
  effect on elution behavior of Sr-85, 165, 166*f,* 169
  See also Exchange chromatography (EXC)
Complexants. *See* Radioanalytical chemistry
Complexation ligands, technetium with aminocarboxylates, 182
Compton scattering
  collimator-less gamma-ray imaging systems, 58–59
  potential scattering sequence, 64*f*
  See also Gamma-ray imaging systems
Concentrated α-activity, α-contamination of soils, 129–132
Contaminants, detection below ppt levels, 7–8
Contamination
  α-activity in soils, 129–132
  Cold War weapons sites, 14
Contraband
  gamma-ray detection, 54*t*
  neutron backscattering for detection, 94
Copper
  functionalized support for extraction chromatography, 171*t*
  neutron activation analysis (NAA) of ancient, 310
Co-precipitation, separation technique, 25–26
Counting efficiencies
  liquid scintillation (LS), 77
  triple-to-double coincidence ratio (TDCR) method, 77–78

See also Triple-to-double coincidence ratio (TDCR) method
Counting rate equations, TDCR method, 79
Courtroom. *See* Forensic science
CR-39 polymer of polyallyl glycol carbonate
α-autoradiography, 123
design of in situ soil exposure of CR-39 detectors to α-emitters, 125–126
etching and evaluation procedure, 126
photographs of etched CR-39 detectors, 130f, 131f, 132f
Crops, neutron activation analysis (NAA), 314

D

Decay, californium-252, 91
Dementia. *See* Alzheimer's disease (AD)
Department of Energy (DOE), funding for Th separation from U-233, 196
Detection. *See* Flow-cell radiation detection
Detection modes
gamma-ray, 54t
See also Gamma-ray imaging systems
Detection technologies
chemical separation methods, 8
gas-proportional counting, 8
high resolution γ- and α-detection, 8
nuclear and radioanalytical sciences, 7–8
nuclear sciences, 10
scintillation counting, 8
Detectors
chemically selective diode, 260–262
See also Chemically selective diode detectors; Gamma-ray imaging systems

Diacidic diphosphonate, characterizing extractant systems, 147–152
Diet, neutron activation analysis (NAA), 310
Diphonix Resin®, actinide extraction, 274
Discovery, elements, 2–3
Dismantlement, nuclear weapons, 13–14
Disposal, nuclear wastes, 178
Distribution coefficients
definition, 144–145
Tc-99 activity, 180, 181t
Doctoral degrees, universities, 19f
Dose effects, radiation research, 16
Double-sided strip detector (DSSD), gamma-ray imaging, 60–61
Dual functionality, sensor materials, 250–252

E

Earth age, Rutherford, 5
Einstein, theories, 5
Einsteinium, production, 7
Electroplates, californium-252, 91
Element and isotope discovery, nuclear sciences, 9
Elements
chemical separation methods, 7
discovery, 2–3
elution from anion exchange resin, 30f
periodic table in 1900, 3f
periodic table in 21st century, 8f
Emergency EXIT signs, tritium, 223
Emission mode, gamma-ray detection, 53
α-Emitters
existence, 122
See also α-Autoradiography
Energy generation, nuclear sciences, 10

Engineering methods, neutron activation analysis (NAA), 311
Environment
  uranium mining, 15
  See also α-Autoradiography
Environmental monitoring
  air pollution, 311–312
  biological systems, 313–314
  biomonitoring, 312
  neutron activation analysis (NAA), 311–314
  solid waste, 312–313
  wastewater, 313
Environmental radiochemistry applications, 39
  efforts of National Institute of Standards and Technology (NIST), 39–40
Environmental samples, instrumental neutron activation analysis (INAA), 96
Ethylenediaminetetraacetic acid (EDTA), complexation ligands, 182-183
EXIT signs, tritium, 223
Explosives detection, prompt gamma neutron activation analysis (PGNAA), 95
Extraction
  characterization by solvating ligands, 152–155
  See also Radioanalytical chemistry
Extraction chromatography (EXC)
  anchored coatings, 172
  application of crown ether-based EXC resin to strontium, 165, 167–168
  behavior of EXC materials, 163
  capacity of EXC material, 163
  capacity of resins, 170
  chemical and physical stability of EXC materials, 163–164
  comparison of elution behavior of Sr-85 on conventional and Sr-selective EXC material, 167f
  conventional EXC material preparation, 162–163
  coupling with anion exchange, 275
  di-(tert-butylcyclohexano)-18-crown-6 (DtBuCH18C6), 165f
  effect of column washing on elution behavior of Sr-85, 166f, 169
  effect of support chemistry on behavior of EXC materials, 170
  .EXC systems employing functionalized supports, 171t
  extractants and diluents, 164–169
  performance of EXC material, 163–164
  relating water extraction and crown ether concentration, 168
  removal of radiostrontium from acidic nuclear waste streams, 164–165
  retention of extractant by support, 163
  reversed-phase chromatography, 162
  schematic of EXC resin stabilization via polymer encapsulation, 172f
  selectivity of resin for Sr^{2+}, 168
  separation and preconcentration of strontium ion, 164
  separation technique, 29, 31
  simultaneous separation and detection, 111–112
  stabilization of EXC materials by support chemistry, 171–172
  support participation in metal ion extraction process, 172–173
  supports, 170–173
  See also Column extraction chromatography
Extraction constant, standard state, 145

F

Fast neutron (FN) activation, californium user facility, 96
Fecal analysis

analysis for actinides, 273–274
See also Column extraction chromatography
Fire, community building and tritium release, 225
Fissile material, californium-252 neutron sources, 93–94
Fission, inducing, 6
Fission fragment (FF) emission, californium-252, 92
Flow-cell radiation detection
  advantages and disadvantages of methods, 116–117
  alpha and beta chromatograms for supernatant from high activity drain tank, *color insert*
  analysis of sequential separation and detection, 112–113
  analysis of simultaneous separation and detection, 113
  analytical systems, 112–113
  background of sequential separation and detection, 110–111
  background of simultaneous separation and detection, 111–112
  chromatogram with scintillating extraction chromatographic separation of high activity drain tank supernatant, *color insert*
  elemental separation, 106
  extraction chromatography, 111
  high activity drain tank, 114, 115t
  high level waste tank sludge, 115–116
  pulse shape discrimination (PSD), 109–110
  scintillation detector, 109–110
  sequential separation/detection configuration, 106, 107f
  simultaneous separation/detection configuration, 106, 108f
Flow-cell scintillation detector
  design, 109–110
  heterogeneous flow-cell, 109
  homogeneous flow-cell, 109
  pulse shape discrimination (PSD), 109–110
Food, neutron activation analysis (NAA), 310
Forensic analyses, high flux isotope reactor (HFIR), 97
Forensic science
  arsenic determination in human hair, 206
  example dual-irradiation particle-induced X-ray emission (PIXE) spectra, 210f
  gunshot residue (GSR) analysis of jeans, 209, 212–213
  high-flux neutron activation analysis for detection of elements, 206–207
  Nobel Prize in Chemistry, 206
  PIXE, 207–209
  PIXE for bovine liver, 211t
  PIXE for naval brass, 211t
  schematic of PIXE system, 208f
  suspect coin market, 215f
  suspect Spanish gold coins, 213–216
Fusion, postulation, 6

G

Gain factors, collimator-less gamma-ray imaging systems, 70–71
Gamma-ray energy tracking array (GRETA)
  comparison to state-of-art Gammasphere, 66, 67f
  gamma-ray imaging, 62–63
  prototype detector, 62f
Gamma-ray imaging systems
  active interrogation procedures, 53–54
  advances in detection of gamma radiation, 53
  advances in detector technology, 65
  advances in manufacture of two-dimensional segmented semi-conductor detectors, 54–55

advantage of double-sided strip detector (DSSD), 60–61
algorithms, 64
applications and imaging goals, 55t
calculated geometrical efficiencies and detector count rates, 72f
coincidence requirement, 61–62
collimator-based, 55–57
collimator-less, 58–65
collimator-less hybrid imager for medical radioisotopes, 71–73
Compton scattering, 58–59
detection modes and corresponding applications, 54t
efficiency values for conventional cameras, 57, 58f
emission mode, 53
evolution of resolving power for gamma-ray detectors and detector arrays, 69f
gain factors for collimator-less, 70–71
Gamma-Ray Energy Tracking Array (GRETA), 62–63
gamma-ray imaging detectors, 60–65
gamma-ray tracking, 59–65
gamma-ray tracking for nuclear physics experiments, 65–66
geometrical efficiencies and gains as function of angular resolution, 70f
high purity germanium detectors (HPGe), 60
high-purity silicon (HPSi) detector, 61
illustration of collimator-based, 56f
improvements of collimator-less over collimator-based, 73
incidence of photons on HPSi detector, 61
incident gamma ray interacting by Compton scattering and absorption by photoelectric effect, 60f
inter- and multidisciplinary research, 53
limitations and trade-offs of collimator-based, 57

lithium-drifted silicon (Si(Li)) detector, 61
nonproliferation applications, 68–71
nuclear medicine as example, 55t
nuclear nonproliferation as example, 55t
particle-induced emission mode, 53
positions of two interactions, 63f
potential advantages of Compton camera or collimator-less, 59
potential scattering sequence of two Compton scatterings and one photoelectrical effect, 64f
scattering mode, 54
schematic diagram of front part of imaging system, 61f
semi-conductor detectors, 65
sensitivity in detection of gamma radiation, 66
state-of-art Gammasphere vs. GRETA, 66, 67f
transmission mode, 54
Gammasphere
comparison to gamma-ray energy tracking array (GRETA), 66, 67f
state-of-art detector array, 65–66
Gaseous tritium light sources (GTLS), radioluminescent devices, 223–224
Gas-proportional counting, detection, 8
Geochemistry, neutron activation analysis (NAA), 309–310, 315–316
Gold, neutron activation analysis (NAA), 310
Gold coins, analysis of suspect, 213–216
Greenhouse effect, nuclear power and, 15
Groundwater, age-dating
copper tube samples for analyzing dissolved noble gases, 241f
detection limits as function of sample size and accumulation time, 240f
dissolved gas procedure, 240–241
^{3}H–^{3}He methods for basins, 241–243

helium in-growth, age-dating
  technique, 236–237
inferred initial tritium vs. mean age
  for Livermore Valley wells, 244f
laboratory procedures, 237–241
Livermore Valley groundwater basin,
  242–243
map showing tritium ages in
  Livermore Valley, 243f
photo of vessel for helium
  accumulation, 238f
plot of initial tritium vs. mean age,
  244f
samples being chilled prior to
  degassing, 239f
tritium procedure, 237–239
Ground zero
  modeling water flow and tritium
    removal, 228–230
  See also Tritium; World Trade
    Center (WTC)
Gunshot residue (GSR)
  analysis of jeans, 209, 212–213
  See also Forensic science

H

Hahnium, aqueous chemistry, 35–36
Half-life property, radioactivity, 4–5
Hanford waste, Tc-99 conversion to
  lower oxidation state complexes,
  178
Hassium, characterization, 36
Heavy element (HE)
  production, 89, 90f
  See also Californium-252
Helium-3
  application of ^{3}H–^{3}He dating method
    to groundwater basins, 241–243
  copper tube for analyzing dissolved
    noble gases in groundwater
    samples, 241f
  dissolved gas procedure, 240–241
  laboratory procedures, 237–239
  noble gas isotope, 236

photo of vessel for He accumulation,
  238f
sum of tritium and, concentrations,
  236–237
See also Groundwater, age-dating
Heterogeneous flow-cell, scintillation
  detector, 109
Hg, functionalized support for
  extraction chromatography, 171t
High flux isotope reactor (HFIR)
  comparison of neutron activation
    analysis (NAA) detection limits,
    97–98
  instrumental neutron activation
    analysis (INAA), 97–99
  laser-induced breakdown
    spectroscopy (LIBS) with INAA,
    98–99
  trace element analyses, 97
High purity germanium (HPGe)
  detectors, gamma-ray imaging, 60
High purity silicon (HPSi) detectors,
  gamma-ray imaging, 61
High resolution γ- and α-detection,
  semi-conductor techniques, 8
Holdback carriers, co-precipitation,
  26
Homogeneous flow-cell, scintillation
  detector, 109
Horwitz, Dr. E. Philip, column
  extraction chromatography, 272–
  273
Hot spots
  α-contamination of soils, 129–132
  gunshot-like structures, 134
  rod-shaped structures, 134
  star-like structures, 134
HTO. See Tritiated water (HTO)

I

Imaging systems. See Gamma-ray
  imaging systems
Inductively coupled plasma mass
  spectroscopy (ICP–MS)

analysis of longer lived radionuclides, 247
automated separation for, analysis of actinides, 266–268
total ^{99}Tc analysis, 263
Innovations, neutron activation analysis (NAA), 317–318
Instrumental neutron activation analysis (INAA)
analysis of environmental and vitrified samples, 96
californium-based, 95–96
comparison of NAA detection limits, 97–98
high flux isotope reactor (HFIR), 97–99
INAA at californium user facility (CUF), 96
laser-induced breakdown spectroscopy (LIBS) with INAA, 98–99
trace element analyses at HFIR, 97
International Atomic Energy Agency (IAEA), national survey of U.S. universities, 19–20
International monitoring, nuclear nonproliferation agreements, 14–15
Iodine isotopes, nuclear medicine, 34
Ion chromatography, separating radionuclides, 106
Ion exchange
elution of lanthanide and actinide ions on Dowex-50, 28f
radiochemical separation, 27–29
Iridium-192, brachytherapy, 92–93
Iron, functionalized support for extraction chromatography, 171t
Island of Stability, nuclear stability, 7
Isotope and element discovery, nuclear sciences, 9
Isotopes
alpha, beta, and gamma emitting, 142–143
alpha spectrometry of alpha emitting thorium, 291–293
californium-252, 91
high flux isotope reactor (HFIR), 97
identity and purity, 143
inductively coupled plasma mass spectrometer (ICP–MS), 110–111
Nobel Prize in Chemistry, 206
radioactive, characterizing new complexing agents, 142
radiotracers, 36–37
term phrased, 5

J

Jeans, gunshot residue (GSR), 209, 212–213

L

Land mine detection
neutron backscattering for detection, 94
prompt gamma neutron activation analysis (PGNAA), 95
Lanthanides, elution on Dowex-50, 28f
Large volume samples, neutron activation analysis (NAA), 317
Laser-induced breakdown spectroscopy (LIBS)
carbon in soils, 99
instrumental neutron activation analysis (INAA) with, 98–99
Lead
functionalized support for extraction chromatography, 171t
gunshot residue (GSR) analysis of jeans, 209, 212–213
Ligand design, process, 141
Liquid scintillation (LS) counting
CIEMAT (Centro de Investigaciones Energéticas Medioambientales y Tecnológicas)/NIST (National Institute of Standards and Technology) efficiency tracing method, 77

preferred method, 77
See also Triple-to-double coincidence ratio (TDCR) method
Lithium-drifted silicon (Si(Li)) detector, gamma-ray imaging, 61
Livermore Valley. See Groundwater, age-dating
Loan program, californium-252, 101–102
Los Alamos National Laboratory (LANL)
analysis of Hanford tank samples, 178
radiochemical method, 179, 180f
See also Non-pertechnetate (^{99}Tc) species
Low activity waste (LAW)
chemical processing, 262–263
testing analyzer instruments, 265–266

M

Magnets, permanent, radiation damage testing, 100
Manganese, functionalized support for extraction chromatography, 171t
Manhattan project, nuclear weapons, 6–7
Mass spectrometry, accelerated (AMS), brain analysis of Alzheimer's patients, 304
Medical radioisotopes, collimator-less hybrid imager, 71–73
Medical studies, neutron activation analysis (NAA), 315
Medicine
nuclear, 13
See also Nuclear medicine; Thorium-229
Medieval ceramics, neutron activation analysis (NAA), 310
Mercury, functionalized support for extraction chromatography, 171t
Metal ion extraction
support participation, 172–173
See also Extraction chromatography (EXC)
Middle ages, alchemy, 2–3
Mining, uranium and environmental problem, 15
Model, water flow and tritium removal from ground zero, 228–230
Monitoring
nuclear nonproliferation agreements, 14–15
on-line process, prompt gamma neutron activation analysis (PGNAA), 94–95
Munitions, prompt gamma neutron activation analysis (PGNAA), 95
Mutation of seeds, fast-neutron-induced, 101

N

National Institute of Standards and Technology (NIST)
1999 master solution verification measurement results, 47t
challenges to address, 44–45
CIEMAT (Centro de Investigaciones Energéticas Medioambientales y Tecnológicas)/NIST efficiency tracing method, 77
Ionizing Radiation Division (IRD), 43–44
measurement assurance programs (MAPs), 43
natural-matrix radionuclide standard reference materials (SRMs), 40–43
NIST Radiochemistry Intercomparison Program (NRIP), 44
performance of laboratories participation in NRIP, 48t
process of developing and verifying NIST evaluation samples, 44–45
program characteristics, 45t
radiochemical analysis, 39–40

SRM measurement discrepancies, 42t
traceability evaluation, 43–46
traceability evaluation program, 44
triple-to-double coincidence ratio (TDCR) system, 79–80
National metrology institutes (NMIs), liquid scintillation counting, 77
Naval brass, particle-induced X-ray emission (PIXE), 211t
Neptunium
 identification, 7
 stacked cartridge approach, 277
Neurofibrillary tangles (NFT)
 time course in AD, 302–305
 See also Alzheimer's disease (AD)
Neutron activation analysis (NAA)
 air pollution, 311–312
 archaeology, 309–310
 biological systems, 313–314
 biomonitoring, 312
 characterizations, 317–318
 Compton suppression, 308
 determination of ^{232}Th by PCNAA (pre-concentration NAA) and pre-concentration radiochemical NAA (PCRNAA), 293–295
 determination of ^{232}Th by PCRNAA, 290–291
 determination of thorium, 287
 discovery, 307
 engineering materials, 311
 environmental monitoring, 311–314
 epithermal NAA, 308
 facilities, 316–318
 food and diet, 310
 geochemistry, 315–316
 innovations, 317–318
 large volume samples, 317
 medical and clinical, 315
 multielemental techniques, 287–288, 308
 nuclear and radioanalytical sciences, 7
 plants and crops, 314

 prompt gamma NAA (PGNAA), 316–317
 reference materials, 308–309
 software, 318–319
 solid waste, 312–313
 statistics, 319
 total quality management (TQM), 319
 upgrades, 317–318
 wastewater, 313
 See also Thorium
Neutron radiography, californium-252, 93
Neutron science, californium user facility, 99–101
Nevada Applied Ecology Group (NAEG)
 fate and transport of actinides, 122–123
 investigation results documenting actinide occurrence, 124t
 Nevada test site and pertinent NAEG results, 123–125
 See also α-Autoradiography
Nevada test site (NTS)
 description, 128f
 field exposure experiments, 128–129
 topographic map, 127f
 See also α-Autoradiography
New Jersey. See Tritium
New York state
 tritium analysis, 222t
 See also Tritium
Nickel, functionalized support for extraction chromatography, 171t
Nineteenth (19th) century, chemistry, 3–4
Nitrilotriacetic acid (NTA), complexation ligands, 182
Nobel Prize in Chemistry, isotopes as tracers, 206
Noble metal catalysts, technetium speciation, 186–187
Non-aqueous processes, radioactive waste, 15–16
Non-pertechnetate (^{99}Tc) species

analysis method, 179
analysis using X-ray absorption near edge spectroscopy (XANES), 181
anion exchange and ^{99}Tc NMR underestimating Tc concentration, 180
anion exchange and chromatography data for soluble complexes, 189t
apparent ^{99}Tc K_d of Tc(IV)–EDTA complex, 183f
autoreduction of TcO$_4^-$, 189f
behavior of Tc-gluconate mixtures, 187
bleaching of 500-nm absorption of Tc(IV)–EDTA dimer after dilution and pH adjustment, 182-183, 183f
calculating approximate distribution coefficient for reduced complexes, 188
catalytic reduction of TcO$_4^-$ to TcO$_2$, 187f
challenge in radiochemical analysis of ^{99}Tc in tank waste, 179
determination of total ^{99}Tc concentrations in 101-SY and 103-SY samples, 179, 180t
discovery of problem, 178–181
distribution coefficients, 180, 181t
distribution coefficients and pH, 182-183
effect of noble metal catalysts on technetium speciation, 186–187
effect of reducing agent SnCl$_2$ on apparent ^{99}Tc K_d, 185f
effect of Sn(II) on ^{99}Tc K_d, 184–185
experimental, 190–191
gamma-emitting tracer ^{95m}Tc, 179
genesis in alkaline solution, 186–190
Hanford tank samples (101-SY and 103-SY) of supernate, 178–179
investigating complex stability, 184
organic compounds and reaction conditions supporting TcO$_4^-$ reduction, 188, 190
paper chromatography separating TcO$_4^-$, 188
radiochemical method for total ^{99}Tc analysis, 180f
rate of bleaching, 182-183, 183f
reactivity vs. radiolysis, 186
screening catalytic reactivity, 186–187
simulant formulations, 184–185
stability in alkaline solution, 182–185
UV-vis spectroscopy monitoring Tc speciation changes, 186
X-ray absorption spectroscopy, 181
Nonproliferation. *See* Nuclear nonproliferation
Nuclear/astrophysics, gamma-ray detection, 54t
Nuclear chemistry, radiochemistry, 22
Nuclear magnetic resonance (NMR) ^{99}Tc, 180
Nuclear Materials Identification System (NMIS), fissile material and nuclear waste characterization, 93–94
Nuclear medicine
  application of radiochemistry, 34–35
  collimator-less hybrid imager, 71–73
  examples, 13
  gamma-ray collimators, 56
  gamma-ray detection, 54t
  imaging goals, 55t
  need for scientists, 17
  nuclear sciences, 9
  See also Thorium-229
Nuclear nonproliferation
  aim, 68
  angular resolution, 68, 70
  evolution of resolving power for gamma-ray detectors and detector arrays, 69f
  gain factors, 70–71
  gamma-ray detection, 54t
  gamma-ray imaging, 68–71
  geometrical efficiencies and gains as function of angular resolution, 70f
  imaging goals, 55t

international monitoring of agreements, 14–15
Nuclear phenomena, chemists and study, 12–13
Nuclear physics, gamma-ray tracking, 65–66
Nuclear power
  accidents, 15
  Greenhouse effect, 15
  non-aqueous processes, 15–16
  radioactive wastes, 15–16
Nuclear sciences
  age dating, 9
  detection techniques, 10
  development of new field, 4
  developments of scientific equipment, 6
  element and isotope discovery, 9
  energy generation, 10
  evolution, 13
  future opportunities, 10
  new analytical tools, 9–10
  nuclear medicine, 9
  positive and negative aspects of applications, 10
Nuclear scientists, areas needing, 17
Nuclear waste
  disposal, 178
  low activity waste (LAW), 262–263
  storage, 247
  X-ray absorption spectroscopy, 181
Nuclear waste characterization, californium-252 neutron sources, 93–94
Nuclear waste reactivity, radiolysis, 186
Nuclear waste streams, removal of radiostrontium, 164–165
Nuclear weapons
  dismantling and disposition, 13–14
  Manhattan project, 6–7
  production, 247
  public health and safety concerns, 162
Nucleus, atomic mass, 5

Nuestra Señora de Atocha, suspect gold coins, 213–216

O

Oak Ridge National Laboratory (ORNL)
  actinium-225 sales, 202f
  heavy element (HE) production, 89, 90f
  Radiochemical Engineering Development Center (REDC), 89
  separation of thorium-229, 203
  Transuranium Element Processing Program (TEPP), 89, 90f
  ^{233}U storage, 195
  See also Californium-252; Thorium-229
Obsidian, neutron activation analysis (NAA), 310
On-line process monitoring, prompt gamma neutron activation analysis (PGNAA), 94–95

P

Paper chromatography, separating TcO$_4^-$, 188
Particle-induced emission mode, gamma-ray detection, 53
Particle induced X-ray emission (PIXE)
  advantage, 209
  analysis system, 208–209
  bovine liver, 211t
  case of suspect coins, 213–216
  disadvantage, 207
  dual-irradiation spectra, 210f
  elemental analysis, 300
  gunshot residue (GSR) analysis of jeans, 209, 212–213
  micro-PIXE of Alzheimer's disease (AD) subjects, 300–301
  naval brass, 211t

radioanalytical technique, 33
schematic, 208f
Spanish gold coins, 216t
technique, 207
See also Forensic science
Particle, wave nature, 5–6
Particle size cartridges, smaller, column extraction chromatography, 273
Passivated ion implanted silicon (PIPS)
alpha spectroscopy, 247
modern PIPS diode detectors, 261
Pb, functionalized support for extraction chromatography, 171t
Performance tests
choice of statistical distribution, 82
counting experiment, 82
experimental and theoretical counting efficiencies, 83f, 84f, 85f
measuring activity of solutions, 85–86
schematic of acquisition hardware of triple-to-double coincidence ratio (TDCR) detection system, 81f
TDCR counting results of National Institute of Standards and Technology (NIST) standard reference material, 86t
TDCR system, 80, 82
See also Triple-to-double coincidence ratio (TDCR) method
Periodic table
additions of 20th century, 9
elements in 1900, 3f
elements in 21st century, 8f
Permanent magnets, radiation damage testing, 100
Personnel needs
International Atomic Energy Agency (IAEA), 19–20
National Academy Report, 17
number of doctoral degrees in nuclear/radiochemistry by U.S. universities, 19f
number of faculty and departments offering nuclear/radiochemistry graduate programs, 18f
studies, 17–20
Pertechnetate
separation, 264
uptake properties of scintillating sensor material vs. nitric acid concentration, 251f
Phase transfer reaction, analysis for stoichiometries, 145–146
Physics
gamma-ray tracking for basic nuclear physics experiments, 65–66
quantum theory, 4–5
See also Nuclear physics
Plant materials, actinide uptake by, 133
Plants, neutron activation analysis (NAA), 314
Plutonium
stacked cartridge approach, 277
valence adjustment, 201
Plutonium Valley, Nevada test site. See α-Autoradiography
Poisoning, arsenic in human hair, 206
Polonium, radiochemistry, 3
Polymer encapsulation, extraction chromatography (EXC) resin stabilization, 172–173
Port Authority of New York and New Jersey (PANYNJ)
World Trade Center, 219
See also Tritium
Positron emission tomography (PET), nuclear medicine, 34–35
Precipitation, separation technique, 25–26
Preconcentration. See Exchange chromatography (EXC)
Pre-concentration neutron activation analysis (PCNAA)
comparison with PCRNAA, 295f

detection limits for thorium isotopes, 296t
determination of ^{232}Th, 293–295
graph of ^{232}Th by PCNAA vs. expected, 294f
See also Neutron activation analysis (NAA)

Pre-concentration radiochemical neutron activation analysis (PCRNAA)
comparison with PCNAA, 295f
detection limits for thorium isotopes, 296t
determination method for ^{232}Th, 290–291
determination of ^{232}Th, 293–296
graph of ^{232}Th by PCRNAA vs. expected, 296f
See also Neutron activation analysis (NAA)

Process monitoring, on-line, prompt gamma neutron activation analysis (PGNAA), 94–95

Prompt gamma neutron activation analysis (PGNAA)
californium-252, 92
carbon in soils, 99
disadvantage, 316
explosive and land mine detection and munitions identification, 95
facilities, 316–317
on-line process monitoring and borehole logging, 94–95
samples, 317

Pumice, neutron activation analysis (NAA), 310

Purification methods. See Thorium-229

Q

Quantum theory, physics, 4–5
Quenching factor, triple-to-double coincidence ratio (TDCR) method, 78

R

Radiation, dose effects research, 16
Radiation damage
testing permanent magnets, 100
testing semiconductor detectors, 100
Radiation detection. See Flow-cell radiation detection
Radioactivity
artificially inducing, 6
half-life property, 4–5
natural matter property, 4
nuclear power production, 15–16
Radioanalytical chemistry
acid dependence in extraction of Am^{3+}, 148–149
activation analysis, 32–33
advancement of science, 5–6
advantages, 141
aggregation of extractant molecules, 150–151
alpha, beta, and gamma emitting isotopes, 142–143
apparent stoichiometries, 150
aqueous anion concentrations for extraction and back extraction, 146–147
basic measurements and methods, 144–147
biphasic chemical equilibrium, 145
calculating distribution coefficient or ratio, 144–145
characterization of diacidic diphosphonate extractant systems, 147–152
characterization of extraction by solvating ligands, 152–155
characterization of new compounds, 141
characterization of water-soluble chelating agents, 156–158
chemical equilibrium, 143
colligative properties of extractant solutions, 149
dependencies of Am^{3+} extraction by tetra(2-ethylhexyl)pyridine-2,6-

bis(methylphosphine oxides) [TEH(NOPOPO)], 155, 156f
detection technologies, 7–8
development of detection technologies, 7–8
distribution experiments elucidating stoichiometry and stability constants, 157f
element synthesis, 6–7
experimental, 141–143
extraction constant defining standard state, 145
extraction curve of Am^{3+} by TEH(NOPOPO), 153–154
historical overview, 3–5
identity and purity of isotopes, 143
ligand and acid dependencies for Am^{3+} extraction vs. temperature, 152f
ligand rigidity and extractant aggregation, 151
liquid cation exchanger extractants, 146
mass balance, 143
need for new chelating agents, 143–144
particle induced X-ray emission (PIXE), 33
phase transfer reaction, 145–146
possible competing reactions, 147
present day technologies and scientific spin-offs, 9–10
production of tranuranium elements, 144
proposed predominant extraction reaction, 152
PUREX process, 144
radioactive isotope characterizing new complexing agents, 142
radioimmunoassay (RIA), 33–34
stoichiometry of extracted Am^{3+} complex, 154–155
techniques, 32–34
thermodynamic equilibrium expressions for extraction reactions, 150

See also Forensic science
Radiochemical Engineering Development Center (REDC)
description, 89
heavy element production, 89, 90f
transuranium element processing program, 89
Radiochemistry
activation analysis, 32–33
anion exchange behavior, 29, 30f
applications, 32–37
bohrium, 36
calculating number of nuclei, 23
chemical effects accompanying specific activities, 24
conventional analytical techniques, 24
definition, 22
distribution ratio D, 26–27
elution of elements from anion exchange resin, 30f
elution of tripositive lanthanide and actinide ions on Dowex-50, 28f
environmental, 39
extraction chromatography, 29, 31
hahnium, 35–36
hassium, 36
integration of separation chemistries with radiation detectors, 268
ion exchange, 27–29
needing upgraded methods at Savannah River Site, 273
nuclear chemistry alliance, 22
nuclear medicine, 34–35
number of atoms and concentration of solutions, 23
particle induced X-ray emission (PIXE), 33
physics and chemistry of heaviest elements, 35–36
polonium, radium, and actinium, 3
precipitation, 25–26
radioanalytical techniques, 32–34
radioimmunoassay (RIA), 33–34
radiotracers, 36–37

rapid radiochemical separations, 31–32
redox properties as function of time, 25
rutherfordium, 35
seaborgium, 36
sensitivity, 24
separation factor, 27
separation of thorium from samples, 289
separation techniques, 25–32
solvent extraction, 26–27
uniqueness, 23–25
See also Academic radiochemistry; Automation of radiochemical analysis
Radiochromatography, applications, 110
Radiography, neutron, californium-252, 93
Radioimmunoassay (RIA), radioanalytical technique, 33–34
Radioisotopes
  chemical separation, 247–248
  collimator-less hybrid imager, 71–73
Radioluminescent devices, tritium, 223–224
Radiolysis, nuclear waste reactivity, 186
Radionuclides
  chromatographic separation, 106
  nuclear medicine, 34–35
  uses, 13
Radionuclide selective sensors
  composite bed sensors, 253
  count rate after sensor equilibration, 255
  detector signals and calibration curve for equilibration sensing approach, 256f
  dual functionality sensor materials, 250–252
  feasibility of ^{90}Sr sensing via ^{90}Y Cherenkov radiation, 258–260

functional requirements and general approach, 248–249
loading and regeneration of Tc(VII) selective composite bed sensor, 254f
pertechnetate uptake properties of scintillating sensor material vs. nitric acid concentration, 251f
preconcentrating minicolumn sensors, 250–260
quantification approaches, 254–257
reagentless ^{99}Tc(VII) sensing in Hanford groundwater, 257–258
response of ^{99}Tc-sensor to sample containing target analyte and unretained species, 252f
schematic, 249f
stopped-flow counting interval, 255
uptake properties of scintillating sensor material vs. nitric acid concentration,
See also Automation of radiochemical analysis
Radionuclide separation and preconcentration. *See* Exchange chromatography (EXC)
Radiostrontium, removal from nuclear waste streams, 164–165
Radiotracers, application of radiochemistry, 36–37
Radium, radiochemistry, 3
Rapid radiochemical separations, technique, 31–32
Rare-earth catalyst, dual-irradiation particle-induced X-ray emission (PIXE), 210f
Reference materials, neutron activation analysis (NAA), 308–309
Research, areas needing nuclear scientists, 17
Rice genes, fast-neutron-induced mutation, 101
Rutherford, age of earth, 5
Rutherfordium, aqueous chemistry, 35

S

Saturnian model, atom, 5
Savannah River Site (SRS)
  accuracy using TEVA–TRU Resin method on urine, 281
  Diphonix Resin®, 274
  electrodeposition preparation, 280
  needing upgraded method for fecal analysis, 273–274
  needing upgraded radiochemical methods, 273
  new urine method, 274
  Pu-236 tracer recoveries on TEVA Resin, 282f
  Pu and Np results on spiked urine samples, 282f
  SRS stacked cartridge approach, 277–278
  SRS urine method protocol and testing, 278–283
  strontium analysis, 280
  test performance using TEVA Resin, 280–281
  TEVA–TRU Resin stacked column method, 283–284
  tracer recoveries using TEVA Resin, 281–282
  valence adjustment of Pu and Np, 279–280
  See also Column extraction chromatography
Savannah River Technology Center (SRTC)
  high activity drain tank, 114, 115t
  high level waste tank sludge, 115–116
  See also Flow-cell radiation detection
Scattering mode, gamma-ray detection, 54
Scientific equipment, nuclear science, 6
Scientists, need for nuclear medicine, 17
Scintillation detector
  counting, 8
  flow-cell, 109–110
  on-line, 112–113
Screening, catalytic reactivity, 186–187
Seaborgium, chemistry, 36
Seeds, fast-neutron-induced mutation, 101
Selectivity, resin for Sr^{2+}, 168
Selenium, neutron activation analysis (NAA) of food and diet, 310
Semi-conductor detectors
  advances, 54–55
  radiation damage testing, 100
Senile plaque (SP)
  average and standard deviation of zinc concentration in SP of AD subjects, 302t
  time course in AD, 302–305
  See also Alzheimer's disease (AD)
Sensitive material, gamma-ray detection, 54t
Sensitivity
  gamma-ray tracking, 66
  radiochemistry, 24
Sensors. See Automation of radiochemical analysis; Radionuclide selective sensors
Separation factor, solvent extraction, 27
Separation methods
  advances in radiochemical, 272
  detection techniques, 8
  extraction chromatography, 29, 31
  ion exchange, 27–29
  precipitation, 25–26
  radioisotopes, 247–248
  rapid radiochemical separations, 31–32
  solvent extraction, 26–27
  spent fuel treatment, 16
  See also Exchange chromatography (EXC); Thorium-229
Sequential injection (SI), sequential injection, 266
Sequential separation and detection

advantages and disadvantages, 116–117
analytical system, 112–113
background, 110–111
on-line scintillation detector, 112–113
schematic, 106, 107f
See also Flow-cell radiation detection
Silicon detection devices, transactinides, 7
Simultaneous separation and detection
advantages and disadvantages, 116–117
analytical system, 113
background, 111–112
on-line detection efficiencies, 113
schematic, 106, 108f
See also Flow-cell radiation detection
Sn(II) reducing agent, effect on ^{99}Tc K_d, 184–185
Software, neutron activation analysis (NAA), 318–319
Solid state nuclear track detectors (SSNTD), α-autoradiography, 123
Solid waste, neutron activation analysis (NAA), 312–313
Solvating ligands, characterization of extraction by, 152–155
Solvent extraction
distribution ratio, 26–27
new complexing agents, 141
separation factor, 27
separation technique, 26–27
Spanish gold coins, analysis, 213–216
Spent fuel treatment, efficient separations, 16
Stability, non-pertechnetate species in alkaline solution, 182–185
Stabilization, extraction chromatography (EXC) resin, 172–173
Stacked cartridge approach, Savannah River Site (SRS), 277–278
Stacked column method

TEVA–TRU Resin, 283
See also Column extraction chromatography
Standard addition approach, ^{99}Tc quantification, 264
Standard Reference Materials (SRMs)
measurement discrepancies, 42t
National Institute of Standards and Technology (NIST) natural-matrix radionuclide SRMs, 41t
natural matrix radionuclide SRMs, 40–43
triple-to-double coincidence ratio (TDCR) counting results of NIST SRMs, 86t
Standard state, extraction constant, 145
Statistics, neutron activation analysis (NAA), 319
Stockpile stewardship, gamma-ray detection, 54t
Strontium
application of crow ether-based extraction chromatographic (EXC) resin, 165, 167–168
comparing elution behavior on conventional vs. Sr-selective EXC material, 167f
effect of column washing on elution behavior of Sr-85, 166f, 169
elution behavior on EXC resin, 167f, 168
evolution of EXC materials, 164
relating water extraction and crown ether concentration, 168
removal of radiostrontium from acidic nuclear waste streams, 164–165
selectivity of resin for Sr^{2+}, 168
stacked cartridge approach, 277
urine analysis, 280
See also Extraction chromatography (EXC)
Styrene-divinylbenzene copolymers, preparation of macroporous, 162
Superheavy elements, production, 7

Support chemistry
  capacity of resin, 170
  choice of conventional EXC materials, 170
  extraction chromatography (EXC), 170–173
  functionalized systems, 171t
  participation in metal ion extraction, 172–173
  schematic of stabilization via polymer encapsulation, 172f
  stabilization of EXC materials, 171–172

T

Table. *See* Periodic table
Technetium-99
  automated total ^{99}Tc analyzer instrument, 262–266
  chemical evolution of long-lived fission product, 178
  composite bed sensor, 253, 254f
  conversion to lower oxidation state complexes, 178
  detector signals and calibration curve for equilibration sensing approach, 256f
  inductively coupled plasma mass spectroscopy (ICP–MS), 263
  methods for analysis, 263
  radionuclide speciation, 263
  reagentless sensing in Hanford groundwater, 257–258
  response of ^{99}Tc-sensor to sample containing target analyte and unretained species, 252f
  See also Automation of radiochemical analysis; Non-pertechnetate (^{99}Tc) species
Terrorist attack
  World Trade Center, 220
  See also Tritium; World Trade Center (WTC)
Thorium
  alpha spectra, 292f
  alpha spectrometry of alpha-emitting thorium isotopes, 291–293
  comparing pre-concentration neutron activation analysis (PCNAA) and pre-concentration radiochemical NAA (PCRNAA), 295f
  comparison of detection limits for important isotopes, 296t
  determination by alpha spectrometry, 289
  determination of ^{232}Th by PCNAA and PCRNAA, 293–295
  determination of ^{232}Th by PCRNAA, 295–296
  determination of ^{232}Th by PCRNAA (ion exchange method), 290–291
  distribution in human body, 287
  electrodeposition of Pa, 291
  elution profile of protactinium from anion exchange column material, 290f
  experimental, 288–291
  neutron activation analysis (NAA), 287–288
  plot of relative alpha spectrometry vs. expected values, 292f
  preparation of reagents and irradiation vials, 288
  radiochemical separation of thorium from samples, 289
  sample preparation, 288
  spectra for determination of ^{232}Th by PCNAA and PCRNAA, 293f
  ^{232}Th by PCNAA vs. expected, 294f
  ^{232}Th by PCRNAA vs. expected, 296f
Thorium-229
  cancer cells, 194
  composition of ten batches of processed ^{233}U (2000–2002), 199–200
  decay chart, 194f
  decay chart for Th-228, 195f
  flow chart for Th purification process, 198f

flow chart for Th separation process, 197f
inventory of ^{229}Th and ^{225}Ac supply and demand, 202f
microporous (AG-1) vs. macroporous (MP-1) resin, 202–203
money for, by U.S. Department of Energy (DOE), 196
ORNL actinium-225 sales, 202f
pathways for obtaining, 194–195
plutonium valence adjustment, 201
process of separation and purification, 196–199
production by irradiating ^{226}Ra targets, 195–196
results of Th separation process, 200, 201t
separation from ^{233}U, 195–196
separation of, at ORNL, 203
supply at Oak Ridge National Laboratory (ORNL), 194
thorium separation process, 196–197
Th purification process, 198–199
Th purification process alternative, 199
See also Nuclear medicine
Three-Mile Island, nuclear power accidents, 15
Timing devices, tritium, 224
Total quality management (TQM), neutron activation analysis (NAA), 319
Traceability evaluations
 definition of traceability, 43–44
 National Institute of Standards and Technology (NIST), 44
 process of developing and verifying NIST evaluation samples, 44–45
 radiochemistry, 43–46
Trace analysis, separation approach, 267
Trace element analyses, high flux isotope reactor (HFIR), 97
Trace metals, Alzheimer's disease, 299–302

Trans-actinides
 identification, 8
 physics and chemistry, 35–36
 silicon detection devices, 7
Transmission mode, gamma-ray detection, 54
Transplutonium elements, separation procedures, 143–144
Transuranium actinide element (Np), need for chelating agents, 143
Transuranium element processing program
 heavy element production, 89, 90f
 Radiochemical Engineering Development Center (REDC), 89
 See also Californium-252
Transuranium elements, large scale production, 144
Triple-to-double coincidence ratio (TDCR) method
 choice of statistical distribution, 82
 counting efficiency, 78
 counting experiment, 82
 counting rate equations, 79
 detection efficiency of system, 77–78
 experimental and theoretical counting efficiencies, 83f, 84f, 85f
 investigating performance, 80, 82
 measuring activity of solution, 85–86
 model, 78–79
 National Institute of Standards and Technology (NIST) TDCR system, 79–80
 optimum operating characteristics, 80
 parameter K, 78–79
 performance tests, 80–86
 quenching factor, 78
 schematic of acquisition hardware of TDCR system at NIST, 81f
 TDCR counting results of NIST standard reference material solutions, 86t
Tritiated water (HTO)
 equation describing HTO concentration, 229

total HTO activity, 229–230
See also Tritium
Tritium
  airplanes as source, 225
  analysis in New York state, 222*t*
  bomb-pulse, 236
  C-124 airplane fire, 226
  concentration in groundwater, 236
  description, 236
  detection limits as function of sample size and accumulation time, 240*f*
  emergency EXIT signs, 223
  fire in Alaska community building, 225
  gaseous tritium light sources (GTLS), 223–224
  half-life, 236
  incorporation into polymer, 223
  inferred initial tritium vs. mean age, 244*f*
  initial tritium vs. mean age, 244*f*
  interest at World Trade Center (WTC), 221
  laboratory procedures, 237–241
  measurements, 221, 223
  mechanism of release from weapons or watches, 227
  modeling of water flow and tritium removal from ground zero, 228–230
  procedure, 237–239
  radioluminescent devices, 223–224
  release scenario from airplanes at WTC, 225–226
  sources and fate at WTC, 224–227
  timing devices, 224
  tritium unit (TU), 236
  watches, 224, 227
  water leak into Bathtub at WTC, 226
  weaponry as source, 226–227
  See also Groundwater, age-dating
Tritium Unit (TU), 236
Twentieth (21st) century, periodic table of elements, 8*f*
Twin Towers
  collapse, 220

World Trade Center, 219
See also World Trade Center (WTC)

U

Ultraviolet-visible spectroscopy, technetium speciation, 186
United States, nuclear waste repository, 14
University. *See* Academic radiochemistry
University Loan Program, californium-252, 101–102
Upgrades, neutron activation analysis (NAA), 317–318
Uranium
  Environmental problem in mining, 15
  extraction chromatography resin, 276
  functionalized support for extraction chromatography, 171*t*
  stacked cartridge approach, 277
Uranium-233
  composition of batches processed (2000–2002), 199–200
  separation of thorium-229 from, 195–196
  See also Thorium-229
Urine
  accuracy of TEVA–TRU Resin method, 281
  adjusting valence of plutonium and neptunium, 279
  electrodeposition preparation, 280
  monitoring actinides, 274–275
  new resin cartridge technology, 274
  protocol and testing, 278–283
  stacked column method, 278
  strontium analysis, 280
  test performance using TEVA Resin only, 280–281
  tracer recoveries, 281–283
  See also Column extraction chromatography

UTEVA Resin®, uranium separation, 276

V

Vacuum boxes, column extraction chromatography, 273
Vitrified samples, instrumental neutron activation analysis (INAA), 96
Volume samples, large, neutron activation analysis (NAA), 317

W

Waste Isolation Pilot Plant, nuclear waste repository, 14
Waste management, gamma-ray detection, 54t
Wastewater, neutron activation analysis (NAA), 313
Watches
  mechanism of tritium release, 227
  tritium, 224, 227
Water
  leak into Bathtub, 226
  modeling water flow and tritium removal from ground zero, 228–230
  See also Automation of radiochemical analysis; Tritiated water (HTO)
Water-soluble chelating agents, characterization, 156–158
Wave nature, particles, 5–6
Weapons
  mechanism of tritium release, 227
  source of tritium, 226–227

World Trade Center (WTC)
  Bathtub, 220
  modeling water flow and tritium removal from ground zero, 228–230
  operation, 219
  Port Authority of New York and New Jersey (PANYNJ), 219
  Slurry Wall, 220
  sources and fate of tritium at WTC, 224–227
  terrorist attack, 220
  tritium analysis in New York state, 222t
  tritium at WTC, 221, 223
  Twin Towers, 219
  water leak into Bathtub, 226
  See also Tritium
Wright–Patterson Air Force Base, C-124 airplane fire and tritium, 226

X

X-ray absorption spectroscopy, tank waste samples, 181

Z

Zinc
  Alzheimer's disease (AD), 299–302
  brain, 300
  disruption of, in homeostasis in AD, 301–302
  functionalized support for extraction chromatography, 171t
  implication in AD, 299–300
  neutron activation analysis (NAA) of food and diet, 310